D1075391

Empire
of the Air

Empire
of the Air

AVIATION AND THE
AMERICAN ASCENDANCY

Jenifer Van Vleck

Harvard University Press

Cambridge, Massachusetts, and London, England

2013

Cataloging-in-Publication Data available from the Library of Congress.

Library of Congress catalog card number: 2012051096

ISBN: 978-0-674-05094-5

For my parents,

Richard and Caren Van Vleck

Contents

Introduction:
The Logic of the Air

On any given day in the summer of 1939, thousands of Americans at Pan American Airways' Dinner Key Terminal in Miami, Florida, were watching the world turn. As newspaper headlines and newsreels warned of impending war in Europe, visitors gazed at a gigantic, rotating globe—one of the largest ever built in the United States, weighing 3.25 tons and measuring 10 feet in diameter—and reflected on the state of the world. Perhaps they marveled at how modern communications and transportation technologies had seemed to shrink distance and time. Perhaps, in this context, they worried about the United States' neutrality and national security, as the expansionist ambitions of Nazi Germany and imperial Japan had once again brought the world to the brink of war. Or perhaps they were simply imagining distant lands that they would like to visit, once the U.S. economy (and their own bank accounts) fully recovered from the ravages of the Depression. But whatever they were thinking or feeling, these thousands of Americans could perceive the world as within their reach, both literally and figuratively.[1]

Built in 1933–1934 by architects William Adams Delano and Chester Holmes Aldrich, who had previously earned their reputation designing estates and social clubs for families such as the Astors, Rockefellers, and

Whitneys, Dinner Key Terminal was the largest marine air facility in the world and a masterpiece of Streamline Moderne architecture, replete with bronze doors and friezes of winged globes (Pan American's logo), suns, and eagles. Now Miami's City Hall, the terminal immediately became a local landmark, serving 50,000 passengers per year during the 1930s and 1940s in addition to an average of 30,000 visitors per month (and as many as 100,000 during the winter months). These enthralled spectators came to watch Pan American's magnificent *Clipper* flying boats take off for Havana, Rio, and beyond—and, indeed, they came to see the globe. Pan American's public relations department invited children from fifty-six Miami public schools to tour the terminal and study the globe, and it even served as the background for wedding pictures when a couple from Havana married at Dinner Key Terminal. As they gazed at the great revolving sphere, passengers and visitors alike could imagine themselves as citizens of the world.[2]

Meanwhile, in his spacious office high atop Manhattan's Chanin Building, an Art Deco skyscraper at the corner of Lexington Avenue and Forty-Second Street, Pan American Airways president Juan Terry Trippe was also looking at a globe. Founded in 1927, Pan American (or "Pan Am," as it would later be known) operated as the United States' exclusive international airline through the end of World War II. From an initial 90-mile route between Key West and Havana, it had grown at an astonishing pace during its first decade, thanks to generous Post Office subsidies and diplomatic support from the State Department. After first expanding throughout the Caribbean and Central and South America (not only by opening its own routes but by buying out local carriers in the region), in 1935 Pan Am inaugurated flights across the Pacific Ocean to the Philippines and China; in the spring of 1939, it began offering transatlantic service to England, France, and Portugal. That summer, even as war threatened to engulf the European continent, Trippe planned further expansion. Using pieces of string, of which his secretary kept a constant supply, he stood before his globe and measured out possible routes. Though smaller than the one at Dinner Key Terminal, this, too, was no ordinary globe. A family heirloom that dated from the 1840s, Trippe's globe proved too large to fit through the Chanin Building's doorway and had to be hoisted by crane through a window. Once installed, though, the globe served Trippe's purposes very well. The most famous image of him, a 1941 photo from *Life* magazine, shows him peering down at his miniature world, absorbed in contemplation or calculation. Other Pan Am executives reportedly kept globes in their own offices—although these, naturally, were not as storied or as magnificent as the one belonging to the airline's president.[3]

As Trippe plotted world air routes on his personal globe, and as thousands of Americans viewed the Dinner Key globe or the many other cartographic images that pervaded aviation culture during the mid-twentieth century—in local newspapers and national magazines such as *Life;* in airline advertisements, timetables, and route maps; and in the dozens of popular films on aviation produced by Hollywood studios—they were learning to see the world, and the place of the United States within it, in new ways. Although air travel (and especially international air travel) remained prohibitively expensive for most people until well into the jet age, popular fascination with the airplane, along with the sheer pervasiveness of aviation discourse and imagery in American culture, made flight an especially powerful metaphor for the United States' rise as a global superpower. The airplane, of course, was also a conduit of power—military, economic, and ideological—and the nation's worldwide infrastructure of commercial air routes, along with its hundreds of overseas air bases, proved critical to its twentieth-century ascendancy.[4]

To underscore the central role of aviation in facilitating the United States' expanding global power and influence, in April 1943 *Fortune* magazine published a lengthy feature article entitled "The Logic of the Air." In stark contrast to the headlines of World War II, which daily reminded Americans of aviation's destructive capacities, *Fortune* reassured its readers that the airplane could also be a catalyst of world peace and prosperity. The "logic of the air," the article explained, "is centered in the fact that the air is a blue-water ocean to which every nation potentially has access for trade and high strategy in all directions. Under its intoxicating implications the ancient ideas of a world divided by land and sea seem to be as outmoded as the Chinese wall."

In *Fortune*'s view, the "intoxicating implications" of aviation derived not only from its ability to expedite global commerce and augment geopolitical power. The "logic of the air" had, in fact, created "a new geography—a geography unbound from the flat Mercator projection that has dominated human thinking for centuries." The airplane, *Fortune* noted, placed all of the world's cities within twenty-four hours' reach of one another, making national borders increasingly irrelevant to the worldwide movement of products and peoples. Indeed, the "logic of the air" appeared to diminish the political and cultural significance of territory itself. In the air-age world, *Fortune* argued, "the traditional boundaries of political sovereignties seem to be on the verge of extinction." Traditional forms of diplomacy and geopolitics also seemed obsolescent. To illustrate the point, *Fortune* asked its readers to imagine that "every officer and employee of the State Department were to be fired tomorrow and replaced by men and women who were flyers or long and expertly acquainted with the flying business." The resulting "airman's foreign policy" could be "disastrous" or "magnificent," but in either case, it would be "wholly different from anything we can now expect."[5]

Published during a world war in which airplanes unleashed destruction on an unprecedented scale, *Fortune*'s utopian vision of the air age may now appear absurd, even delusional. But in 1943, this vision was far from marginal or unusual. *Fortune* was the brainchild of Henry R. Luce, the publishing magnate who also helmed *Life* and *Time*, and its oversized, lavishly illustrated pages broadcast views that were widely held among America's corporate and political elites. In particular, *Fortune* and the other Luce publications played a key role in elaborating and popularizing a new way of seeing the world—an optic through which Americans came to envision their nation as a global power unlike any other in history. This optic, as "The Logic of the Air" suggests, was an aerial perspective, which looked down on the world as a unified, unbounded sphere. Yet this way of seeing the world also yielded visions of national greatness, which depicted the world as a unified, unbounded sphere of American influence. Just as the airplane facilitated global interconnectedness, the "logic" of the air helped Americans to understand and literally to visualize the United States' global ascendancy.

In imagining aviation as the optic and catalyst of a new world order, *Fortune* was articulating what had become, by 1943, a prevailing view among U.S. government and business leaders. In fact, "The Logic of the Air" received enthusiastic endorsement from Assistant Secretary of State

Adolf A. Berle Jr., the Roosevelt administration's point man for international aviation policy, who opined that the article offered "an intelligent and balanced presentation of the problem and one that should be a most helpful contribution to a proper understanding of it." Aviation industry leaders made similar arguments about the airplane's effects on international relations. "We are entering the Air Age—the Age of Flight—when not just a single nation or a single continent, but the entire globe will soon be one neighborhood," proclaimed Juan Trippe, a former Yale classmate and personal friend of Henry Luce.[6]

Discourse on aviation thus framed the policies, strategies, and ideas that propelled the United States' ascendance as a global power. Invented on U.S. soil, the airplane promised not only to make the world more interconnected but also to extend the nation's "manifest destiny" to the skyways. As airplanes transported American people and products to places that were no longer distant, the "logic of the air" became the commonsense logic of a type of American empire—an empire based, primarily, not on the direct control of territory but on access to markets, on the influence of culture and ideology, and on frequent military interventions in other countries, all of which the airplane facilitated. Aviation both created and legitimized a new international order in which power itself was increasingly defined in extraterritorial terms, consistent with U.S. foreign policy objectives. The air age thus inaugurated what Henry Luce famously called the "American Century"—a vision of the world united under the aegis of U.S. hegemony.[7]

Empire of the Air reframes the global history of the American Century by following the course of U.S. aviation from the early 1900s through the 1970s, from the pioneering flights of the Wright brothers through the mass journeys of the jet age. Aviation accelerated the worldwide transmission of American capital and consumer goods, popular culture, technical expertise, and weaponry. As the U.S. military established air bases around the world, as airlines such as Pan Am and TWA dispatched jetloads of tourists on foreign soil, and as Boeing, Douglas, and Lockheed became the world's leading suppliers of both military and civilian aircraft, few places remained distant from Wall Street and Washington. Discourse on aviation, meanwhile, gave rise to visions of the entire world as the United States' appropriate sphere of influence, reconciling expansionist ambitions *[handwritten margin notes:] Airplane facilitated a new American empire in commerce + markets under U.S. aviation*

with Americans' traditional antipathy to imperialism. By creating an ostensibly extraterritorial "empire of the air," the airplane appeared to offer the United States an empire without imperialism—an empire for the American Century, based on markets rather than colonies, commerce rather than conquest.

This book's central protagonist is Pan American Airways, the United States' exclusive international airline through 1945 and the nation's flagship airline until its slow demise that began during the early 1970s with financial crisis and leadership turmoil. Yet this is not simply a corporate history. Rather, Pan Am's dramatic rise and fall serves as a lens onto the broader historical developments that *Empire of the Air* explains: the critical significance of aviation (and technology more generally) in facilitating and legitimizing the United States' ascendancy; the inextricable interconnections between the history of civil and military aviation; shifting patterns of state-corporate cooperation and the important role of private contractors in implementing Washington's strategic and diplomatic objectives; aviation's extraordinary cultural salience as a symbol and catalyst of both nationalism and globalism; and, not least, foreign responses to the United States' expanding global presence. Interwoven with Pan Am's story are those of an eclectic cast of other characters, from the Wright brothers and Charles Lindbergh to West African mechanics, Afghan technocrats, and "ordinary" American air travelers and aviation enthusiasts.

Pan Am's changing corporate logo neatly captures the chronological arc of this book. The original logo, from 1927, portrayed a winged Western Hemisphere, reflecting the airline's initial expansion throughout the Americas. During World War II, this had become a winged Atlantic world that also encompassed Europe and Africa, in light of Pan Am's transatlantic service and wartime trans-African military supply route. In 1957, Pan Am introduced its iconic jet-age logo, which suggested its worldwide reach: an abstract, sky-blue globe depicting no continents, only white horizontal lines indicating latitudes and "PAN AM" emblazoned in boldface type across the equator.[8]

In the 1910s and 1920s, the development of aviation reanimated older nationalist narratives about Americans' special aptitude for technological innovation, individual self-sufficiency, and the conquest of frontiers. In spite of aviation's transnational origins, by the late 1920s the accomplishments of the Wrights and Charles Lindbergh had transformed the airplane into an American icon, identified with ostensibly unique national characteristics. During the interwar years, aviation expanded the United States' power in its traditional sphere of influence, the Western Hemisphere. Commercial aviation proved particularly conducive to U.S. objectives in the region, as it elevated the nation's economic, political, and cultural status without the overt military trappings of "Yankee imperialism." It was during World War II, however, that the United States began to project its power on a truly global scale. This period also marked the coherence of American globalism as such—and ideas about aviation, as well as its striking visual iconography, played a key role in clarifying the United States' role in the war and the postwar international order. The international aviation policies formulated at the end of the war enabled the United States to extend its "open door" free trade doctrine to the commercial airways, and subsequently the skies became the circuitry of U.S. global power. As the Cold War national security state funneled a steady stream of federal funds to the aerospace industry, the airways facilitated worldwide flows of American capital, consumer goods, tourists, technical advisors, and weaponry.

The 1950s and 1960s have been characterized as the golden years of air travel, when sleek Boeing jets, soaring airport architecture, and glamorous stewardesses embodied the confidence and optimism of the ascendant American Century. Even at its zenith, however, America's empire of the air grappled with unexpected limits and unintended consequences. Jet lag weakened individual bodies; foreign travel by U.S. citizens exacerbated the nation's balance-of-payments problem; and, by the 1970s, the airlines' investments in expensive jet fleets had plunged them into deep debt. Complicating notions of American exceptionalism, the jet age also gave rise to transnational identities and communities, such as the cosmopolitan "jet set," and aviation globalized the United States itself by expediting immigration and imports. The jet age represented the triumph of the American imperium but also the ineluctable reality of imperial overreach.

The story of how airlines such as Pan American Airways, in partnership with Washington, imagined, wielded, and marketed a particular

vision of global power thus sheds fresh light on the origins and implications of the United States' ascendancy. The cultural "logic of the air"
seemed to justify American dominance of the world, and aviation, in
turn, implemented and sustained the infrastructure of American empire. Importantly, though, other nations did not passively submit to U.S.
hegemony, either in the air or on the ground. Using case studies from
Latin America, West Africa, Afghanistan, and both sides of iron-curtained
Europe, the following chapters show how other governments used aviation to further their own national agendas, which often conflicted with
those of the United States. By the early 1970s, moreover, the state of the
airline industry itself indicated that U.S. political and economic power
was not nearly as coherent, as confident, or as uncontested as leaders such
as Luce and Trippe had once believed. The declining financial health of
flagship airlines including Pan Am and TWA, as well as the new threat of
hijacking, suggested that the aviation industry's onward-and-upward ambitions might be unsustainable. Aviation, then, not only elucidates the
ascent of the American Century but also reveals its turbulence and possible decline.

Looking Down on the World

Aviation's profound, seemingly universal cultural allure rendered it particularly compelling as a site for the formation and refashioning of power.
During the first half century of the airplane's existence, people around
the world were fascinated by flight, whether they welcomed the airplane
as an instrument of peaceful commerce or feared it as an agent of mass
destruction. Commentary on aviation could be read or heard in political
speeches, advertisements, newsreels, magazines and newspapers, film and
radio productions, and museum exhibitions. Airplanes inspired religious
sermons, novels, plays, poetry, paintings, songs, board games, clothing
styles, and countless household objects from "Lucky Lindy" cigarette
lighters to aerodynamic toasters. This rich and pervasive culture of aviation brought the airplane into the daily lives of millions, even though only
a small percentage of the world's population had actually flown by the
mid-twentieth century. As Robert Wohl has written, "The irony is that in
comparison with other technologies, such as electricity, the telephone,

the automobile, the cinema, or the radio, the airplane had little or no immediate or direct impact on the way that most people lived their lives; yet its invention nonetheless inspired an extraordinary outpouring of feeling and gave rise to utopian hopes or gnawing fears."[9]

In the United States, the "logic of the air" inspired Americans to reimagine the world and the place of their nation within it. Aviation catalyzed the formation and the expression of a new global imaginary, a literal worldview that allowed Americans to conceive the entire globe as an object of analysis and sphere of influence. As Benedict Anderson has argued, in the early modern era the emerging culture of print capitalism—inexpensive, mass-produced books and newspapers written in vernacular languages—"made it possible to 'think' the nation," to imagine the populations of nation-states as naturally homogeneous and bounded communities. During the twentieth century, similarly, the culture of aviation made it possible to "think" the world, to imagine the globe as a single social entity. "There are no distant places any longer; the world is small and the world is one," wrote politician Wendell Willkie, a prominent Republican internationalist, whose 1942 round-the-world goodwill flight inspired his best-selling book *One World* (1943). President Franklin D. Roosevelt, noting that Americans "are today closer to other continents than Boston was to Mount Vernon in Washington's day," believed that his "Four Freedoms" should apply not just to U.S. citizens but to people "everywhere in the world." When adviser Harry Hopkins questioned the scope of the president's aims, stating, "I don't know how interested Americans are going to be in the people of Java," FDR replied, "I'm afraid they'll have to be. . . . The world is getting so small that even the people in Java are getting to be our neighbors now." Popular culture, and particularly print and visual culture, disseminated such globalist sentiments far beyond the elite ranks of airline passengers. As they read about air travel in *Life* and *Time*, marveled at the feats of Charles Lindbergh, or watched an American man marry a Brazilian woman aboard a Pan Am Clipper in the Hollywood musical *Flying Down to Rio* (1933), Americans perceived the world as increasingly small and interconnected.[10]

Yet if the airplane fostered global unity, it was also an instrument and symbol of national power, which could be deployed to lethal effect. Even before the atomic age, the use of strategic bombing in both world wars, as well as in numerous "smaller" conflicts (especially in the colonized world), emphatically demonstrated aviation's destructive capacities. Nations competed for dominance of the commercial skyways, too, and after 1919, international law designated the air itself as a form of sovereign national property with de facto borders corresponding to the territories below. Aviation augmented the military and economic power of certain nations over others and that of the United States above all. Invented by Americans and widely viewed as a symbol of national greatness, the airplane promised to extend the United States' frontiers "to infinity," as Trippe was fond of saying. By 1945, air routes mapped the geography of a nascent American empire—not a colonial empire but an empire of the air, a "market empire" (to use Victoria de Grazia's phrase) based on commerce and capital and backed by a formidable military arsenal. The "logic of the air," which rejected the significance of territoriality, became the logic of a world order marked by the simultaneous decline of Europe's colonial empires and the ascent of the United States.[11]

It is no coincidence, then, that Henry Luce's "American Century" editorial—perhaps the most influential midcentury formulation of American globalism, published in *Life* on February 17, 1941—used the airplane as a metaphor for the American Century. After World War II, Luce argued, the United States stood to become "the most powerful and vital nation in the world," which therefore had a right "to exert upon the world the full impact of our influence, for such purposes as we see fit and by such means as we see fit"—including "the right to go with our ships and our ocean-going airplanes where we wish, when we wish, and as we wish." Consequently, Luce called upon U.S. citizens to embrace "a truly *American* internationalism," an internationalism that would become "as natural to us in our time as the airplane or the radio." However, he emphasized, U.S. global power would derive not from conquering some "vastly distant geography" but from "big words like Democracy and Freedom and Justice." The air routes and air waves, meanwhile, would disseminate those "big words" to foreign lands, transforming them in the United States' image without colonizing so much as a grain of sand.[12]

As "The American Century" suggested, the vision of "one world" proved eminently compatible with visions of American hegemony. Nationalism and globalism, often seen as opposing forces in political culture, were, in fact, deeply enmeshed. Luce's idea of the American Century can best be described as a form of what John Fousek has fittingly termed "nationalist globalism": a global imaginary that represented the world as one but also endowed the United States with exceptional national characteristics and unique entitlements to global power. Nationalist globalism conflated the world's interests with the United States' interests. Its discourse cloaked American exceptionalism in a rhetoric of universality; its grand strategy combined overt multilateralism and commitments to international institutions with covert (and sometimes not-so-covert) unilateralism. As the dominant ideology of U.S. foreign policy during the twentieth century, nationalist globalism enabled the United States to attain a preponderance of global power even as Americans denied imperialist intentions. Nationalist globalism eschewed the label of empire yet simultaneously authorized the projection of nearly limitless extensions of influence over the earth's surface.[13]

The airplane allowed the United States to look down on the world both literally and figuratively. On the one hand, the aerial optic could deepen feelings of connection with other lands and peoples, as Wendell Willkie discovered during his 1942 *One World* flight: seen through an airplane window, the earth was unified by a "navigable ocean of air." On the other hand, this lofty vantage point could foster a sense of aloofness and superiority, as writer and aviator Anne Morrow Lindbergh emphasized in a 1948 *Harper's* essay chronicling a commercial flight to Europe. "It occurs to me that America looks on Europe as an air-observer looks on the earth below," she wrote. "He is near it—yes; he can see it spread out at his feet . . . but he is insulated from it, by his power and his freedom, by his mechanical and material genius, by the infinite resources at his ground base, which keep him safe above it, at this great height, from which he can observe so well—and need feel so little." Exemplifying what Mary Louise Pratt has termed "planetary consciousness," the aerial optic seemed to offer objective knowledge of the world but did not necessarily encourage, or even allow for, identification with it. Miles above messy on-the-ground realities, the airborne American could remain "comfortable, well-fed, aloof, and superior," as Lindbergh wrote. Cultural theorist Bruce Robbins has similarly identified the aerial optic as a position of power—the subject position favored by the architects of the American Century. The "assumed mobility of the view from above," Robbins argues, gave rise to "a set of expectations about the openness and submissiveness of the world," which in turn seemed to justify (and even necessitate) "a period of unprecedented American hegemony over the rest of the planet." Historian Michael S. Sherry also attributes the global violence of the air age to distance: the bomber pilot's physical distance from the earth prevented him from seeing his targets as human beings, and the analogous intellectual distance of science and grand strategy allowed aeronautical engineers and policymakers to deny the moral consequences of the lethal technologies they invented and unleashed. In Anne Morrow Lindbergh's words: "That is the terrible thing—the curious illusion of superiority bred by height. . . . The illusion of irresponsible power."[14]

Significantly, the view from above was not merely metaphorical. The corporate, government, and military leaders who helmed the United States' twentieth-century ascendancy—men including Henry Luce, Juan Trippe, and Adolf Berle—were frequent fliers and citizens of what Walter Kirn, in his novel *Up in the Air* (2002), has memorably dubbed "Airworld." Comfortably ensconced in cockpits or cabins, shuttled to and from airports in limousines, waited on by flight attendants, huddled over papers in hotel bars and conference rooms, these men (and air travelers were, for much of the twentieth century, predominantly male) inhabited lofty social realms that were remarkably insulated from the world that their decisions so dramatically affected. The experiential gestalt of commercial flight goes far to explain its cultural and political salience. Replicating the subjective experience of air travel, the "moral geography" of the American Century placed the United States above and apart from the world while simultaneously expanding its access to the world.[15]

However, this emphasis on distance and insularity can easily be exaggerated. Indeed, the very notion of the United States as an "empire of the air" risks obscuring the extent to which its commercial and military airpower required the control of territory and resources—air bases and airports, factories, oil, metals, navigation and communications facilities— and involved diverse types of "on-the-ground" international interactions. Relations of power, after all, emanate from proximity as well as distance, from international contact and integration. Moreover, although Americans claimed to eschew territorial conquest, the United States' ever-expanding global footprint—in the form of military bases, diplomatic missions, non-governmental organizations, tourism, and commerce—involved everyday close encounters between U.S. and foreign citizens, both official and unofficial, formal and informal. The worldwide expansion of U.S. aviation both necessitated and facilitated such encounters. And although bomber pilots and scientists may have remained distant from those affected by their work, the United States' global aviation infrastructure brought people from all walks of life, in hundreds of nations and colonies, into intimate contact with Americans and their airplanes. The "empire of the air" was, in significant ways, deeply grounded: rooted in human interaction (often in the context of highly unequal relations of socioeconomic and geopolitical power) and implanted on the earth. The sense of superiority felt by the American "air-observer," Anne Morrow Lindbergh noted, derived not only from his lofty perspective but also from "the infinite resources at his ground base."[16]

The history of aviation thus unveils the dialectic of distance and proximity from which U.S. global power ultimately emanated. The airplane imposed distance between Americans and others, giving rise to what Lindbergh described as "the curious illusion of superiority by height." Yet the airplane was also an inherently globalizing technology that appeared to reduce the significance of territorial boundaries, creating a world with "no distant places." And in this world, no place remained too far away to be touched by U.S. influence.

The Airplane, the United States, and the World

Aviation was hardly the first technology to facilitate expansionist projects. Roads were the arteries of the Roman Empire; innovations in shipbuilding brought the galleons of Spain, Portugal, and England to the New World; and, of course, Great Britain's unrivaled nineteenth-century naval power enabled it to acquire and control an empire on which the sun, famously or infamously, never set. As with the airplane, previous advances in transportation technologies had inspired visions of global unity. After steamships shrunk the voyage between New York and London from five weeks to ten days, commentators on both sides of the Atlantic predicted that "education conducted by steam" would lead to "the adoption by one nation of such customs as were found worthy from another—the amalgamation of languages—the spread of knowledge—the exchanges of commerce," as New York magazine *The Knickerbocker* stated in 1835. Like air travelers of later generations, steamship passengers claimed that their journeys fostered cosmopolitan sensibilities. "The ship is a little world, where individuals of different nations are thrown together in such close and familiar relationship," wrote an American traveler in 1852. Travel writers of the steamship era imagined the ocean, as twentieth-century travel writers would imagine the air, as a neutral space, where geopolitical disputes would dissolve or evaporate. "How natural it is for men to feel kindly toward one another when there is no miserable competition of earthly interests to set them in conflict!" exclaimed an American minister in his account of a voyage to Europe in 1836.[17]

Even in its own time, the airplane was not unique in its ability to influence how Americans imagined the world. At the turn of the twentieth century, an efflorescence of new transportation and communications technologies revolutionized perceptions of time and distance. Wireless telegraphy and the telephone made possible rapid communication across vast spatial expanses; radio brought distant voices into living rooms; cinema screens revealed close-up glimpses of faraway places. Nor was technology the only catalyst of American globalism during the so-called air age. In the first few decades of the twentieth century, a "vogue" for fashions and handicrafts from Asia and Latin America brought traces of the exotic into provincial households. Progressives, Wilsonians, and Marxists of the Popular Front era forged connections with social reformers in other countries, resulting in a robust, transnational cross-current of ideas about how to improve modern society. What W. E. B. DuBois famously called "the problem of the color line" was transnational in scope, and throughout the twentieth century, black intellectuals, activists, and artists—including DuBois himself—linked the African American freedom struggle to anticolonial movements in Asia and Africa. And a plethora of nongovernmental organizations, as well as new international institutions such as the United Nations, formally committed the United States to a leadership role in an increasingly global economic and legal system.[18]

Yet if aviation was not unique in its ability to inspire global consciousness, it nonetheless did so in distinctive ways. Unlike radio or film, the airplane was also a weapon of mass destruction, and as such it evoked the implications of global interconnectedness with particular starkness and urgency. Aviation's ability to serve as an instrument of both "hard" and "soft" power also rendered it a particularly effective instrument of U.S. global hegemony. The commercial airplane's benign cultural connotations enabled Americans to reconcile themselves with, and sometimes to ignore, the military airplane's destructive capacities. Ironically, even as the airplane enabled the United States to practice increasingly destructive methods of warfare, discourse on aviation did important cultural work to sustain conceptions of national innocence and benevolence, including the long-standing notion that the United States fights only defensive wars.[19]

Americans were not alone, of course, in their fascination with the airplane as a symbol of national identity and a means of global influence. In Germany during the Weimar era, the Zeppelin dirigible became a nationalist icon that the Nazis later appropriated; in the Soviet Union, aviation served as a barometer of both industrial progress and independence from the capitalist great powers; in Peru, elites believed that the airplane would civilize the Amazonian jungle and elevate their nation to the ranks of the world's modern great powers. By World War II, Western Europe's major colonial powers—Britain, France, Belgium, and the Netherlands—had all become "empires of the air" in their own right, with government-subsidized national airlines (Imperial Airways, Air France, Sabena, and KLM, respectively) that linked metropolitan capitals to far-flung colonial outposts in Asia and Africa. Some of the earliest uses of strategic bombing also occurred in colonial settings, as imperial powers used airplanes to suppress insurrections and intimidate local populations. And as vividly revealed by a 2007 exhibition of international airline posters at the Smithsonian Institution's National Air and Space Museum, the culture and visual iconography of Western European aviation blended nationalist and globalist imagery, symbolism, and rhetoric, as did American representations of air travel.[20]

Importantly, however, just as aviation was distinct from other globalizing forces, American fascination with the airplane had distinctive meanings and consequences. In both cases, the distinguishing factor was power: the power of the airplane to destroy; the power of the United States to implement its particular vision of the air age on a worldwide scale. After 1945, the United States' unrivaled economic affluence, cultural influence, and military strength lent unique weight to American interpretations of the "logic of the air." In other words, although Americans were not alone in envisioning a global air age, U.S. material and ideological preponderance during the mid- to late twentieth century ensured that the air age would indeed be an "American Century." The United States built and maintained the world's largest network of overseas military bases (which numbered 666 in 2012, according to the Department of Defense's annual Base Structure Report), sent more tourists abroad than any other nation, flooded foreign markets with American-made consumer goods and popular culture, and crafted international institutions, laws, policies, and programs that favored U.S. interests—including those of airlines and aircraft manufacturers, as well as the national security state on whose patronage the aerospace industry depended. Just as the airplane's destructive powers gave it special potency as an agent of globalism, the United States' preponderant power during the latter part of the twentieth century enabled it to shape the air age in its own image. Furthermore, American cultural attitudes toward aviation have been distinctively characterized by optimism, national self-confidence, and even evangelical fervor—as Joseph Corn has argued in his appropriately titled book *The Winged Gospel* (1983). In sharp contrast to most of Europe and many nations in Latin America, Asia, and Africa, the continental United States has never experienced aerial bombardment, and even the Japanese attack on Pearl Harbor in December 1941 did not diminish Americans' romance with the airplane. Widely perceived as a technology that the United States had triumphantly invented and dominated, aviation rarely evoked, in American culture, the anxieties and dystopian visions that characterized its reception elsewhere in the world.[21]

The history of aviation thus offers a privileged view—like the aerial view itself—of the origins, character, and effects of the American ascendancy, as well as its challenges and limits. The airplane's meanings and uses embodied the defining modalities of twentieth-century U.S. foreign policy: nationalism and globalism, hard and soft power, security and commerce, distance and proximity. But aviation matters to twentieth-century history not merely because it projected U.S. power—sustaining the technocratic projects of expansion that Michael Adas has fittingly described as an "engineers' imperialism"—but also because it shaped other nations' responses to that power, which ranged from willing consent to selective adaptation to outright resistance. Aviation, in short, was never simply a tool of American interests. Signifying universalist conceptions of modernity, progress, and power, the airplane gave other nations a means of asserting their own interests even within a context of overall U.S. hegemony. John Krige has defined this dynamic as "co-production," a term that underscores multidirectional agency and creativity in relations of power.[22]

Viewing technology as central to the history of international relations in the twentieth century, this book integrates aviation into a growing body of scholarship on what has been called "the new global history," or history after "the transnational turn." This work has rightly critiqued, and worked to rescind, what Thomas Bender described as "the unexamined assumption that the nation was the natural container and carrier of history." Yet in advocating the globalization of history without historicizing globalization, historians risk simply replacing the nation with the world as history's new naturalized container. For the global is not the blank canvas upon which history unfolds but a historically constituted category in its own right. Starting from that premise, the project of this book is to explain how the global was conceived in American culture as a "thinkable" spatial register in the era of the United States' ascendance as a great power. Explaining how Americans came to envision and to desire the world at a time when the airplane made such visions and desires seem attainable, this is a history of "the world," as Americans have imagined it, as well as of the United States' actions and role in the international arena.[23]

1

The Americanization
of the Airplane

On December 17, 1903, from atop the windswept dunes of Kitty Hawk, North Carolina, two bicycle mechanics from Dayton, Ohio, accomplished the world's first sustained, controlled flight in a machine-powered airplane. The flight lasted for twelve seconds and covered a distance of 120 feet. Later that day, Orville and Wilbur Wright made three more flights, including one of fifty-seven seconds. The brothers' brief telegram to their father—"Success four flights Thursday morning"—scarcely conveyed the magnitude of their achievement. Only five observers witnessed the event: three men from the nearby Kill Devil Hills lifesaving station and two residents of nearby towns. Several weeks later, the Wrights released a statement to the Associated Press, but concerned about patent infringement, they divulged few details about how their flying machine actually worked. Lacking precise information, newspapers that reported on the Wrights' claims published stories rife with misinformation and rumors. The U.S. government, meanwhile, largely ignored the Wrights; believing them to be cranks, military leaders repeatedly rejected their bids to sell their flying machine to the War Department. For the next five years, then, few Americans had any inkling about what the events of December 17, 1903, would mean for their nation and the world.[1]

In Europe, meanwhile, scientists made important advances in aeronautics, and governments actively invested in the development of both military and commercial aviation. As a result, during World War I, the bombers and pursuit planes used by England, France, Germany, and Italy were considerably more advanced than aircraft produced in the United States (whose fledgling air service flew French- and British-made aircraft during the war). And by 1925, European commercial airlines, mostly government-operated, were flying mail and passengers between the continent's major cities and even to colonies in Africa and South America, whereas the United States had yet to develop a single international airline.[2]

By the end of the 1920s, however, the United States had become the world's leading aerial power. By 1929, U.S. airlines had carried more passengers than the airlines of England, France, Italy, and the Netherlands combined, airmail routes stretched from coast to coast, aviation stocks were soaring, and the American public had gone wild for the airplane, resulting in countless aviation-themed popular songs, films, books, magazines, and collectibles. "America has once again used her natural advantages to zoom ahead of her European competitors in the race for the empyrean of the aerial age," observed an editorial in the British trade journal *Airways*.[3]

Europe's postwar social and economic devastation, the comparative strength of the U.S. economy, and the Air Mail Act of 1925, which brought federal revenue to the nation's airlines, all contributed to the United States' rapid ascent into the skyways. Two events, however, proved decisive in catalyzing public and government interest in aviation: the U.S. Army Air Service's 1924 round-the-world flight and Charles Lindbergh's nonstop transatlantic flight of May 1927. Discourse surrounding these two history-making flights ultimately resulted in the national adoption of the airplane as a "child of America," as the trade journal *Aeronautical Digest* stated in 1922. Aviation had developed simultaneously in several nations; its most important innovations, after the Wright brothers' initial invention, came from Europe; and the airplane itself was a globalizing technology that appeared to diminish the salience of national borders. Yet in the American cultural imagination, aviation tapped into nationalist narratives about frontier conquest, manifest destiny, and American exceptionalism (the idea that the United States possessed a unique character and mission). The accomplishments of the Wright brothers, the U.S. Army Air Service, and especially Lindbergh appeared to prove historian Frederick Jackson Turner's 1898 "frontier thesis," which attributed American democracy and industriousness to the nation's pioneer history. But whereas Turner had pessimistically concluded that Americans had no more territorial frontiers to conquer, the airplane opened an entirely new frontier—a frontier of the air, filled with virtually unlimited promise.[4]

To be sure, aviation animated nationalist narratives in every country that developed the technology.[5] While the peaceful implications of commercial air travel encouraged internationalist or cosmopolitan sensibilities, the destructive capacities of military airpower, amply demonstrated in World War I, raised concerns about national security. Aviation's political economy, moreover, developed within the institutional framework of the nation-state: aeronautical laboratories were typically government funded, most of the world's airlines originated as state-owned utilities, and even in the United States, the aviation industry would not have gotten off the ground without federal support for research, manufacturing, and airmail carriage. Yet if the link between aviation and nationalism was not unique to the United States, the transnational culture of the airplane increasingly looked and sounded American. By the end of the 1920s, people around the world had witnessed the 1924 U.S. Army flight, danced the "Lindy hop," and marveled at Hollywood films such as Frank Capra's *Flight* (1929).

The Americanization of the airplane, however, was far from inevitable. That the Wright brothers had first flown on the sands of North Carolina did not necessarily mean that the United States was destined to win "the race for the empyrean of the aerial age." In fact, for the first two decades of the century, it seemed rather unlikely to do so.

The Wright Brothers and the Early Air Age
in the United States and Europe

The first American experiments in machine-powered aviation yielded highly discouraging results. In 1898, Samuel Pierpont Langley, a renowned astronomer and the secretary of the Smithsonian Institution in

Washington, DC, secured a $50,000 grant from the War Department to design a flying machine. Widely publicized, his experiments led many to believe that the ages-old dream of human flight was about to be realized. But in October 1903, when Langley catapulted his "Aerodrome" from the roof of a houseboat docked in the Potomac River, the ungainly machine plummeted directly into the icy water. A second attempted flight ended the same way, ending also Langley's career as an aeronautical inventor. Newspapers around the country made great sport of mocking Langley— the *Boston Herald* suggested that he try designing submarines, since his inventions seemed to have a penchant for water—and the War Department decided to eschew further investment in aviation.[6]

Thus, when the Wright brothers announced their own flight in December 1903, the same month that the Aerodrome made its second and final plunge into the Potomac, the American public had good reason to be skeptical about their claims. Orville and Wilbur's fears of patent infringement exacerbated such skepticism, as the brothers refused to divulge photographs or any information that might allow rival inventors to understand how their flying machine worked. And in 1905, when the Wrights invited a group of journalists to observe a flight, mechanical problems and inclement weather forced them to cancel the event. In the absence of reliable information, reporters resorted to speculation (the *New York Herald* described the Wright Flyer as a balloon) and, in one notable instance, outright forgery (*The Independent* magazine assigned Wilbur Wright's name to a 1904 article, "The Experiments of a Flying Man," which contained excerpts from two lectures that Wilbur had given but had not actually been written by him).[7]

The first eyewitness account of the Wright Flyer appeared on January 1, 1905, in a rather unlikely place: pages 36 through 39 of *Gleanings in Bee Culture*. The journal's publisher, beekeeper Amos Root of Medina, Ohio, had observed the Wrights flying at Huffman Prairie, a field near their Dayton home, and his report appeared alongside such articles as "Judging Honey at Fairs." Root had "recognized at once," he wrote, that the Wrights were "really *scientific explorers* who were serving the world in much the same way that Columbus did when he discovered America." Beyond his fellow apiarian enthusiasts, however, Root's testimony remained in obscurity. Meanwhile, leading scientific journals continued to discount the Wrights' claims. According to *Scientific American*, for example:

> Unfortunately, the Wright brothers are hardly disposed to publish any substantiation or make public experiments, for reasons best known to themselves. If such sensational and tremendously important experiments are being conducted in a not very remote part of the country, on a subject in which almost everybody feels the most profound interest, is it possible to believe that the enterprising American reporter . . . would not have ascertained all about them and published them long ago?

As the *New York Herald* put it, the Wrights "are in fact either fliers or liars. It is difficult to fly. It is easy to say 'We have flown.' "[8]

Though War Department officials stopped short of calling the Wrights "liars," they too were skeptical and dismissive. In December 1904, the Wrights met with their congressman, Representative Robert Nevin (R-Ohio), and informed him that they wished to sell airplanes to the government. Nevin instructed them to write a letter stating their intentions and promised that he would bring it to the secretary of war, William Howard Taft, and arrange an appointment for the brothers to meet with him. Illness, however, prevented Nevin from delivering the letter to Taft personally, and the War Department summarily rejected the Wright brothers' offer to "furnish machines of 'agreed specifications at a contract price.' " The department's $50,000 investment in Langley's Aerodrome debacle had been a public embarrassment; for years, moreover, it had received hundreds of letters from crackpots claiming to have invented flying machines. As aviation historian Tom Crouch has noted, "Yet another letter, this one from two 'inventors' who claimed to have solved the problem of the ages in the back room of a bicycle shop, was not calculated to impress."[9]

After a second rejection from the War Department in October 1905, the Wrights turned elsewhere—to interested parties from Britain and France. "It is no pleasant thought to us that any foreign country should take from America any share of the glory of having conquered the flying problem," Wilbur wrote, "but we feel that we have done our full share toward making this an American invention. . . . We have taken pains to see that 'Opportunity' gave a good clear knock on the War Department door." Since the War Department refused to "reopen the door it had slammed in our faces," in April 1906 the Wrights sold their first flying machine to a French syndicate, acting as a government proxy, for the sum of one million francs (about $200,000).[10]

Even in France, the Wright brothers encountered skepticism and hostility. Prior to World War I, Paris had been the world's capital of aviation, and the French maintained considerable national pride in their history of aeronautical innovation—which dated to 1783, when the Montgolfier brothers discovered how to fly a hot-air-filled balloon. Thus, when Wilbur Wright arrived in Paris in June 1908 to perform demonstration flights, he was widely perceived as a dubious foreign interloper. French aviators believed that the Wrights' flying machine would prove incapable of fulfilling the terms of their contract with their buyers, which required the aircraft to "ris[e] by itself from a hard ground," to make two flights of no less than 50 kilometers in length and one hour in duration, and to land "without damages."[11]

Initial French skepticism, however, would quickly turn to wild enthusiasm. In August of 1908, after spending two months reassembling his aircraft (which had been shipped overseas in parts), Wilbur Wright began flying at Le Mans, a field on the outskirts of Paris. Disproving his critics, the flights were sensationally successful. Spectators "nearly went wild with excitement," he reported in a letter to Orville. "You never saw anything like the complete reversal of position that took place after two or three little flights of less than two minutes each." The French aviators who had initially scorned the Wright brothers now sang their praises, and the press conceded that the United States now rivaled France as the world capital of the air age. "Our supremacy in flying is no longer questioned over here," Wilbur wrote. The Aéro Club awarded the brothers a gold medal, and *Les Sports*, a sportsmen's journal that had previously been critical of the Wrights, "is now one of our strongest boosters." Within a week, Wilbur Wright had become a celebrity in France. Thousands flocked to Le Mans, from millionaires and royals to peasants who traveled 50 miles by bicycle to get a glimpse of the flying machine. Wilbur was celebrated in song ("Il Vol," or "He Flies,"), verse, and portrait; images of the Wright Flyer, both realistic and caricatured, appeared in commemorative pamphlets and postcards sold by street vendors. "The furor has been so great as to be troublesome," Wilbur wrote his sister Katharine. "I cannot even take a bath without having a hundred or two people peeking at me."[12]

Shortly after Wilbur Wright's success in France, Orville achieved equivalent prestige in the United States. In February 1908, the Army signed a contract to purchase a single Wright B Flyer for the sum of $25,000. To demonstrate that the aircraft could meet the Army's specifications, in September 1908 Orville conducted a series of public flights at Fort Myer, Virginia, across the Potomac from the nation's capital. As at Le Mans, the flying machine attracted thousands of awestruck spectators, and once-skeptical newspapers and magazines now published effusive paeans to the Wrights. In an editorial dated September 11, 1908, the day Orville set a world endurance record by remaining airborne for seventy minutes and twenty-four seconds, the *Washington Post* described the sight of the airplane as "a delicious thrill," inspiring one of those "moments in every man's life when he is a poet." It was, the *Post* concluded, "a wonderful achievement for the man and a glorious triumph for the race of men, which has conquered the air, as it won the earth and the sea, even to the uttermost horizon." Yet triumph was soon marred by tragedy: on September 17, the Wright Flyer crashed, gravely injuring Orville (he would be hospitalized for seven weeks) and killing his passenger, Army lieutenant Thomas Selfridge, in the world's first airplane fatality.[13]

Even after the crash, however, the Wrights' popularity and acclaim continued to grow in both Europe and the United States. Wilbur remained in France during the fall and winter of 1908–1909, winning several awards including the prestigious Michelin Cup (for a flight of over two hours). In January, Orville and Katharine Wright joined their brother in the town of Pau, where over the next three months Wilbur trained French pilots to operate the Wright Flyer, and then for a tour of Europe and more successful demonstration flights in Germany, Italy, and England. When the Wrights returned to the United States in the spring of 1909, they arrived as national heroes. Initially ignored or mocked, their invention was now celebrated as a great patriotic achievement, the latest evidence of national pioneering and ingenuity. The brothers themselves were construed as quintessential Americans, embodying the small-town virtues of hard work and humility: "they are the same imperturbable 'men from home,'" wrote Arthur Ruhl in *Collier's*, who "still live in the little side street across the river from the main part of Dayton."[14]

Although the French had appreciated the Wrights' achievement more quickly than had their own fellow citizens, Americans now represented their flights as evidence of the United States' superiority over Europe in aviation. According to an article in *The World's Work*, flight in Europe had been merely "parade-ground experimental" before Wilbur Wright arrived, and the French "added nothing essential to the principles that the Wrights were practicing." The Wrights' own publicity materials contributed to this nationalist narrative by emphasizing that their airplanes were products of American resources and labor. "We send our lumber expert regularly into the West Virginia forests and there select our stock piece by piece. . . . Our cloth is made by the most reliable mills in New England," stated a Wright Company brochure from 1911. Ironically, the machine that had to cross the Atlantic to attain recognition had become a national icon.[15]

Indeed, the Wright brothers' celebrity catalyzed public excitement about all things aeronautical. When an airplane flew over Chicago in 1910, it attracted over a million spectators. By 1911, *Scientific American* reported, at least a dozen U.S. companies were manufacturing airplanes; fifty others were producing parts and supplies. During the 1910s, Americans began to see the airplane as not simply a technological marvel but a symbol of social progress. Aviation affirmed the Progressive faith in rationality and science as tools of social improvement. If humans could fly through the air, it stood to reason, they could certainly perfect their living conditions on earth. The airplane also inspired religious faith: as historian Joseph Corn has argued, Americans often wrote and talked about aviation in spiritual terms. The airplane, ascending into the heavens, could be a conduit to the divine—a "winged gospel" that brought humanity closer to God.[16]

During the 1910s, stunt flying and air races further stimulated public interest in aviation. In cities across the United States, crowds numbering in the tens of thousands converged on airfields to watch pilots compete for speed and endurance records or perform daredevil feats such as wing walking and "looping the loop." Sensational and carnivalesque, these events involved more showmanship than science, but they did test the technical capabilities of airplanes, producing some important milestones in aviation history. In a 1911 contest, pilot Calbraith Rodgers made the first flight across the United States in a modified Wright B Flyer named the *Vin Fiz*, after the soft drink company that sponsored his endeavor. For "birdmen" like Rodgers, aerial exhibitions offered glory, riches (Rodgers won a $50,000 prize donated by publisher William Randolph Hearst), and the affections of women. Flying was often lethal in this era, and although there were several well-known American female pilots—notably Harriet Quimby, who in 1912 became the first woman to fly across the English Channel, and Ruth Law, famous for her stunts and endurance records—the culture of aviation was highly masculinized, defined by daredevil machismo and hard living. Aviators' lives tended to be short: Rodgers died in a crash just five months after his record-setting *Vin Fiz* flight; Quimby, who replaced Rodgers as spokesperson for the Vin Fiz company, perished at a Boston air meet three months after her flight across the English Channel.[17]

A growing chorus of critics began to argue that stunt flying was harming the progress of "legitimate" aviation by associating it with spectacle and danger. In 1905, a group of wealthy sportsmen formed the Aero Club of America with the goal of making aviation safer and more socially respectable. During the next decade, the Aero Club licensed pilots, gathered statistics, sponsored competitions, lobbied Congress to fund aeronautical research, and awarded the prestigious Collier Trophy (which to this day annually recognizes "the greatest achievement in aviation in America"). Though stunt flying increased aviation's mass popularity, the Aero Club lent it greater legitimacy within the upper echelons of American society. Lobbying by Aero Club members also led to the creation of an important institution: the National Advisory Committee on Aeronautics (NACA), established by Congress as part of the Naval Appropriations Act of 1915. The epicenter of aviation research in the United States, the organization would become better known as NASA, the National Air and Space Administration, the name it adopted in 1958. Its original membership included professors from Columbia, Stanford, Johns Hopkins, and Northwestern, along with the secretary of the Smithsonian, the commanders of Army and Navy aviation, the chief of the Weather Bureau, and the director of the Bureau of Standards.[18]

Yet in spite of aviation's growing popularity and institutional backing, the pace of its development in the United States continued to lag behind that of Europe throughout the 1910s. In 1914, the United States had only sixteen aviation manufacturing firms, which produced a total of forty-nine planes. Compared with the governments of the major European powers, Washington showed little interest in aviation, either military or civil. In 1911 and 1912, for example, Italy spent $2 million to develop air-power (albeit largely due to its war with Turkey); France and Germany, which were not at war, each invested $1 million and Russia $900,000. The U.S. Army, by comparison, received only $125,000 for aviation development. Limited funding for aeronautical research also caused the United States to lag behind Europe. With an initial budget of only $5,000 per year, NACA could not produce research on par with that of the government-funded national laboratories of France, Britain, and Germany, which had all been established long before World War I. Three years after NACA was created, director Joseph Ames reported that the organization had neither "an established program" nor plans "to encourage the design of any type of airplane." He gloomily concluded, "our work is 99 percent clerical."[19]

World War I revealed how far Europe had surpassed the United States in the manufacturing and development of airplanes. In 1914, France possessed 1,400 military aircraft; Germany, 1,000; Russia, 800; and Great Britain, 400. The U.S. military, by contrast, had just those forty-nine planes. "It was quite a shock to look helplessly at Europe that year and remember that ours was the country of the Wright brothers," recalled Air Force General Henry H. "Hap" Arnold in his autobiography. To be sure, European governments had immediate incentives to build up their air forces. The continent had a long and recent history of conflict—the Franco-Prussian War had ended in 1871—and its geography enabled nations to use airplanes to bomb one another in a future war. After Louis Blériot flew across the English Channel in 1908, even Great Britain could no longer count on its "splendid isolation" from a continental war. The United States, in contrast, faced no aerial threats; neither Canada nor Mexico had developed air forces, and airplanes were not yet capable of crossing oceans. U.S. military leaders, moreover, had been committed to naval development since 1898, when the United States' resounding victory in the Spanish-American War had vindicated the theories expounded by Navy strategist Alfred Thayer Mahan in *The Influence of Sea Power upon History* (1890). Finally, the Wright brothers' continuing obsession with patent security had resulted in a string of lawsuits against their first real competitor, aircraft maker Glenn Curtiss, which effectively stalled progress in airplane design for years. Thus, when the United States entered the war in April 1917, its aircraft were so "antiquated in design and defective in workmanship" that U.S. pilots ended up flying British and French planes. Most military aviators received at least part of their training in Europe, and many American aviation schools employed Allied officers as instructors.[20]

A wartime boom in aircraft manufacturing proved as brief as the United States' participation in the conflict. In July 1917, Congress approved $640 million for Army aviation, then the largest specific-purpose appropriation to date; that year's appropriations for Navy aviation totaled $45 million. Also in July, NACA's Subcommittee on Patents produced a cross-licensing agreement that settled the Wright-Curtiss dispute by allowing manufacturers to share technical innovations. Freed from patent lawsuits and flush with military contracts, the U.S. aviation industry briefly thrived. By the November 1918 armistice, firms had manufactured 13,894 aircraft and 41,953 engines; the industry had grown from sixteen companies with 168 employees (in 1914) to over 300 companies employing 175,000 workers. Some 10,000 U.S. servicemen had qualified as pilots (out of 23,000 who entered flight school during the war), and many had proved their mettle in combat, flying with the Royal Air Force and France's elite Lafayette Escadrille as well as with the U.S. Army and Navy. During demobilization, however, the U.S. government canceled $100 million of contracts to aircraft manufacturers, reducing the industry to 10 percent of its wartime size. In 1922, just 263 planes were produced in the United States, compared with 14,000 in 1918. Many decommissioned pilots became "barnstormers" who purchased cheap surplus planes, toured the country, and earned a living by performing stunts and taking passengers for rides. Giving thousands of Americans a firsthand glimpse of flight or even a ride in the cockpit, the barnstorming phenomenon left enduring legacies: those who flew for the first time with the barnstormers included future TWA chief Howard Hughes, test pilot Tex Johnston, and *Enola Gay* pilot Paul Tibbets. But although barnstorming helped to further publicize aviation, it did little to resolve the industry's financial crisis.[21]

Criticism of federal aviation policy, or the lack thereof, steadily mounted. According to NACA chairman Joseph Ames, the cancellation of military contracts had struck a "suicidal" blow to aeronautical research and manufacturing. NACA had finally completed construction of its laboratory at Langley Field in Virginia, but as one observer reported in 1919, due to inadequate funding work "seems to be completely stopped," the facility "merely a storage place, in a state of great disorder." Lieutenant Colonel Billy Mitchell, deputy director of the Army Air Service and a decorated World War I combat pilot, became the most prominent and vociferous critic of the state of American aviation. In widely read books and magazine articles, he advocated an aggressive buildup of airpower, a new doctrine of warfare centered on strategic bombing, and an independent air force equal in stature with the Army and the Navy. In July 1921, Mitchell staged a spectacular demonstration of airpower when his First Provisional Air Brigade destroyed a captured German battleship during training exercises off the coast of Virginia. Yet the Army and Navy Board's official report on the bombing trials refused to concede Mitchell's argument that aviation had fundamentally transformed the nature of warfare, concluding instead that "the battleship is still the backbone of the fleet." Undeterred, Mitchell continued to proselytize for airpower, pressing his case to Congress and to the public even at the expense of his career. In 1926, he resigned from duty after being court-martialed for insubordination. But Mitchell had succeeded in winning key military allies—notably Henry H. "Hap" Arnold, commander of the U.S. Army Air Forces during World War II—who subsequently took up his cause. Aviation trade journals, too, reiterated Mitchell's argument that the underdeveloped state of U.S. aviation posed a threat to national security. "Even the children in all parts of Russia are building toy planes with rubber-band motors," *Aero Digest* derisively noted.[22]

For the United States to have become "the world's tail-enders" in aviation, as *Aero Digest* put it, struck many critics as all the more humiliating since Americans were "fellow citizens of the Wrights." The idea that the airplane was America's birthright—"a distinct advance in civilization given to the world by America," stated a 1918 NACA report—pervaded discourse on aviation during the early 1920s. To aviation promoters, the development of the technology became nothing less than a patriotic duty. The National Aeronautic Association (NAA), incorporated in 1922 as the successor to the Aero Club, dedicated itself to promoting "a patriotic and progressive aeronautical development program" that would enable "America, which was first in the air, back there in 1903, [to] be First in the Air in the 1920s." According to industrialist Daniel Guggenheim, a prominent aviation enthusiast, "America has been the creator of the airplane. She must not allow herself to be diverted from its full development." Aviation, "the most entirely and characteristically American of arts," exemplified the "peculiarly energetic and progressive psychology of American youth," Guggenheim argued. J. Rowland Bibbins, head of the U.S. Chamber of Commerce's Department of Transportation and Communications, similarly described the United States as a "speed-loving Nation . . . of immense production and population" whose future prosperity would depend on its ability to "develop long-haul transport at minimum cost." And *Aero Digest* attributed the airplane to a national "spirit of enterprise and progress," which had previously "produced the great telegraph, telephone, and railroad systems."[23]

If the airplane was, in fact, America's birthright, the nation was failing in its patrimonial obligations. The Wright brothers' 1908 flights at Le Mans and Fort Myer had galvanized enthusiasm for U.S. aviation on both sides of the Atlantic. After fifteen years of relative neglect, however, the nation's aviation industry would need more than the Wright brothers' past accomplishments to once again become "first in the air." It would need something bold and unprecedented: a flight around the world.

Around the World with the U.S. Army Air Service

Worldwide flights had been imagined long before they were possible. The cover of the first issue of *Aerial Age Weekly*, from 1915, depicted an airplane superimposed on a globe; its lead editorial proclaimed, "The prospects of American aeronautics are world-wide and wonderful." Harry F.

Guggenheim, the son of Daniel Guggenheim and a prominent aviation advocate, believed that air routes would become the conduits of a truly global world order, enabling "the economic development not of any particular empire, but of the entire world." Billy Mitchell kept a globe in his Washington office, which he used to demonstrate how airpower made possible (and, in his view, required) an expansion of the United States' defensive perimeter. Mitchell would lead visitors to the globe, Hap Arnold recalled, and then "run his finger from Alaska down to Natal, off the coast of Brazil, then up to the islands at the mouth of the St. Lawrence River [near New Brunswick, Canada]. He would tell all who would listen to him or read what he wrote, that air power made these locations 'the three-pronged suspension of our real national defense.'" Mitchell, Guggenheim, and *Aerial Age* all envisioned aviation as, at once, a catalyst of globalization and an instrument of American power and expansion.[24]

With such powerful advocates, it was not long before global flight became a reality. In the early 1920s, the Army and the Navy both began considering plans for an around-the-world flight. This would not be the first time that the U.S. military had toured the world. In 1907, President Theodore Roosevelt ordered the Navy to embark on a worldwide cruise that would display the nation's strength and expand its influence—particularly in the Pacific, where the acquisition of the Philippines in 1898, after the Spanish-American War, had rendered the United States a bona fide imperial power. For fourteen months, the Navy's "Great White Fleet" of sixteen battleships and some 14,000 sailors circumnavigated the globe. The flotilla announced to the world that the United States' naval power was second to none and that the nation had become a global actor of the first rank. Now military leaders believed that a round-the-world flight would accomplish similar goals. Army Air Service Major Herbert Dargue first proposed the idea in 1922, but in spite of endorsements from Billy Mitchell, Air Service Chief Mason Patrick, and Navy Aeronautics Chief Rear Admiral W. A. Moffett, the General Staff decided that the flight would be too costly. However, when aviators from Britain, Portugal, and several other nations announced their own intentions to fly around the world, the military brass realized that the United States would gain special prestige from being the first nation to do so. By the end of 1923, the worldwide flight was a go.[25]

The Army identified four objectives for the flight: to "gain for the Air
Service added experience in long distance flying, particularly the supply
problems connected therewith; to complete an airplane flight round-the-
world in the shortest practicable time; to demonstrate the feasibility of
establishing an airway around-the-globe; and, not incidentally, to secure
for the United States, the birthplace of aeronautics, the honor of being
the first country to encircle the world entirely by air." Yet there was
nothing "incidental" about the nationalist significance of the flight. Its
insignia, depicting two eagles appearing to swoop toward an image of
the globe, strongly suggested such connotations. And although the
Army downplayed the flight's military implications, choosing instead to
construe it as a stimulus to commercial aviation, the knowledge and experience gained as a result would clearly benefit U.S. aviation in any future war.[26]

Planning for the flight, which involved daunting logistical challenges,
began during the summer of 1923. The proposed route traversed the
earth's climatic and topographical extremes, from mountainous Alaska
and frozen Greenland to tropical Indochina and the deserts of the Middle
East. With the cooperation of the Navy and the Coast Guard, the Army
dispatched officers around the world to map potential routes, find suitable
landing areas, set up lines of radio communication, gather meteorological
data, and enlist local agents to assist the fliers. Once landing places were
selected, supplies—gasoline, tools, lumber, several hundred types of aircraft hardware, and rations for the men—were sent ahead on Navy ships.
An equivalent effort went into the selection of the planes and the men
who would fly them. The Army commissioned aircraft manufacturer
Donald Douglas, a thirty-two-year-old prodigy who had previously built
bombers for the Navy, to design four planes specifically for the world
flight. Rugged and simple to operate, the "Douglas World Cruisers" were
single-engine biplanes, with twin cockpits in which a pilot and copilot-
mechanic each had his own set of controls. Detachable pontoons permitted the planes to land on water; tanks capable of carrying 644 gallons of
fuel gave them a range of nearly 2,200 miles. The Army christened the
four planes the *Seattle*, the *Chicago*, the *Boston*, and the *New Orleans*, as if
the flight would represent all four regions of the United States to the
world. As pilots, Army officials selected four men from among 110 Air

Service officers rated as "superior": Major Frederick Martin (flight commander), Lieutenant Lowell Smith, Lieutenant Leigh Wade, and Lieutenant Erik Nelson. Four mechanics would accompany them: Staff Sergeant Alva Harvey, First Lieutenant Leslie Arnold, Staff Sergeant Henry Ogden, and Lieutenant Jack Harding, respectively.[27]

Diplomatic planning for the flight, which would make fifty-two stops on foreign soil, involved as many challenges as its logistics. In most cases, foreign governments granted permission for the Army fliers to land, but there were a few notable exceptions. The Chinese government initially refused to allow the Douglas World Cruisers to land on its territory because the planes belonged to a foreign army; China granted permission only after receiving State Department assurance that the flight's purpose was commercial and not military. The Turkish government issued similar objections and placed restrictions on where the planes could fly and land. The case of Japan created special hurdles. As the State Department sought permission for the Army fliers to land on Japanese territory, Congress was debating a bill to restrict Japanese immigration, and leaders in Tokyo saw no reason to let the American fliers in while the United States was planning to keep the Japanese out. Negotiations between Washington and Tokyo dragged on through the fall of 1923. In trying to explain Japan's intransigence on the issue, some State Department officials invoked the same racially charged tropes that had fueled the campaign to restrict Asian immigration. "This matter of foreign aircraft in Japan is a matter of such a degree of suspicion and chauvinistic nationalism as can scarcely be understood by those who have not had experience of the abnormal psychology of the Japanese in regard to questions of 'national defense,'" wrote one official from the Division of Far Eastern Affairs. The Japanese government ultimately did grant landing permissions for the Army fliers—but in April 1924, just after the flight began, Congress passed the Asian Exclusion Act, which barred Japanese from entering the United States.[28]

Adding to the irony of the United States sending airplanes to Japan at the same time that it closed its doors to the Japanese, the Army made the decision to circumnavigate the world from west to east, such that Japan would be the first non-American country visited by the Douglas World Cruisers. The other pilots attempting round-the-world flights in 1924 all planned to take the east-to-west route from Europe to Asia to North America. Army planners believed, however, that weather conditions would favor the opposite direction. By flying through northern Canada and Alaska in the spring rather than fall, the fliers would encounter milder temperatures, and they would presumably arrive in southeast Asia before the typhoon and monsoon season began. The final itinerary had the flight beginning in Seattle in April and proceeding up the Pacific Coast, with refueling stops in British Columbia and Alaska. The four planes would then cross the Bering Strait, overflying the Soviet Union (where, because the United States and the USSR lacked diplomatic relations, they could not land) and touching down on Japan's Kurile Islands, the first of several stops on Japanese territory. Proceeding west across Asia and Europe, the Douglas World Cruisers would visit China, French Indochina, Siam, Burma, India, Persia, Mesopotamia, Syria, Turkey, Romania, Hungary, Austria, and France. Then, to cross the Atlantic, they would fly northwest to England, the Orkney Islands, Iceland, Greenland, Labrador, Newfoundland, and Nova Scotia. According to plan, the four planes would return to their native soil in the early fall, landing in Boston and then flying across the United States back to Seattle.[29]

In spite of its careful planning, however, the beginnings of the Army's world flight were far from auspicious. Just two weeks after departing Seattle on April 6, 1924, amid a blizzard and thick fog, the *Seattle* crashed into a mountain near Chignik, Alaska. Pilot Martin and mechanic Harvey survived the crash and a harrowing week stranded in the Alaskan wilderness, but their wrecked plane was irreparable. The remaining three World Cruisers forged on, only to encounter their own share of near disasters during the next five months: a dicey emergency landing on Russian territory; overheated radiators, broken radios, and seventeen burned-out engines; punishing extremes of heat and cold that taxed both the planes and the pilots; and in the case of Lowell Smith, who took command of the flight after Martin's crash, a debilitating bout of dysentery. During one of the final legs of the flight, between the Faroe and Orkney Islands, the *Boston* plummeted into the frigid waters of the northern Atlantic Ocean after a sudden drop in oil pressure. Its crew (Wade and Ogden) emerged unharmed, but the *Boston* was destroyed in the process of being hoisted onto a Navy ship for repair. (The original Douglas World

Cruiser prototype, the *Boston II*, was flown to Nova Scotia so that Wade and Harding could finish the flight.)[30]

Still, the U.S. Army fliers fared better than their foreign competitors. By September 1924, aviators from Britain, Portugal, France, Italy, and Argentina had all failed in their attempts to circumnavigate the world, due to crashes or inadequate equipment and supplies. And on September 28, 175 days after they had departed, the *Chicago*, the *New Orleans*, and the *Boston II* triumphantly returned to Seattle before a cheering crowd of 50,000 people. In spite of losing two of its original planes, the flight had fulfilled the Army's goals of demonstrating "the feasibility of establishing an airway around-the-globe"—and bestowing on the United States "the honor of being the first country to encircle the world entirely by air." The two original World Cruisers had covered 26,445 miles in 363 to 366 hours of flying time, setting world records that included the first aerial crossings of the Pacific Ocean, the Atlantic Ocean from east to west, and the Yellow Sea.[31]

Around the world, the flight also succeeded in stimulating public interest in aviation. According to reporters, diplomats, and the fliers themselves—whose testimonies were compiled by journalist Lowell Thomas in *The First World Flight* (1925)—the event had "increased activity in aeronautical circles the world over" (to quote *Aero Digest*). At each port of call, curious spectators, many of whom had never seen an airplane, thronged the harbors and airfields where the Army planes landed. In Calcutta, recalled pilot Lowell Smith, "Flocks of natives gathered . . . to watch us and it took fifty policemen to hold back the mob." The southeastern Chinese port city of Amoy, where the Douglas World Cruisers were the first airplanes ever to appear, was "covered with Chinese people in every conceivable place to witness the flight . . . the harbor was jammed with sightseeing sampans and junks literally filled with Chinese," according to the commander of a U.S. Navy vessel. Invoking a well-worn stereotype, the U.S. consul in Amoy added, "there was a thrill in this extraordinary event which even stirred the Chinese out of their customary stolid indifference."[32]

Diplomatic and military officials also believed that the flight had helped improve foreign perceptions of the United States. According to a U.S. vice-consul in Persia, it was "of great value . . . in giving the Persians a better idea of the power and ability of the American nation." Since the Douglas World Cruisers' arrival, he claimed, Persians had begun to "show more respect for our government." In Bangkok, a Navy commander reported, the flight "increased to a considerable extent . . . the admiration with which the American people are regarded by the Siamese." An official from the U.S. legation in Vienna described it as "excellent propaganda for the United States." Indeed, the flight seems to have influenced perceptions of the United States even in nations that the Army did not visit. The U.S. legation in Bogotá, Colombia, for example, cited a laudatory editorial in the newspaper *El Tiempo* as evidence "of a more friendly attitude toward American achievements than the press of Bogota has at times exhibited."[33]

Even in Japan, objections to the Asian Exclusion Act deterred neither the government nor thousands of people from warmly welcoming the Army aviators. "Popular interest is keen," stated the U.S. ambassador in Tokyo, "and the marked cordiality of [the fliers'] reception . . . is occasioning considerable comment by the press which ironically contrasts their welcome here with the action of Congress in excluding Japanese from America." The contrast could not have been sharper. In the southwestern coastal city of Kagoshima, where the American pilots first landed on Japanese territory, a "committee of welcome" had spent weeks organizing a lavish reception for them. When they arrived on the evening of June 2, they were greeted by some forty thousand people, including "numerous officials of every rank" and students from local public schools, who waved American flags and sang "My Country 'Tis of Thee" in English. It was, copilot Leslie Arnold recalled, "one of the most impressive receptions we had anywhere in the world." In the other Japanese cities the fliers visited, residents "fairly overwhelmed us with courtesy and hospitality," Arnold said. In Minato, a crowd estimated at twenty to thirty thousand held up "huge 'welcome' signs"; in Tokyo, geishas pinned American flags to their kimonos, the entire Japanese diplomatic corps greeted the aviators at the Imperial Hotel, and the government arranged so many official banquets that the fliers remained there for two weeks. Arnold noted that the flight reached Japan "at a time when feeling was running high against Americans" but claimed that "we saw not the slightest evidence of it." The U.S. ambassador reported, "the Japanese have carefully avoided making any reference to the impending exclusion legislation and as a matter of fact appear to regard it as a matter of pride to make a distinction between the two matters." Japanese officials' magnanimous treatment of the Army aviators could very well have been a strategic way of underscoring the unfairness of the U.S. legislation by contrasting it to their own government's openness.[34]

The circumnavigation of the globe by air thus appeared to bring nations and peoples closer together, inspiring predictions that the air age would bring about world peace. The U.S. Army fliers themselves endorsed such ideas. "The speed with which men will fly from continent to continent will bring all peoples into such intimate contact that war will be as out of date as the cuneiform inscriptions of the ancient civilizations that the shadows of our Cruisers were passing over," declared copilot Erik Nelson. According to Army Air Service Chief Mason Patrick, aviation promised to bring about "better understanding among nations." Though airplanes were "terrific engines of war," they could also "do much to promote the peace of the world." Newspaper headlines, meanwhile, called the Army fliers "Modern Magellans" and "Magellans of the Air" (a comparison, incidentally, that the men disliked, as Portuguese explorer Ferdinand Magellan had died during his attempt to circumnavigate the globe in 1521). Journalists also predicted that the airplane would soon allow ordinary people to travel to other countries. In the words of a *Washington Post* editorial, published just after the Douglas World Cruisers returned to the United States: "Think of the pleasure, doubtless to be enjoyed by almost anybody before long, of getting aboard the family air flivver on a pleasant day, with wife and children, and making 'hops' over a few thousand miles of territory in the course of a week's or a fortnight's vacation!" Such articles often included images of maps and globes that graphically showed how the airplane would make the world "smaller" and more unified. The cover of *Aeronautical Digest* from January 1924, for example, showed the earth as if seen from above; on another *Aero Digest* cover, airplanes encircled the earth. These images of global unity suggested that the aerial perspective, a rarified view from above, would enable people to perceive the essential interconnectedness of all countries.[35]

Internationalist discourse and images, however, proved entirely compatible with nationalist arguments that interpreted the Army flight as evidence of the United States' distinctive national greatness. "America is the birthplace of aviation. It was first 'in the air' and it should remain first in the air," stated an editorial in the *Washington Post* from February 1924. According to the *New York Times*, "the United States gave the airplane to mankind ... and now it will enjoy the supreme triumph of being first round the world in the air." This idea that the United States deserved to be "first in the air," expressed in many articles on the Army flight, did not simply reflect the historical fact that the Wright brothers had invented the airplane. It also derived from, and built upon, older cultural narratives about frontiers and pioneering. As the *Post* pointed out, the U.S. Army had previously enabled the nation's westward expansion by "guard[ing] the trails for the covered wagons, the roads for the pony express, and the routes for the railroads." A century later, its round-the-world flight had proven that the Army "is still the pioneer in the newest method of transportation." By drawing a straight line of progress from pioneer wagons through the pony express and railroads to airplanes, the *Post* made it seem inevitable that the United States would conquer the new "frontier" of the air just as it had previously vanquished territorial frontiers. The article suggested that the Army's victory in the race around the world was preordained, the product of Manifest Destiny rather than careful planning and providential contingency.[36]

Expedited access to foreign lands, moreover, did not necessarily mean that Americans would accept cultures different from their own. The Army fliers' experiences, rather than changing their preexisting beliefs about foreign cultures, seem in some ways to have strengthened those beliefs—particularly those concerning Asia. In describing people and places they encountered in China, Japan, India, Siam, and French Indochina, the men invoked and reproduced cultural tropes that construed Asia as simultaneously decadent and primitive, beautiful and barbarous, desirable and frightening. Exoticized spectacles dominated their recollections. In Siam, the fliers reported seeing "white elephants," "sampans filled with naked people," and "20,000 Buddhist temples, golden-spired pagodas, and teak monasteries." The king invited them to witness a beheading; they declined. In Burma, they spent their evenings at nightclubs "where the European throws off the 'White Man's Burden'" in order to indulge in "a life of languid hours, whispered scandals, waving punkas, and clinking glasses." In Calcutta, they witnessed "ash-covered fakirs" and people "smeared with cow-dung and diabolic-looking red-and-white Hindoo symbols on their foreheads and bodies," whom they described as "the most hideous specimens of humanity that we encountered on the Flight." In Persia, they discovered "that most picturesque of men, the desert Arab." Similar tropes filled the pages of U.S. newspapers, which sensationalized the "Thrilling Adventures of 'Round the World Flyers" (to quote a *Hartford Courant* headline) as they entered "jungles where no white person had ever been" and explored "the most romantic countries in the whole of the mysterious continent of Asia." During the 1920s, such stylized images of Asia were much in vogue among middle- and upper-class white Americans, particularly women, who filled their homes with imported Chinese porcelains, Japanese screens, and Persian carpets. Discourse on the Army flight, infused with richly detailed descriptions of the exotic East, both reflected and reinforced American orientalism.[37]

Americans' views of foreign peoples and places, then, did not necessarily change after the airplane allowed them to see and to know the foreign firsthand. In the case of the Army pilots, the diplomatic and military officials who assisted them, and the journalists who covered their travels, encounters with the foreign tended to confirm preexisting beliefs. The airplane itself, moreover, could underscore conceptions of cultural difference. The language that U.S. officials used in describing Chinese and Indian reactions to the Douglas World Cruisers—"flocks of natives," "the mob," "jammed with sightseeing sampans," "covered with Chinese"— evoked older stereotypes of the "Asiatic horde," which the U.S. government had used to justify exclusionist immigration policies. Because individuation was widely understood to be a prerequisite for citizenship in a democratic society, the image of Asians as an undifferentiated mass proved highly effective in mobilizing opposition to Chinese and Japanese immigration.[38] In accounts of the Douglas World Cruisers, the six sovereign individuals operating the flying machines stood out against the "flocks" and "mobs" on the ground. Accounts of local reactions to the airplanes thereby served to distinguish the foreign from the American. "Intimate contact" with other nations did not necessarily mean that airborne Americans would come to view the world's peoples as equals. Nor did internationalist discourse on aviation replace claims that the United States was and should be "first in the air." Rather, new feelings of international intimacy coexisted with older nationalist sentiments.

The peaceful nature of the flight, furthermore, did not diminish its undeniable military significance. Not only did the endeavor demonstrate the U.S. Army's superiority in the air; its very logistics—the surveying of the route, the transportation of equipment, the construction of communications relays—were made possible by the naval supremacy that Teddy Roosevelt's Great White Fleet had exhibited seven years earlier. Showing how airpower could augment sea power, the flight served to reconcile Alfred Thayer Mahan's theories with Billy Mitchell's. Although the Douglas World Cruisers carried no bombs, their successful world flight suggested how the airplane and the battleship could work together to ensure American victory in a future war. The relationship between airpower and sea power would, for decades, remain a subject of great contention among U.S. military strategists, but the U.S. Army World Flight nonetheless foreshadowed how both the Army and the Navy would increasingly rely on airpower.

During the late 1920s, several key developments further strengthened the U.S. aviation industry, reanimating the airplane's symbolic associations with national greatness. In 1926, Daniel Guggenheim created the Guggenheim Fund for the Promotion of Aeronautics, which provided much-needed financial support for aviation education and research. By 1930, the fund had granted nearly $2.7 million for the establishment of aeronautical engineering programs at eight leading universities, an aviation law institute, and educational initiatives for elementary and secondary schools. The federal government also took unprecedented steps to stimulate aviation development. In September 1925, President Calvin Coolidge, acting on the advice of Secretary of Commerce Herbert Hoover, convened a nine-man President's Aircraft Board, chaired by financier Dwight Morrow, to study and make recommendations for all aspects of the nation's military and commercial aviation. Following the Morrow Board's recommendations, both the Army and the Navy issued five-year aircraft procurement plans, and federal expenditures on aviation doubled from $6 million in 1922 to $12 million in 1926.[39]

Most important for the development of commercial aviation, in February 1925 Congress passed the landmark Air Mail Act (also known as the Kelly Act), which authorized the Post Office to contract with privately owned airlines to deliver mail, paying them fixed rates for each pound of airmail. Although the Morrow Board had not recommended direct government subsidies to the aviation industry, as was the common practice in Europe, military and postal contracts channeled a steady stream of federal revenues to the nation's struggling aircraft manufacturers and airlines. By reducing financial risk within the U.S. aviation industry, these measures proved crucial to the early growth of companies including Boeing, Douglas, United Airlines, and Pan American Airways. And to meet the requirements of carrying mail, manufacturers came up with innovative new designs such as the Ford Trimotor, which could carry ten passengers along with mail sacks. Airmail, in turn, stimulated the U.S. economy by expediting the transit of mail and goods, by facilitating communication between a company's national headquarters and its local branches, and by reducing interest charges (by lessening the time required for checks to clear). In 1924, the assistant postmaster general estimated that airmail would save the New York Federal Reserve Bank $809,589 in annual interest charges on its transactions.[40]

The Morrow Board also recommended that the government regulate civil aviation by creating a federal agency to license pilots, enforce safety standards, maintain the nation's aerial infrastructure (such as lighting and navigational aids), and implement nationally consistent air traffic control procedures. Aviation insiders, including NACA and the NAA, airline executives, and trade magazines, had in fact been calling for such regulation for years. Although it may seem odd for an industry to demand its own government oversight, aviation leaders believed that consistent, enforceable laws and standards—on a federal level, since flights regularly crossed state borders—would increase public confidence in flying, in turn increasing the industry's profitability. Congress had considered aviation regulation bills every year since 1918, and in May 1926 it finally passed the Air Commerce Act, which authorized the Department of Commerce to create an Aeronautics Branch to regulate the technical aspects of civilian flying within the United States. Industry reactions were overwhelmingly favorable, and the *New York Times*, declaring that the legislation marked a "red-letter day" in aviation history, concurred with Herbert Hoover's prediction that within three years, the United States would have "the most complete air service in the world."[41]

Hoover was not far off the mark. The Air Mail Act and the Air Commerce Act together created the institutional foundations that enabled the U.S. aviation industry to take off in the late 1920s, just as the U.S. Army Air Service's world flight had rallied public enthusiasm for aviation and demonstrated the capabilities of American airplanes and pilots. For the United States to truly become first in the air, however, an additional and unexpected factor proved decisive: a twenty-three-year-old, St. Louis–based airmail pilot and stunt flier named Charles A. Lindbergh.

Airplane Diplomacy and the "Lone Eagle"

"We specialize in Fair and Carnival Exhibition Work, Offering Plane Change in Midair, Wing Walking, Parachute Jumping, Breakaways, Night Fireworks, Smoke Trails, and Deaf Flights," announced Lindbergh's business card from early 1927. But the young aviator aspired to accomplish something even more risky than the exploits of the barnstorming circuit: a nonstop flight across the Atlantic Ocean. Although pilots had crossed the Atlantic over one hundred times prior to 1927, most had flown dirigibles rather than airplanes, and all had made midway stops in Newfoundland and the British Isles. In 1919, Raymond Orteig, a French-born Manhattan hotelier and aviation enthusiast, offered a prize of $25,000 to the first aviator to fly without stopping between New York and Paris. Seven years later, the prize remained unclaimed, although several men had lost their lives in the attempt. Undeterred, Lindbergh not only rose to Orteig's challenge but upped the ante by proposing to fly solo across the Atlantic—a feat that had never even been attempted—in order to increase speed and maximize the amount of fuel that his airplane could carry. To many of his fellow aviators, the plan seemed suicidal, earning him the nickname "Flying Fool." Still, Lindbergh succeeded in obtaining financial backing from several prominent St. Louis businessmen (who viewed the flight as an opportunity to promote their city); Ryan Aeronautical, a startup based in San Diego, agreed to build him a plane capable of crossing the Atlantic.[42]

At 7:51 a.m. on May 20, 1927, Lindbergh and his single-engine Ryan monoplane, christened the *Spirit of St. Louis*, took off from Long Island's Roosevelt Field amid a steady drizzle of rain. Ascending above a large crowd of journalists, aviation enthusiasts (including a recent Yale graduate named Juan Trippe), and local residents, the tiny plane soon vanished into the clouds. Lindbergh's death-defying endeavor had by then been highly publicized, and throughout the day and night, much of the U.S. population waited by their radios for any news of the flight. It would ultimately take thirty-three and a half hours, passed mostly in darkness over open water. Lindbergh later said that he had remained awake only due to the periodic rolling and pitching of his plane. Although he had brought five sandwiches and a jug of water, he scarcely ate or drank; at one point, he experienced hallucinations of ghostly forms moving through the cockpit. It was a test of physical and psychological endurance as much as flying skill—a test that Lindbergh survived, he said, by imagining himself and his plane as symbiotic beings. Indeed, he consistently used the plural first person when speaking of the flight, titling his memoir *We*.[43]

When the *Spirit of St. Louis* finally landed at Le Bourget airfield outside Paris, an estimated 150,000 people formed a "human tidal wave" (as Lindbergh described it), rushing forth and literally ripping the plane apart in a quest for souvenirs. They pulled Lindbergh out of the cockpit, tore his helmet from his head, and carried him aloft until two sympathetic French aviators came to his rescue. Exhausted but in good health, Lindbergh had landed precisely where he intended, using only rudimentary navigational instruments, and the *Spirit of St. Louis* suffered more damage at Le Bourget than it had during its journey. In the United States, reactions to Lindbergh's flight reached levels of hysteria. "LINDBERGH DOES IT!" exclaimed the usually staid *New York Times*, which devoted its entire front page and the following four pages to the flight. Lindbergh had not only won the coveted Orteig Prize; overnight he became a global media sensation. The New York papers alone published nearly 300,000 stories about him during the first twelve days after the flight; motion picture cameras filmed some 7,430,000 feet of newsreel footage (two million more than for the Prince of Wales, previously the most documented newsreel subject); the *New York Times* received over two thousand reader-submitted poems; and songwriters commemorated the event in hundreds of popular ditties, including "Lindbergh (The Eagle of the U.S.A.)," "Columbus of the Air," and "Eagle of Liberty." The following summer, an estimated thirty million Americans—a quarter of the country's population—had the privilege of seeing their idol in the flesh during his eighty-two-city, cross-country goodwill flight sponsored by the Guggenheim Fund.[44]

More than any other single event, Lindbergh's solo transatlantic flight identified aviation with the United States. Widely interpreted as not just an individual feat but as a national triumph, the event inspired outpourings of patriotic sentiment. The *Spirit of St. Louis* "represented American genius and industry," stated President Coolidge (in a characteristically terse comment on the occasion). "As air-minded Americans we are proud that the New York to Paris air trail has been blazed by an American pilot in an American airplane," proclaimed *Aero Digest*. Among the hundreds of letters that Lindbergh received from ordinary U.S. citizens, many expressed nationalistic pride in his achievement: "I am glad that an AMERICAN under the STARS and STRIPES has accomplished this," wrote Ross H. Rohrer, a manufacturer of flour and cornmeal from Quarryville, Pennsylvania. A boy from Northampton, Massachusetts, wrote that Lindbergh's picture hung on his family's wall alongside portraits of Washington and Lincoln, in a pantheon of "our three great Americans."[45]

Why did Lindbergh's flight so beguile the American imagination? It was, to be sure, audacious and unprecedented, but so too were the Wright brothers' first flights and the Army world flight, neither of which elicited the same intensity of emotion. According to biographer A. Scott Berg, Lindbergh fulfilled a cultural need for a national hero—a need that Americans acutely felt in 1927. The handsome, clean-cut young pilot, who abstained from alcohol and tobacco, publicly expressed love for his mother, and appeared to live in celibacy, "stood in sharp contrast to the rest of the current newsmakers—bootleggers, racketeers, and millionaire playboys." On the day Lindbergh departed for Paris, U.S. newspapers carried grim headlines about the Teapot Dome scandal and an anarchist who had blown up an elementary school in Michigan, killing thirty-eight children and six adults. As historian John Ward has argued, Lindbergh's flight "came at the end of a decade marked by social and political corruption and by a sense of moral loss." In American political culture, the idealism and engagement of the Progressive years had given way to cynicism, malaise, and materialism. The "war to end all wars" had failed to produce the harmonious international order envisioned by Woodrow Wilson, the Harding and Coolidge administrations seemed to revere profit over ideals, and Prohibition had only increased lawlessness and violence. In this context, Ward wrote, Lindbergh "gave the American people a glimpse of what they liked to think themselves to be at a time when they feared they had deserted their own vision of themselves." Furthermore, in the context of post–World War I anxieties about the destructive consequences of modern technology, Lindbergh's feat represented a harmonious merger of man and machine, implying that technological innovation was not incompatible with individual heroism. In contrast to the haunting images of mechanized mass slaughter that had emerged during the recent world war, his solo flight seemed to indicate that individuals could still control the meanings and uses of technology—and harness it for peaceful purposes.[46]

Lindbergh's flight also resonated with older nationalist narratives. Its significance had much to do with the fact that Lindbergh flew alone. The "Lone Eagle" appeared to embody the virtues of Yankee individualism. Thousands of news stories and dozens of popular biographies portrayed Lindbergh as self-sufficient, thrifty, industrious, stoic, modest, and morally upstanding. In the words of one typically hagiographic biography, *Lindbergh: The Lone Eagle* (1927): "He saved what he earned, he neither drank nor smoked, he kept to himself." James West, leader of the Boy Scouts of America, authored *The Lone Scout of the Sky* (1927), which praised Lindbergh for upholding the dual purposes of his organization: "complete self-reliance, and at the same time, the desire, the will to share in the work of the community's welfare." Such representations of Lindbergh were deeply nostalgic. Although his transatlantic flight had depended entirely on modern technology, the "Lone Eagle" image evoked an imagined past in which the American character had been molded by hardship and hard work. Linking the nation's history to its future, Lindbergh's legend bridged the pioneer age and the machine age. Again, the air was imagined as a new frontier, such that aviation would steer Americans away from modern decadence and back to their pioneer roots.[47]

Journalists and biographers explicitly likened Lindbergh to the pioneers. He had grown up in rural Little Falls, Minnesota, on "country not
long since the hunting-ground of Indians," noted one popular biography.
Another biography described how Lindbergh's paternal grandfather, a
Swedish émigré, had settled in Minnesota and "built a log cabin on a pre-
emption of land." In the words of Boy Scout leader West, "The lone Path-
finder, blazing a trail through the arch of the sky, *called to the blood of the
pioneer in every American boy*" (emphasis in original). Lindbergh's own
boyhood, as such writers imagined it, conjured the legends of Tom Saw-
yer and Huck Finn. In Little Falls, he had reportedly spent his youth
hunting, fishing, and hiking in the woods with his dog. Young Charles's
father, "bred of hard pioneering stock," taught him how to use a rifle and
revolver and once took him boating up the Mississippi River. In reality,
though, his childhood was not exactly hardscrabble. His father, Charles
Lindbergh Sr., served as a U.S. congressman from 1907 to 1917, and the
family split their time between Little Falls and Washington, where
Charles attended the elite Sidwell Friends School. In his own autobiogra-
phies and memoirs, however, Lindbergh lent credence to his legend by
emphasizing his Scandinavian heritage, rural boyhood, and love of nature. "A wilderness lies beneath my wings," he wrote in a 1953 account of
his transatlantic flight. "My grandfather must have found a country like
this when he immigrated to America from Skåne in the southern part of
Sweden."[48]

Lindbergh's race and ethnicity added further layers of meaning to his
legend. While scholars have typically interpreted him as an American
national icon, he also served as an idealized racial type—an embodiment
of transnational whiteness. Coming at the end of a decade marked by immigration restriction, rising white supremacism, and cultural concerns
about the very boundaries of whiteness, Lindbergh's flight seemed, to
many observers, to confirm the vitality of the white race. Contemporary
journalists and biographers made much of his ethnic heritage: "he is half
Swedish and for the rest of him, English, Irish, French"; "he is a lanky,
six-foot blond Viking type"; he "seemed to be composed of a mixture of
the traits of his Swedish and English grandfathers." According to Richard
Beamish, author of *The Boys' Story of Lindbergh, the Lone Eagle* (1928), "it is
a fact worthy of note that the young American who, in this twentieth cen-

tury, was the first to fly from New York to Paris was the son of a Norseman; and the first voyager to cross the Atlantic, back in the eleventh century, was a Norseman." Transportation, Beamish stated, was "synonymous with the development of civilization," and like Leif Ericson, Lindbergh belonged to a long line of transportation pioneers that had begun with "the ancient Briton and Gaul padd[ling] his 'coracle' of skin stretched on wickerwork." Other writers called attention to his mother's French ancestry, and Lindbergh himself reportedly said that this ethnic heritage enabled him to understand and relate to the French people.[49]

At stake in Americans' interest in Lindbergh's genealogy was an imagined white identity that transcended national borders, bridging the old world and the new. His transatlantic flight appeared to perform a kind of racial rapprochement, healing recent political tensions between the United States and several European nations, notably Britain and France. Bringing "the Old World and the New within less than a day and a half's reach of each other," wrote Beamish, the flight would "bring our peoples together, nearer in understanding and in friendship than they ever have been." Similarly, a letter sent to Lindbergh from the employees of the post office in Wauseon, Ohio, described his achievement as "the discovery of the old world by the new." In this interpretation, Lindbergh not only brought the United States and Europe closer together; he also reversed the currents of progress, such that they now moved from new world to old. His flight seemed to confirm that the United States, not Europe, now stood on the leading edge of modernity, civilization, and technological innovation.[50]

Lindbergh flew to France, moreover, at a time when the United States desired rapprochement. Relations between Paris and Washington had recently soured because of the French government's debts to the United States (resulting from loans floated during the world war), the United States' increasingly restrictive tariff policies, and what the French perceived as American softness toward Germany's rearmament. Aviation, ironically, compounded these tensions. Two prominent French pilots, Charles Nungesser and François Coli, had disappeared over the Atlantic in their own attempt to capture the Orteig Prize just two weeks before Lindbergh's flight; in sensationalist (and inaccurate) reports, French newspapers claimed that the United States Weather Bureau had caused their disappearance by not providing adequate meteorological information. The U.S. ambassador to France, Myron Herrick, thus described the spring of 1927 as "one of those periods of petulant nagging and quarreling between the French and ourselves which have flared up and died down more than once since the Armistice." Herrick also warned the State Department that Lindbergh would only exacerbate French hostility toward the United States if he proceeded with his flight while Nungesser and Coli were missing and presumed dead.[51]

Yet the effects of Lindbergh's achievement were precisely the opposite of what Herrick had feared. Later the ambassador proclaimed that the flight had created "an entirely new atmosphere in the relationship between the countries." The aviator, he wrote, carried "the spirit of America in a manner in which it could never be brought in a diplomatic sack." The French Foreign Office saluted Lindbergh by flying the American flag, and thousands of French citizens wrote letters to Lindbergh expressing admiration for the United States. "I am full of admiration for your wonderful performance and as a French girl—great lover of America—I hope it will be a tie between our two nations!" exclaimed one admirer. Lindbergh himself proved adept at diplomacy. During his stay in France (during which time he resided at the U.S. Embassy, upon Herrick's personal invitation), he conspicuously refrained from gloating over his achievement, instead making conscious efforts to recognize France's aviation legacy. Lindbergh immediately paid a visit to Nungesser's mother, informing her that his own flight had been easy compared with the more dangerous east-to-west route taken by the French aviators. He also described Louis Blériot as his "mentor," visited Napoleon's tomb, auctioned his autograph to benefit French veterans, and appeared on balconies waving both American and French flags. By presenting himself as a member of a transnational aviation community and not just as a citizen of the United States, Lindbergh may have done more for U.S.-French relations than if he had acted as an agent of Washington. As historian Brooke Blower has written, Lindbergh's celebrity construed him "as a figure who stood above politics," capable of bringing France and the United States together in spite of their disputes over diplomatic and economic matters.[52]

The effects of Lindbergh's "airplane diplomacy" reverberated far beyond France. He received more than 15,000 gifts from sixty-nine countries, along with letters from citizens of Germany, Holland, Belgium, Switzerland, Romania, Panama, Suriname, and many more nations. In Colombia—which had a thriving commercial airline and strong public interest in aviation—a group of air-minded young men formed a social club called Salon Lindbergh, and a hotelier rechristened his property Hotel Lindbergh. Some letter writers implored Lindbergh to send them money or to help them secure employment. A twenty-four-year-old French woman, the wife of a plumber and the mother of an infant son, requested money for her family to purchase a house; an Armenian Turk asked for help finding engineering work in the United States; a sixty-year-old Russian woman, exiled in Rome, informed Lindbergh that the Bolsheviks had massacred most of her family and wondered whether he could find her surviving son a job.[53]

Responses to Lindbergh's transatlantic flight indicate that the Lone Eagle was not simply an icon of American greatness, as his fellow citizens tended to see him. In other countries, Lindbergh was also viewed as a world citizen, epitomizing transnational human abilities and values. A Hindi newspaper in India, for example, described his transatlantic flight as "a matter of glory, not only for his own countrymen, but for the entire human race."[54] Although many of Lindbergh's admirers professed to be "great lover[s] of America," their affection for the man did not necessarily imply affection for the United States or Americans generally. The letters that Lindbergh received suggest that many people related to him as a universal hero—or as an imagined personal friend or lover—and not simply as an American. The meanings of Lindbergh's legend, then, cannot be contained within a national frame. Rendered in discourses of universalism as well as nationalism, Lindbergh's feat took on different meanings in different cultural contexts.

Even within the United States, not everyone imagined Lindbergh in purely nationalistic terms. A series of articles in the prominent black newspaper *New York Amsterdam News*, for example, demonstrated that his flight could have very different meanings for African Americans. In an editorial entitled "Lindbergh and the Negro Problem," published in June 1927, writer Sybil Bryant Poston addressed "the thousands of Negroes [who] joined a million or more white people in acclaiming Charles Lindbergh as our hero." It was "fitting and proper" that African Americans pay tribute to Lindbergh, she argued, as they would do well to follow his example of daring to "go it alone":

> Too often as a racial group, in an effort to solve our problems, we have been afraid to GO IT ALONE. Too often we have followed the advice and accepted the support of those who knew not our problems and whose hearts did not vibrate sympathetically with ours. We have been afraid to go it alone politically. We have been afraid to go it alone religiously. We have been afraid to go it alone economically. We have failed to manifest that confidence in self which was so evident in the character of Lindbergh. Too long has the Negro hobbled through life on the crutch of philanthropy. Too long has he followed beaten paths by mimicking the social, moral, political, and economic standards set by others. . . . [W]e will never be a Lindbergh and receive the acclaim of the world until we have become pioneers, trail blazers, with confidence in ourselves and a belief in the ultimate triumph of our race, and last, the will to GO IT ALONE!

Poston thus mobilized Lindbergh's legend to advocate for black self-sufficiency. Indeed, her call to "go it alone" resonated with the black nationalist and pan-African politics of the 1920s, manifested most notably in the Universal Negro Improvement Association led by Marcus Garvey. Favorable letters in response to Poston's piece suggested that her arguments touched a chord among readers: "No people is as badly in need of 'The Spirit of Lindbergh' as the Negro," wrote George Wallace Hunter. His letter appeared in the same issue of the *Amsterdam News* as another Lindbergh-related article, "What Will the Negro Contribute to Aviation?" Because two African Americans, Bessie Coleman and Herbert Julian, had already won fame as aviators, and "the Negro's pioneering prowess" had been proven in other fields, the writer expressed hope that "some Negro youth, inspired by this feat of Lindbergh's, will begin a serious apprenticeship in aeronautics"—and that this youth would "show the same stamina in the face of ridicule . . . as did Lindbergh." In Paris, the famed African American performer Josephine Baker was also captivated by Lindbergh, interrupting her popular cabaret show at the Folies Bergère to announce his arrival to the audience. In her excitement, Baker later recalled, she had forgotten that Lindbergh was based in the segregated southern city of St. Louis, where she had been born. "I forgot that Lindbergh was a white man and that he came from St. Louis and might not have liked Negroes," she stated. "I only remembered that he was an American and that he had done something great for the progress of the world."[55]

Like the cinema screens that showed his image, "Lucky Lindy" was a canvas on which people projected their imagined conceptions of the national and the global, as well as their own racial or ethnic identities. Lindbergh's legend could sustain a multiplicity of cultural narratives and political positions, from transatlantic whiteness to self-sufficient blackness to universalistic conceptions of "the entire human race." As a popular icon, "Lucky Lindy" was a shape-shifter who could variously embody a race, a nation, a region, or all of humanity. Lindbergh's legend, like the airplane itself, also illustrates the compatibility of nationalism and globalism. Signifying both American greatness and universal human progress, he was an ideal ambassador for a nation that imagined itself as a model for all nations.

Prior to the stock market crash of October 1929, the late 1920s were boom years for the U.S. aviation industry. In 1927, NACA reported, "aeronautical progress during the past year has surpassed the hopes of a year ago." Most strikingly, "aviation is being accepted by the people as a means of transportation and as a business in which industrial capital is being invested." Production of civilian aircraft quadrupled between 1927 and 1929, from 1,374 to 5,516 planes; production of military aircraft, during the same two years, increased from 1,995 to 6,193. In 1928 alone, twenty-three new airlines began offering passenger service in the United States. By mid-1929, the nation had sixty-one passenger airlines, forty-seven airmail lines, and thirty-two air cargo lines. The number of airline passengers increased nearly thirtyfold, from 5,782 in 1926 to 173,405 in 1929. While most of these passengers were affluent businessmen, they represented a surprising range of geographic diversity: in early 1930, one airline found that its clientele hailed from forty-two states and 407 cities. Investment in aviation also soared. Between March 1928 and December 1929, $1 billion in aviation securities was traded on the New York Stock

Exchange. "I tell you there's money in it. . . . In the air," says the lead character in John Dos Passos's 1929 play *Airways, Inc.*, in which a pilot-turned-airline-entrepreneur betrays his radical working-class roots to grab the gold of the skyways.[56]

Americans had become *Plane Crazy*, to quote the title of a 1928 Walt Disney film featuring Mickey Mouse as a wannabe Lindbergh. As of 1928, some twenty aviation magazines were published in the United States. In one New York newspaper, coverage of aviation news increased from eighty columns and three pages of pictures in 1925 to 450 columns and thirty-six pages of pictures in 1928. Accomplished pilots, including Amelia Earhart and several other women, became national celebrities. Lindbergh, in particular, continued to be an object of obsession. In the United States and abroad, his name endorsed everything from cigars and ladies' shoes to bread and canned fruit (despite his lawyers' best efforts to suppress unendorsed uses of his image), and the U.S. Patent Office received hundreds of applications for patents on objects designed to resemble the *Spirit of St. Louis*, from lamps to letter openers.[57]

Commentators described the ascendance of U.S. aviation in the late 1920s as a "Lindbergh boom." The "transoceanic flights of Lindbergh and others," concluded NACA, "led to the awakening of the American people to the possibilities of aviation."[58] But if Lindbergh "awakened" Americans to aviation, why had they remained "asleep" for so long? Why did the great aerial awakening occur in 1927 and not in 1908, with the Wright brothers' celebrated public flights, or in 1924, with the U.S. Army Service world tour? The answer lies in the mutable meanings of Lindbergh's legend—his ability to animate both nationalist and universalist narratives. There is yet another reason, however. The airplane proved exceptionally well suited for the types of international involvements pursued by the Coolidge and Hoover administrations, namely, the expansion of foreign commerce and investment, backed up by periodic military invasion and occupation. Between 1927 and 1941, the United States would begin using commercial aviation as a chosen instrument of foreign policy—starting in its traditional sphere of influence, the Western Hemisphere.

2

Good Neighbors
Are Close Neighbors

In the climactic scene of the hit Hollywood musical *Flying Down to Rio*, released by RKO Studios in December 1933, a bandleader from the United States (Gene Raymond) marries a Brazilian woman (Dolores del Rio) aboard a Pan American Airways Clipper, the silver-hulled flying boat that made flying synonymous with glamour during the 1930s. That same month, at a diplomatic conference in Montevideo, Uruguay, Secretary of State Cordell Hull pledged U.S. aid for the development of Latin American transportation and communications facilities. His offer followed the "Good Neighbor" policy that President Franklin D. Roosevelt had proclaimed in his March 1933 inaugural address, which promised that the United States would curtail military interventions in Latin American countries and would instead aim to improve diplomatic, economic, and cultural ties with its southern neighbors. *Flying Down to Rio* staged the Good Neighbor policy as a romance—a marriage between a U.S. man and a Brazilian woman—and it is no coincidence that the union took place aboard an airplane.[1]

It was in Latin America that the United States first experimented with the airplane as an instrument of diplomacy, commerce, and warfare. This resulted, in part, from technological constraints: the airplanes of the time could fly only about 500 miles before needing to refuel; aircraft capable of traversing the Atlantic and Pacific Oceans would not be developed until the mid-1930s. Geopolitics and history, however, also charted the southward course of U.S. aviation. In 1823, President James Monroe declared that the U.S. government would henceforth oppose, with military force if necessary, any attempt by European powers to colonize territory in the Americas or otherwise interfere in the region's internal affairs. The Monroe Doctrine, which pronounced the Western Hemisphere as the United States' exclusive sphere of influence, acquired additional potency in 1904, when President Theodore Roosevelt added a corollary provision stipulating that the United States had the right to intervene in Latin America whenever its own political or economic interests were threatened. Subsequently, the region served as a laboratory of U.S. global power, where strategies of political, economic, and military influence were invented and tested. This political and economic context made it both possible and desirable for the United States to use Latin America as its international aviation laboratory.[2]

During the interwar years, the airplane extended the Monroe Doctrine to the skyways, enabling soldiers, sailors, and what one observer described as "the small army of American business men traveling in Latin America" to perform their jobs more quickly and efficiently. Consistent with its long history of military interventions in the region, the United States first used airplanes in a military conflict during the Mexican Revolution. The Navy employed reconnaissance planes in its 1914 occupation of Vera Cruz, and the Army used them again two years later in its "Pershing Punitive Expedition" against revolutionary leader Pancho Villa. In the era of the Good Neighbor policy, however, aviation also helped to reinvent—or indeed to rebrand—U.S. regional hegemony by facilitating more neighborly activities. As Pan American Airways developed air routes throughout the Americas during the late 1920s and 1930s, its advertisements claimed to "make good neighbors close neighbors." Ultimately, the development of commercial aviation did important work to make the U.S. presence in Latin America appear more benign while also bringing the region within closer reach of Washington and Wall Street.[3]

The cultural meanings of aviation embodied the contradictory impulses of the Good Neighbor policy, which eschewed military intervention but embraced other means of power and influence. During the Good Neighbor era, gunboat diplomacy largely gave way to more peaceful activities— commerce, tourism, philanthropy, educational exchanges, public health initiatives, audiovisual propaganda—but the United States retained, and in many ways strengthened, its longtime regional preponderance. By earning Latin Americans' friendship, the Roosevelt administration believed, the United States could obtain its political and economic objectives without the costs and controversy of military invasions. And aviation, signifying both national power and international goodwill, promised to be an ideal instrument of Good Neighbor diplomacy.[4]

Importantly, however, the history of aviation also reveals significant Latin American resistance to U.S. hegemony. During the 1920s and 1930s, governments tried to curtail the expansion of U.S. commercial airlines in the region either by restricting their rights to operate or by forming national airlines to compete against them. Throughout Latin America, aviation promoters advanced nationalist agendas that often conflicted with, or at least diverged from, the interests of Washington and Wall Street. The airplane's cultural meanings, moreover, were always shaped by local context; Charles Lindbergh, for example, came to be identified with Latin American or pan-American traits, not those of the *Yanqui*. If aviation made good neighbors close neighbors, it did not always do so precisely in the ways that U.S. leaders envisioned.

"There Must Be Something to the Americans"

In the spring of 1927, as Lindbergh's solo transatlantic flight helped to mollify anti-American sentiment in France, relations between the United States and Mexico were approaching a point of crisis. In 1926, President Plutarco Elías Calles had threatened to nationalize Mexico's oil industry, invoking a provision in the Mexican constitution that defined oil reserves as national property. Mexico was, at this time, the United States' largest supplier of foreign oil, and the Coolidge administration viewed Calles's threat to expropriate foreign companies' assets as a grave danger to the national interest. U.S. leaders were also concerned about Mexican radicalism. In 1926, the Soviet Union had opened an embassy in Mexico City, evoking the specter of Bolshevik saboteurs south of the border.[5]

Ironically, however, the United States' own actions further intensified Latin American radicalism and anti-Americanism. In early 1927, President Coolidge ordered an invasion of Nicaragua, with the intention of stabilizing the pro-Washington Chamorro regime against a peasant insurgency led by Augusto César Sandino. U.S. Marines would remain in Nicaragua for the next six years, embroiled in a brutal guerrilla war that caused extensive casualties on both sides and met with vociferous condemnation throughout the region.[6]

Amid these escalating regional tensions, in October 1927 Coolidge dispatched a new ambassador to Mexico City: Dwight R. Morrow, formerly a banker at J. P. Morgan. Morrow immediately sought to improve U.S.-Mexican relations with his publicity-savvy "ham and eggs" diplomacy, so called because his first meeting with Calles took place over breakfast at the latter's ranch. He also turned to his friend Charles Lindbergh, for whom he had served as a financial adviser. Recalling how Lindbergh's transatlantic flight had bolstered the United States' image in France, Morrow reasoned that a flight to Mexico City would have similar results. What better symbol of international goodwill than "Lucky Lindy," who was idolized throughout the world?[7]

The State Department, after some initial hesitation, endorsed Morrow's proposal to dispatch Lindbergh on a goodwill flight throughout Latin America. Beginning in mid-December, the *Spirit of St. Louis* would visit Mexico, Guatemala, British Honduras, El Salvador, Honduras, Nicaragua, Costa Rica, Panama, Colombia, Venezuela, St. Thomas, Puerto Rico, the Dominican Republic, Haiti, and, finally, Cuba, where Lindbergh would appear at the Pan American Conference in Havana in late February. For the publicity-shy aviator, the tour offered a welcome opportunity to get back in the sky, away from the throngs of fans and reporters that now followed his every move. Having recently flown on a barnstorming tour across the United States, Lindbergh believed that his intrepid little airplane could survive one more international adventure before its impending retirement to the Smithsonian Institution. To increase the symbolic value of the event, he decided to begin the tour by making the first-ever nonstop flight between Washington, DC, and Mexico City.[8]

In the United States, the news media reproduced the State Department's argument that the proposed flight would do wonders for relations between the United States and its southern neighbors. "It is a foregone conclusion that Col. Lindbergh will be assured a resounding reception," stated the *Los Angeles Times*. "There can be no doubt of the excellent impression which his visit will make upon the Mexicans," concurred the *New York Times*, which went on to assert that Mexicans were "an impressionable people" who would undoubtedly be influenced by "one of our very finest types of young manhood." According to the *Washington Post*, Lindbergh's flight would convince Mexicans that "the welfare of Mexico can be promoted best by respecting the rights of foreigners." What promised to be a "red-letter event in the lives of millions of Mexicans" would also be a lesson in proper manhood: "gallant lads . . . will be stirred to emulation of [Lindbergh's] character and deeds." As with his transatlantic flight, Lindbergh's race and ethnicity shaped the cultural meanings of his latest endeavor. His civilized Nordic manliness, U.S. newspapers claimed, would set an example for Mexican "lads," teaching them to be gallant and modest—and teaching them to respect the United States.[9]

After a slight detour due to bad weather and some linguistic confusion (reading train station signs with the word "Caballeros," indicating the location of the men's restrooms, Lindbergh tried to locate his position by finding the nonexistent town of Caballeros on his map), the *Spirit of St. Louis* landed at Mexico City's Valbuena Airport at 3:40 p.m. on December 14, 1927, before a crowd of 150,000 people. President Calles gave Lindbergh the keys to the city and thanked him for creating "closer spiritual and material relations" between the two countries. President Coolidge proclaimed in a wire message that the flight would "materially assist the two countries to cement friendly relations," while the Mexican ambassador to Washington, again appealing to the virtues of manliness, hailed Lindbergh as "a very clear example of the worth of the young manhood of the United States of today." The celebrations continued as Lindbergh remained in Mexico through Christmas, staying with the Morrows. Local elites treated him to banquets, parades, bullfights, and cultural festivals; Lindbergh reciprocated by taking them for airplane rides. The Mexican Regional Federation of Labor, the nation's largest workers' organization, held a massive parade in his honor (reportedly attended by over 100,000), and nearly five hundred towns sent telegrams requesting him to visit. When the aviator visited Tampico, the U.S. consul reported that he received a reception "such as has rarely, if ever, been accorded to an American or other foreigner in this city."[10]

In the United States, representations of the flight suggested that Lindbergh had single-handedly repaired U.S.-Mexican relations. "Cheers for a President of the United States were heard today in the Mexican Congress probably for the first time," reported the *Washington Post*. "Some of the [U.S.] oil interests could afford to pension this young man for life," claimed *Popular Mechanics*. A *New York Times* editorial described Lindbergh as "a great national asset" who could "draw together nations" through his "modesty, his poise, his sound judgment . . . and his remarkable self-control." In a separate article, *Times* aviation correspondent Russell Owen congratulated Lindbergh on making "a definite advance toward a solution of the so-called Mexico problem." "The Gringo has been looked upon instinctively for many generations as the national enemy of this country," Owen wrote, but "Lindbergh has given the populace a new conception of American manhood; they have decided that after all there must be something to the Americans." As Owen's statement suggests, gender ideology continued to shape U.S. interpretations of Lindbergh's visit to Mexico. The aviator's own exemplary "manhood," characterized by modesty and self-reliance, would teach Mexicans to be proper men—men who could see that "there must be something to the Americans."[11]

During the remainder of Lindbergh's Latin American tour, local reactions reprised what had occurred in Mexico. Cheering throngs greeted Lindbergh's every public appearance; government officials feted him with lavish banquets and gifts. Embassies and consulates transmitted glowing dispatches to Washington: "The universal enthusiasm for Colonel Lindbergh [is] unprecedented in this country in the memory of living persons" (Honduras); "an unqualified success from every point of view . . . has added materially to the popularity of Americans in this country" (Venezuela); "never before has such a welcome been given anyone. . . . It proves that Haitians as a whole are not anti-American"; "Nothing of this kind . . . could have accomplished more for the relations between the U.S. and the Dominican Republic." From Colombia, the U.S. consul reported, "Lindbergh's reception was beyond anything we could have hoped for, the people are simply wild over him, he has done a world of good." In Costa

Rica, people "pour[ed] into San Jose from villages, coffee and banana plantations, [by] foot, ox cart, horseback, automobiles and special trains, [the] latter furnished by government gratis." (Informed that Costa Rica's population numbered 500,000, Lindbergh reportedly quipped, "I believe they are all here.")[12]

Without exception, U.S. diplomats believed that Lindbergh's tour had greatly improved local opinion of the United States. According to the consul in Costa Rica, for example, Lindbergh was "acclaimed not only as [a] famous aeronaut but as [a] citizen [of the] United States." Conflating praise for Lindbergh with praise for the United States itself, such statements reprised the nationalist interpretations of his transatlantic flight. Diplomats, like journalists, also invoked conceptions of masculinity in arguing that Lindbergh would teach Latin Americans to respect and to emulate the United States. The *Spirit of St. Louis* was the first airplane to land in Belize City, capital of British Honduras, and according to the U.S. consul there, Lindbergh's "bravery and skill," combined with his "unassuming boyish manner," had proven "very beneficial" in making a good impression on the local population. "A colored chauffeur told me that he wished he had half of Lindbergh's spirit," the consul reported. "Someone here said that manhood glued to the ground was of course helpful to people near it, but that real manhood personified in the air was, like a spirit, practically limitless in its beneficial scope." By exporting "real manhood" across national lines, Lindbergh universalized American gender ideology, offering a model for men of all nations to follow.[13]

Lindbergh's own reports, published in the *New York Times*, affirmed that the goodwill flight brought about "a distinct improvement in the relationships between the nations of the two continents." Aviation, he argued, stirred "fundamental emotions, common to all people." He had nothing but praise for the countries of Latin America: "I can only say that every one of our Latin-American sister republics I have visited makes me regret more that I do not know Spanish to express directly my appreciation of their welcome." Later in life, however, Lindbergh described his tour in very different terms. "I felt the great superiority of our civilization to the north," he recalled in his 1977 autobiography. "There seemed to be an unbridgeable chasm between it and Central American culture—in science, in industry, and in political organization." This shift reflected Lindbergh's increasingly rigid conceptions of racial and ethnic difference—expressed, most notoriously, in his flirtation with Nazism during the late 1930s. But tensions between pan-Americanism and nationalism had been present all along in U.S. accounts of Lindbergh's tour: in journalists' predictions that the aviator's modest manliness would teach Latin American men to respect U.S. interests, in diplomats' easy conflation of praise for Lindbergh with praise for the United States, and in Lindbergh's own rhetoric. Good neighborly appeals to pan-Americanism were not incompatible with assertions of U.S. superiority. In fact, pan-Americanism could justify Yankee nationalism by construing U.S. hegemony in Latin America as a form of benevolent tutelage.[14]

Beyond the United States' borders, however, Latin American leaders crafted very different interpretations of Lindbergh, which often reflected their own national interests and nationalisms rather than (or in addition to) admiration of the United States. Even as regional leaders effusively praised the aviator, their responses to his tour were more complex than the accounts of U.S. diplomats and journalists suggested. At a reception in Lindbergh's honor, for example, a Guatemalan dignitary said that the aviator "gives us a great lesson"—phrasing that echoed the claims of leading U.S. newspapers. But, the official went on to emphasize, rather than teaching Guatemalans to respect and emulate the United States, Lindbergh "teach[es] us that our country, however small, is as great as any of the other countries that he has visited." Lindbergh's visit, in this interpretation, proved that Guatemala was equal to the United States and other "great" powers, not in need of their guidance.[15]

If Lindbergh inspired love of nation, then, the object of that love was not always the United States—or not the United States alone. In the Dominican Republic, politicians and journalists compared Lindbergh to a previous foreign visitor, Christopher Columbus. Emphasizing Santo Domingo's status as the oldest city in the New World, Dominican leaders placed the aviator within a pan-American narrative: in the words of President Horacio Vásquez, Lindbergh was "a product of these American lands," plural. His flight from the United States to France, Vásquez argued, exemplified how the American "new world" had advanced beyond the European "old world." And as members of the new world, Dominicans could rightfully claim Lindbergh as their own. An editorial in the

Honduran newspaper *Renacimiento* likewise construed Lindbergh's legacy as multinational. On its hull, the *Spirit of St. Louis* bore painted images of flags from the countries and colonies that it had visited, the newest being that of Honduras. Thus, when future generations would view the historic plane at the Smithsonian Institution, they would see "the indelible colors of the sister flag of Honduras" bearing "silent witness" that "Honduras remains bonded with the *Spirit of St. Louis.*" The editorial went on to praise Lindbergh's "extraordinary valor," but it also identified valor as a defining trait of the Honduran people, which they consequently knew how to recognize in others.[16]

In African American newspapers, similarly, pan-American interpretations of Lindbergh resonated with transnational constructions of black identity. After Lindbergh's visit to Haiti, the *New York Amsterdam News* likened him to Toussaint-Louverture, leader of the Haitian Revolution: "as our great airman wings his way to those islands, let us wipe out the barriers of prejudice, following his example . . . [and] remember the other man who was also, a century [and] a half ago, an apostle of freedom and good-will." By linking Lindbergh to Toussaint-Louverture, and the airplane's demolition of territorial barriers to the demolition of "barriers of prejudice," the *Amsterdam News* incorporated aviation into a transnational black freedom movement.[17]

When Latin Americans did identify Lindbergh with the United States, moreover, they did not always do so in celebratory ways. In two documented instances, his arrival catalyzed fierce protests of the U.S. invasion of Nicaragua, which at the time was occupied by three thousand Marines and National Guardsmen. In El Salvador, representatives from the Nicaraguan Autonomous Association met Lindbergh at the U.S. legation and gave him a letter imploring him to acknowledge the "victims of the unjustified slaughter of a handful of men who term themselves soldiers of the United States." The legation in El Salvador received another letter, signed by forty-eight university students, which identified Lindbergh with previous agents of imperialism. "You are the hero of the air, but you are a Yankee, your hair is blonde, your eyes are blue, and your language is English. Of the same type were and the same language spoken by Walker, Cole, Knox, Roosevelt, Wilson, Coolidge, Bryan, Jefferson, and Latimer"— statesmen and presidents who had authorized the United States to invade or economically exploit Latin America. Just as Lindbergh physically resembled these men who had committed "sorrowful deeds" in the region, the *Spirit of St. Louis* looked, to the students, like an instrument of conquest: a "hand that writes in the clear sky a sentence of possession . . . announcing the crowd which will one day come to conquer our lands." In the context of the United States' imperial history in the Americas, the students wondered what good the aviator's "goodwill" tour would accomplish. "Is it bringing back gold to her safes which have been looted by your wicked bankers? Does it bring liberty to our Nicaraguan brothers who have been oppressed by the Department of State?" They concluded their letter by urging Lindbergh to boycott Nicaragua: "Do not go there with your glorious airplane bellowing a hymn of power from your mighty land."[18]

But Lindbergh did fly on to Nicaragua, and at the airfield in Managua, a crowd of approximately six thousand cheered his arrival, in spite of the presence of U.S. military planes on the airfield. (On the morning of Lindbergh's landing, a medical plane had just arrived with U.S. troops wounded in recent fighting; a second plane, carrying Marine commander Colonel Louis Mason Gulick, was taking off en route to the port town of Corinto to meet an arriving shipload of Marines.) For U.S. diplomats in Nicaragua, Lindbergh's visit even seemed to confirm the necessity of the invasion. According to the consul in Managua, locals' jubilant reactions to Lindbergh represented the "best evidence that Nicaragua welcomes American aid in terminating the disorders which have recently disturbed [it]." In the United States, meanwhile, leading newspapers argued that Lindbergh's example would inspire Nicaragua to resolve its political differences in a civilized fashion. "Fiery Nicaragua saw a hero today such as it never dreamed could exist," stated the *New York Times*. Lindbergh arrived as "a superman and a demi-god to these warm-souled people of the Tropics . . . a Northern conqueror as different in manner from the exuberant Nicaraguans as can be conceived." Still, "they understood him . . . and loved him." Construing Nicaraguans as emotional, superstitious, and childlike, this language implicitly affirmed that the United States had the right and duty to impose order in Nicaragua and other regional trouble spots. As the Marines suppressed peasant rebellions in the countryside, Lindbergh, embodying cool-headed Nordic self-control, would teach Nicaraguans to tame the "fiery" and "warm-souled" temperaments that produced such political instability. However, because this "Northern Conqueror" was not simply a U.S. citizen but a worldwide hero, his particular brand of conquest would presumably be welcomed with open arms.[19]

The letters from El Salvador, though, reveal that Latin Americans did not always welcome Lindbergh with open arms—and even when they did so, their fondness for "Lucky Lindy" himself did not always silence their criticisms of U.S. foreign policy. Nor did they necessarily view his flight as a symbolic triumph of the United States. Many Latin American leaders placed Lindbergh in the service of their own nationalist claims; others appropriated him as a pan-American hero. From the perspective of the State Department, Lindbergh's tour successfully "developed a real feeling of good will towards us and the countries to the South." But the meanings of this "goodwill" were more complex, more contingent, and more locally determined than the State Department or the U.S. media were willing or able to acknowledge.[20]

And more was at stake than good will. Though the *Spirit of St. Louis* was not the herald of a conquering army, the Salvadoran students accurately perceived affinities between Lindbergh's tour and Washington's military and commercial interests in Latin America. Like the world flight of the U.S. Army three years earlier, Lindbergh's flight served as a demonstration of power, displaying the United States' technological superiority even as it purportedly expressed international goodwill. The idolized aviator had been sent to Latin America to assuage anti-U.S. sentiment, but the United States was simultaneously using airplanes to bolster its invasion of Nicaragua, which had exacerbated such sentiment in the first place. As would be the case throughout aviation history, the military and civilian uses of flight appeared contradictory, yet were closely intertwined.

Furthermore, Lindbergh's Latin American tour had an equally important, though less publicized, secondary purpose: surveying and testing future commercial routes for U.S. airlines. Shortly after Lindbergh returned to the United States, he received a letter from John Hambleton, the vice president of newly formed Pan American Airways, expressing appreciation for "all the help you gave us" in promoting U.S. air transport in Latin America. When the postmaster general placed six foreign airmail routes up for bidding in March 1928, their geography retraced the so-called "Lindbergh Circle" through Central America and the Caribbean.

The airline that won the bids, receiving the maximum postal subsidy of $2 per mile, was Pan American—and its president, Juan T. Trippe, gratefully informed Lindbergh that his company's success "is due directly to your personal demonstrations that long distance flying is possible and safe."[21]

A Monroe Doctrine for the Skies

Commercial aviation interests on both sides of the Atlantic had long viewed Latin America as a lucrative and untapped market. The region's topography—dense jungles, forbidding mountains, and chains of islands—favored aviation over railroads and other forms of ground transportation, and the airplane promised to facilitate the transportation of its rich natural resources. Beginning in the 1910s, U.S. aviation trade journals had proposed that "Hundreds of Aerial Lines Could Be Established Immediately in South and Central America." As it turned out, few U.S. entrepreneurs would risk investing in international air transport until the mid-1920s. But European concerns, backed by government funding, eagerly entered the Latin American market. In 1919, a French company founded the region's first airline, Société des Transports Aériens Guyanais, in the colony of French Guiana. Another French airline, Lignes Latécoère, began flights between Rio de Janeiro and Buenos Aires in 1925. Germany, prohibited from developing military aviation by the Treaty of Versailles, also took an active interest in Latin America's commercial skyways. A group of German émigrés to Colombia founded the Sociedad Colombo-Alemana de Transportes Aéreos (SCADTA), Latin America's first commercially successful airline, in 1919. Under the direction of Austrian engineer Peter Paul von Bauer, by 1925 SCADTA had carried some five thousand passengers along routes that spanned Colombia's mountainous interior. Germans also founded airlines in Argentina (Aero-Lloyd), Bolivia (Lloyd Aéreo Boliviano), and Brazil (Sindicato Condor).[22]

Meanwhile, U.S. military leaders watched Germany's expanding presence in South American aviation with increasing alarm. After an inspection tour of the Panama Canal in early 1924, General Mason Patrick, chief of the U.S. Army Air Service, warned that the German-run airlines posed an imminent threat to the canal's security. Concerns escalated after

April 1925, when SCADTA petitioned the postmaster general for rights
to carry airmail from Colombia to the Canal Zone and Key West. The
Post Office denied SCADTA's petition, but Bauer's bold move, evoking
visions of Junkers flying over the Panama Canal, convinced military leaders that the United States needed to take action to counteract German
influence in South American aviation.[23]

The solution, General Patrick argued, was to establish a U.S.-run airline to initiate airmail service in Latin America. This move would uphold
the Monroe Doctrine by containing European interference in the regional air transport market, and it would help protect U.S. sovereignty
over the Panama Canal, the defense of which "practically demands that
the United States dominate aeronautical employment throughout Central
America," Patrick wrote. He acknowledged that such an airmail service
could not be justified on economic grounds, as the presumably low volume of mail would not warrant its operating expenses. However, he
emphasized, the issue was "not a Post Office problem, but as we see it, a
military problem involving the protection of the Panama Canal from
attack by the air." The commanding general of the Army's Panama Canal Department concurred, stating: "If great corporations controlled
by nationals, other than American, should dominate commercial aviation in this territory, our interests, political, economic, and military,
will suffer."[24]

In early 1927, after the Post Office announced that it would accept bids
for airmail carriage between Miami and Havana, U.S. Army Air Corps
Major Henry "Hap" Arnold and two flying buddies, Carl Spaatz of the
Army and Jack Jouett of the Navy, decided to incorporate an airline. Their
company, called Pan American Airways, received the Miami-Havana
contract in July. From its very inception, the airline that would later be
known to the world as "Pan Am" functioned as a privately owned proxy
for the U.S. government—an agent of expansionism whose stated mission was not only to facilitate commerce but to uphold the United States'
strategic, economic, and political interests. Indeed, Hap Arnold and his
colleagues viewed their airline less as a moneymaking scheme than as a
means of enforcing the Monroe Doctrine and protecting U.S. sovereignty
over the Panama Canal. "In a sense, the formation of Pan American Airways turned out to be the first countermeasure the United States ever

took against Nazi Germany," claimed Arnold, somewhat hyperbolically, in his autobiography (German, and especially Nazi, influence in South American aviation turned out to be vastly overstated). Yet with "not a dime between them," Arnold and his colleagues lacked the resources to turn their fledgling company into a functioning airline. Pan American existed only on paper until October 1927, when an enterprising twenty-eight-year-old Yale graduate with the improbable name of Juan Terry Trippe came to its rescue.[25]

In spite of his Spanish first name, Trippe had only tangential hereditary connections to Latin America: an aunt by marriage was Venezuelan. The son of an investment banker, he grew up amid upper-class WASP surroundings on Manhattan's East Side; as a child, he flew model airplanes in nearby Central Park. After studying engineering at Yale's Sheffield Scientific School—where he became active in the Yale Aeronautical Society and trained as a Naval Air Reserve pilot during World War I—Trippe worked briefly on Wall Street before deciding to combine his two interests, business and flying, by starting an airline. Using his inheritance money, Trippe formed Long Island Airways in 1922 to ferry socialites between Manhattan and the Hamptons; revenues, however, were insufficient to cover operating costs, and the airline quickly went out of business. Trippe had better luck with his second venture, Colonial Air Transport, which received the coveted contract to carry airmail between Boston and New York (and was one of several companies that merged to become American Airlines in 1930).[26]

Trippe's greatest ambition, however, was to establish an international airline. In June 1927 he formed the Aviation Corporation of the Americas, a holding company whose board and financial backers included such scions of fortune as Cornelius Vanderbilt Whitney, William H. Vanderbilt, William A. Rockefeller, and Sherman Fairchild (many of whom had attended Yale during the same years as Trippe). In 1925, during a demonstration flight to Cuba with aircraft builder Anthony Fokker, Trippe had persuaded Cuban president Gerardo Machado to grant him the right to operate air transport between Cuba and the United States. Thus, when the postmaster general awarded the Miami-Havana airmail contract to Pan American Airways, Trippe immediately initiated negotiations to acquire the company. In early October 1927, Arnold and his colleagues sold their cash-strapped airline to Aviation Corporation of the Americas for $10,000 in cash and $45,000 in stock options. As president, Trippe elected to retain Pan American's original name.[27]

On October 28, 1927, Pan Am inaugurated scheduled airmail flights between Key West and Havana. Passenger flights began the following January—precisely the same time that Lindbergh, too, was linking the Americas by air. Pan Am's first seven revenue passengers paid round-trip fares of $100 and arrived in Havana in about two hours, compared with thirteen hours by steamer. In the words of Postmaster Harry New, the airline promised to become "a great public utility" that "will bring about better relations between the countries it is to serve." Across the United States, newspapers and magazines expressed similar arguments. "Trade and friendship follow the airplane," stated a *Washington Post* editorial, which predicted that Pan Am, like Lindbergh, would help the United States to improve its image in Latin America. Importantly, however, articles on Pan Am also emphasized that aviation would strengthen the United States' hegemony in the Western Hemisphere. The *Christian Science Monitor*, in an editorial entitled "The Eagle Extends His Wings," observed that Pan Am would "scarcely fail to add pinions to the Monroe Doctrine." As its founders had intended, the airline served the United States' national interests while facilitating regional trade and friendship.[28]

Within just a few years, Pan Am had expanded far beyond the Miami-Havana "cocktail circuit," so nicknamed because flights catered to wealthy U.S. citizens seeking relief from Prohibition. But Pan Am did not simply carry airmail and rum-swilling high rollers. Its flying Clippers—named after the U.S. merchant vessels that had once ruled the high seas, and piloted by "Masters of Ocean Flying Boats" whose navy blue, gold-braided uniforms resembled those of naval commanders—transported U.S. political and economic power throughout the Americas. The airline's president, like the railroad barons of the nineteenth century, was an "empire builder" (as *Time* labeled him in a 1940 cover story) who pursued his vision of a globe-girdling airline with single-minded intensity: "J. T. T. will *never* delegate authority," wrote one subordinate. Under Trippe's leadership, Pan Am grew with astonishing rapidity. With a steady flow of capital from its wealthy investors, the company purchased the most advanced aircraft available, employed top-notch pilots and technicians (including

Lindbergh, hired as a consultant in 1929), and boasted a 99.678 percent efficiency record, with only one fatal accident, during its first five years. In 1928, it carried 9,500 passengers along 251 miles of routes, operating with eleven airplanes, six airports, 118 employees, and a capital investment of $3.5 million. By 1934—the year that Pan Am opened Dinner Key Terminal in Miami, with its iconic Art Deco architecture and massive, rotating globe—the airline had carried 106,875 passengers and boasted 133 airplanes, 165 airports, and 2,701 employees.[29]

Pan American's success can be attributed to solid business foundations: an ambitious and visionary leader, abundant financial resources, skilled personnel, and the latest technological equipment. Quite literally, however, Trippe's airline would never have gotten off the ground without "the progressive and far-sighted policy of our Government," as its 1929 Annual Report acknowledged. As was the case with all U.S. airlines, federal airmail subsidies effectively underwrote Pan Am's operating expenses during the 1920s and 1930s, as air transport remained far too costly for companies to survive on passenger revenues alone. As an international airline, furthermore, Pan Am needed the support of the State Department. Flying to and from a foreign country required operating permissions both from Washington and from the government in question. And because many governments, fearing that their own airlines would be destroyed by U.S. competition, refused to grant such permissions willingly, diplomatic assistance proved crucial to Pan Am's initial expansion. Keenly aware that his company needed Washington's support, Trippe became a skilled lobbyist who secured government favors for Pan Am by branding it as an agent of the national interest. In negotiations with the State Department, he echoed Hap Arnold in arguing that a "U.S.-owned, U.S.-operated" airline in the vicinity of the Panama Canal would be "of real value to our national security." Air transport would also benefit "our social and cultural relations with foreign countries," Trippe emphasized, by helping to cultivate an "intangible item, sometimes referred to as 'good will.'"[30]

On such points, President Herbert Hoover needed little convincing. An engineer by training, the technocratic Republican had supported commercial aviation since his tenure as secretary of commerce under Coolidge. As president, Hoover granted corporations wide latitude to do business as they wished in other countries, effectively entrusting them to protect the United States' economic and strategic interests. In Latin America particularly, Hoover also sought to curtail military interventions, which had so tarnished the United States' image during the Coolidge and Wilson years. Instead, he promoted trade, foreign direct investment, and the extraction of natural resources as means of U.S. regional hegemony. Pan Am, therefore, was perfectly suited to Hoover's foreign policy objectives.[31]

During his pre-inaugural goodwill tour of South America in November 1928, Hoover met with Pan Am's vice president and promised that his administration would do whatever it could to support the airline's endeavors. Subsequently, the State Department went to extraordinary lengths to help Pan Am obtain operating concessions from Latin American governments. In 1928 and 1929, Assistant Secretary of State Francis White sent nearly sixty communications instructing U.S. consulates in the region to render Pan Am "all possible and proper assistance." Diplomats arranged meetings with local government officials and at times lobbied directly on the airline's behalf. "We have been moving heaven and earth to help Pan American Airways," wrote Stokeley Morgan, of the State Department's American Republics Division, in January 1928. In the Commerce Department, Pan Am had another powerful ally in Clarence M. Young, assistant secretary of commerce for aeronautics, who believed that airmail, by increasing the efficiency of business transactions, would save the United States some $4 million per year in its trade with Latin America. Thus, beginning in the Hoover years, Pan Am became the United States' "chosen instrument" in aviation—a private corporation officially authorized to implement state policy objectives.[32]

Yet even as it received strong support from Washington, the airline faced strong opposition from many Latin American governments. During its first two years, Pan Am received generous and exclusive operating concessions from reliable U.S. allies such as Cuba's Gerardo Machado (whose subordinates escaped to Miami aboard Pan Am planes after Machado was deposed in a 1933 coup) and General Rafael Trujillo of the Dominican Republic. More typically, however, Latin Americans perceived Pan Am as a threat to local airlines and an agent of Yankee imperialism, like other U.S. corporations. In Central America, newspapers linked the airline to the detested United Fruit Company: according to the Hondu-ran newspaper *El Cronista*, "The United Fruit Company is conquering our lands and monopolizing our waters, and its older daughter, the Pan American Airways, is claiming at present our blue sky." In Venezuela, Pan Am was rumored to be in league with U.S. oil interests. Other critics, like the students who protested Lindbergh's visit to Nicaragua, suspected that inter-American air routes would facilitate U.S. military interventions. An editorial in the Santiago, Chile, newspaper *El Mercurio*, for example, warned that commercial airplanes, though typically used for "inoffensive purposes," could "easily be transformed into formidable war machines."[33]

Such concerns were not unfounded. Pan Am did, in fact, work closely with other U.S. corporations in Latin America, including United Fruit, Standard Oil, the Radio Corporation of America, and shipping conglomer-ate W. R. Grace—with whom it formed a subsidiary, Pan American–Grace Airways, nicknamed Panagra, which operated along South America's west coast. Air transport benefited U.S. companies by expediting the transit of goods, correspondence, and businessmen themselves. Traveling frequently within Latin America as well as to and from the United States, the latter were among Pan Am's most loyal customers, and they rewarded the airline by introducing its representatives to local officials, providing insider infor-mation on political and economic conditions, and offering advice on where to construct airfields and seaplane bases. Exemplifying such connections, Pan Am's chief representative in Latin America, John MacGregor, had pre-viously worked for U.S. oil companies in the region.[34]

Moreover, the operating concessions that Pan Am demanded from Latin American governments could be interpreted as imperialistic. The proposed contract that MacGregor presented to the Guatemalan govern-ment in 1928, which typified the airline's demands, requested permissions to operate both foreign and domestic air transportation within Guate-mala, exclusive rights to carry airmail between Guatemala and the United States, exclusive rights to acquire Guatemalan property, free usage of communications infrastructure (including mail, radio, cable, telephone, and telegraph), and exemption from local taxes, duties, and customs—all for a period of twenty-five years. The Guatemalan Council of State sum-marily rejected Pan Am's demands as "unacceptable," and contract nego-tiations continued for two years. Through the early 1930s, this pattern would be repeated in Costa Rica, El Salvador, Colombia, Chile, Bolivia, Peru, and Venezuela.[35]

To counteract Latin American hostility toward Pan Am, in the fall of 1929 Juan Trippe employed a proven strategy: he sent Charles Lindbergh on a goodwill flight. This time, however, the "Lone Eagle" did not fly alone. On the 7,000-mile, twenty-day trip, he was accompanied by his new wife, Anne Morrow Lindbergh—the shy middle daughter of Ambassador Dwight Morrow, whom Lindbergh had met in Mexico City during his first Latin American flight and married in May 1929—along with Trippe and his wife, Betty. Lindbergh remained wildly popular, and state leaders who had rebuffed Pan Am's initial advances now wined and dined the attractive young couples with "the fanfare of a royal honeymoon." The tour, U.S. diplomats reported, proved highly effective in "stirring up . . . enthusiasm" for Pan Am. "I regard Lindbergh's coming as extremely important for success of further Pan American negotiations in this country," wrote the U.S. ambassador to Colombia, echoing statements made by many other diplomats. Indeed, within just six months of the Lindberghs' and Trippes' tour, Pan Am had received all of its desired operating contracts. In El Salvador, for example, the U.S. ambassador reported that the minister of war had been "convinced of the necessity of an international air service" after dining with Trippe and Lindbergh; a month later, his government awarded Pan Am a contract.[36]

When Pan Am reported its first year of profit in 1931, its routes extended from Miami and Brownsville, Texas, through Central America, the Caribbean, and along both coasts of South America. Although its clientele continued to be dominated by businessmen, Pan Am also increasingly attracted tourists. According to a *New York Times* reporter, Latin American hotel and restaurant owners were "accustomed now to seeing a crowd of sun-burned, khaki-clad flying Yankees trooping in to lunch, just in from a long flight." By the mid-1930s, Pan Am had swallowed up most of its competition, acquiring ownership of domestic rival NYRBA (the New York, Rio, and Buenos Aires Line, which had briefly operated flights between the United States, Brazil, and Argentina), the Cuban national airline Cubana de Aviación, Mexico's Compañia Mexicana de Aviación and Aerovías Centrales, Chilean Airways, Peruvian Airways, and Colombia's Urabá, Medellín, and Central Airways. Even the formidable SCADTA succumbed: in February 1930, Trippe negotiated a secret agreement with Bauer to purchase 84.4 percent of the Colombian airline's stock. In many cases, the acquisition of local airlines enabled Pan Am to enjoy economic privileges that it could not obtain by government contract, such as tax exemptions and land expropriation rights.[37]

Still, Latin American governments continued to assert their own national interests in aviation. Most of the contracts that Pan Am received offered far less generous terms than the airline had initially demanded, placing limitations on where, how often, and under what terms it could fly. Even Nicaragua, which was at the time essentially a client state occupied by U.S. military forces, reduced Pan Am's proposed thirty-year contract to ten years and refused to grant it a monopoly on airmail carriage and the use of national airports. In Chile, Panagra bought its contract with a "voluntary donation" of 500,000 pesos. This dynamic, in which foreign nations negotiated with American power instead of merely submitting to it, would continue to characterize international responses to Pan Am, and to the expansion of U.S. commercial aviation generally, during the next several decades.[38]

The push back that Pan Am encountered during its initial contract negotiations (and which it would continue to encounter each time these contracts came up for renegotiation) subsequently led to an important transformation in corporate strategy. In response to Latin American suspicions that Pan Am was a tool of Yankee imperialism, Trippe devoted increasing energy and resources to advertising and public relations. During the 1930s, Pan Am would market itself as a truly pan-American company—not an agent of Yankee imperialism but a conduit and embodiment of regional friendship.

"The Good Neighbor Who Calls Every Day"

Pan American Airways devised a corporate Good Neighbor policy well before President Franklin D. Roosevelt coined that term. During the late 1920s, Pan Am executives promoted their company's services by arguing that air transport would bring the nations of the Western Hemisphere closer together, both literally and figuratively, while increasing U.S. political and economic influence in Latin America. In a series of articles published in trade journals between 1929 and 1931, Pan Am public relations director William Van Dusen argued that commercial aviation would give all Americans "the means at hand to visit their neighbors, to find out who they are and what they are doing." Yet, he emphasized, the airplane would also facilitate U.S. corporations' access to Latin America's "virgin markets" and "rich areas" of natural resources. In *Aero Digest*, Pan Am's vice president Evan E. Young echoed Van Dusen's claims that South America "is more receptive to American goodwill gestures now than at any time in recent years," thanks in large part to Pan Am's presence. Thus, when Roosevelt announced the Good Neighbor policy in 1933, the "small army of businessmen" who read such publications were already well aware of the incentives to develop better relations with their southern "neighbors."[39]

The airplane was a fitting vehicle for the Good Neighbor policy, as it appeared to make the world a smaller "neighborhood" while also expanding the United States' global influence. In Latin America, Charles Lindbergh's 1927 and 1929 tours had proven that aviation could be a potent symbol of international goodwill—and, at the same time, a potent instrument of U.S. political and economic power. Aviation promised to satisfy the dual impulses of FDR's Good Neighbor policy: the improvement of U.S.–Latin American relations and the preservation of U.S. hegemony in the Western Hemisphere.

During the 1930s, Pan American Airways' publicity strategies did important cultural work to promote pan-Americanism both in the United States and abroad. The masthead of the company newsletter *Pan American Air Ways* (whose readership included employees, passengers, and aviation industry insiders) featured an illustration of the Chanin Building, the airline's Manhattan headquarters, flanked by North America on its left and South America on its right. Inside the newsletter, articles emphasized themes of American unity. "Pan American Airways System isn't just busy running an airline," stated a typical editorial. "Its real job is the bringing together into happy, profitable relationships the seventy-five millions of South America and the hundred and fifty millions of North America. . . . We are all equals, serving the greater good of the western hemisphere." To show how Pan Am served "the greater good," *Pan American Air Ways* devoted special attention to its humanitarian activities. Uplifting accounts of airborne medical clinics and disaster relief efforts portrayed the airline as the "largest life-saving institution in the world." The newsletter also publicized Pan Am's educational fellowship program, created in 1937, which enabled Latin American students to attend U.S. universities and vice versa. Though *Pan American Air Ways* was published in English and catered primarily to U.S. readers, Latin American newspapers often re-published its articles in Spanish, thus disseminating Pan Am's good neighborly message beyond U.S. borders.[40]

As it publicized Pan American (and pan-Americanism), *Pan American Air Ways* also publicized Latin America itself, portraying the region as fascinating yet unthreatening, exotic yet accessible. Its trivia column, "Do You Know?," informed readers that Brazil had more "undiscovered terri-tory" than all of Africa, that Mexico City constructed the hemisphere's first sewage system some six thousand years ago, and that La Paz, Bolivia, was the world's highest national capital. The magazine also testified to the airplane's popular appeal in Latin America. Many airports served as "community center[s]," one article reported, where residents brought pic-nic lunches to "watch the big international airliners come and go." An-other article, entitled "Ox-Cart to Air Now Commonplace," described how "men coming down from the interior step down from their burros into a Pan American airliner as if it were the most natural thing in the world." An airport manager from Mexico City recalled a "native man" who had purchased a ticket to visit his ailing mother: "At first the poor man was very upset as he had never seen such a monster which could fly through the skies like the birds of his village. By and by he felt more at home and finally dared to sit down more comfortably in his seat." As the plane ascended above the clouds, "he got down on his knees in the aisle and exclaimed in a loud voice, with his hands folded together, 'Thanks to You, My Lord, for having me granted the favor of this experience in life.'" Such articles represented aviation as a wondrous, even divinely inspired gift that a beneficent United States had bestowed on its grateful southern neighbors.[41]

Pan Am managers, meanwhile, increased their efforts to ensure that employees stationed in Latin America behaved in accordance with the neighborly sentiments expressed in *Pan American Air Ways.* In 1930, R. Leslie Cizek, an aviation insurance agent, flew over Pan Am's entire route network in order to evaluate its equipment and personnel. While Cizek expressed overall satisfaction with Pan Am, he urged management to be more selective in hiring foreign representatives. Airport managers and ticket sales agents, he wrote, "must be picked for their diplomacy, tact, and ability to command the respect of the more prominent citizens in their community." They also should be "thoroughly familiar with Spanish and the people and conditions in their territory." Following Cizek's advice, Pan Am's public relations office used house publications such as *Pan American Air Ways* to transform employees into pan-Americans. At New York headquarters, a group of secretaries became "active Good Neighbor[s]" when they organized a twice weekly Spanish class. Employees even ate like good neighbors: at one 1938 meeting, sales managers dined on Guatemalan olives, Antigua celery, cocktail de Cuba, soup à la Dominicana, Trinidad beef, salad Mexicana, Venezuelan potatoes, Virgin Islands pie, and Puerto Rican coffee.[42]

One employee in particular, Foreign Counsel David E. Grant, became the public face of Pan Am's good neighbor policy. Fluent in Spanish, Grant toured the airline's Latin American destinations, giving speeches at business conventions and airport opening ceremonies on such topics as "Trade and Friendship with Latin America." The United States had exploited Latin America in the past, Grant acknowledged, but in the present and future, Pan Am would serve as a model for a newly respectful, and mutually profitable, relationship. Like Lindbergh, Grant seemed to exemplify an anti-imperial manliness, wielding power through restraint rather than force. In the words of *Aviation* magazine, he was one of "several thousand specimens of our best young manhood" who were "making friends with the peoples of three dozen countries as part and parcel of their job of running an air line."[43]

Juan Trippe, too, cultivated a good neighborly image. Rumor held that he "deliberately played upon the ethnic connotations of his name in order to curry favor with Latin American politicians and businessmen," as *Life* noted in 1941. In fact, Trippe disliked his name, preferring to go by "J. T." or "J. Terry." Still, public relations director Van Dusen encouraged Trippe to use "Juan," especially since he looked vaguely Hispanic, with dark hair, dark eyes, and an olive-toned complexion. And although Trippe expressed reluctance to exploit his name, he embraced Van Dusen's larger strategy of promoting his airline as *simpático* to Latin America. Like David Grant, Trippe frequently gave speeches on pan-American themes. "Air transport will be the catalyst in building One America," he proclaimed in a typical statement. In his view, Pan American Airways did not simply bring the Americas closer; it was a pan-American endeavor that both fostered and depended on regional cooperation. When the Americas Foundation honored Trippe with its first "Americas Award" in 1944, he dedicated the award to "those men and women without whom it could never have been earned: to the faithful weather observer at his high post in the Andes; to the mechanic at the Santos Dumont airport in Rio de Janeiro; to the traffic men in three hundred cities 'south of the border.' " In the United States, meanwhile, Trippe became an icon of pan-Americanism. A Texas radio show, for example, aired a play in which the character of "Juan Trippe" engaged in dialogue with South American liberator Simón Bolívar, who praised the airline president for "binding the Americas in a brotherhood . . . with a silver ribbon of commerce."[44]

That "silver ribbon of commerce" was, indeed, the lifeline of Pan Am's corporate Good Neighbor policy. "Under our wings will flow trade—world trade," predicted *Hearst's International Cosmopolitan* in 1933, and Trippe dedicated his life's work to maximizing the commercial potential of the airplane. The self-titled "America's Merchant Marine of the Air" transported an eclectic array of cargo, including the valuable commodities and natural resources that U.S. firms extracted from Latin America: "the world's chief supply of sugar, of coffee, of nitrate, of bananas, of many cabinet woods . . . rum, spices and dyes, bauxite for aluminum, brazil nuts, rubber, diamonds, abrasive carbon, tannin, hides, preserved meats, copper and silver, Panama hats," as *Hearst's* informed its readers. Bolstering the airplane's cultural associations with commerce, such lists of products became standard features of newspaper and magazine articles on Pan Am. In the *Atlantic Monthly,* an account of a flight to South America enumerated the items stored in the Clipper's cargo hold: "Samples of cold cream going to Buenos Aires. Machine parts tagged for Santiago de Chile. Silk stockings and a lingerie shipment for Bogota. Surgical instruments, a set of X-ray tubes, and drugs for Panama. A feature picture to be shown in the shadow of Pizarro's tomb in Lima. Gramophone records. Typewriter ribbons. Cigars. Telephone diaphragms for Mendoza." Such accounts of Pan Am's cargo business served to make Latin America seem both desirable and available to the United States—richly endowed with minerals, gemstones, and foodstuffs that could be profitably exchanged for U.S. manufactured goods, electronics, pharmaceuticals, and films.[45]

Material culture, too, especially clothing and handicrafts, quite literally showed U.S. citizens what they could gain from better relations with their southern neighbors. Pan Am sponsored window displays of imported artisan-made crafts in major department stores including Bloomingdale's and Macy's, contributing to the Depression-era vogue for all things Latin. A Los Angeles fashion designer, meanwhile, created a series of "Clipper dresses" inspired by the indigenous styles of Central and South America.[46]

After President Roosevelt announced his Good Neighbor policy in March 1933, Pan Am began to directly reference that policy in its advertising. An ad from 1941, which appeared in mass-circulation magazines such as *Time* and the *Saturday Evening Post*, identified Pan Am as "The Good Neighbor Who Calls Every Day." The text informed readers that the nation's security required more than "guns and ships and planes"; it also involved "how we rate in friends." Meanwhile, the image of a gaucho doffing his hat in salute to a Pan Am Clipper implied that air routes linking nearly two hundred cities across the Americas were earning the United States new friends within its "neighborhood." "To untold thousands of these Latin neighbors, Pan American not only reflects Uncle Sam; it *is* Uncle Sam—the chief contact that many of them ever have with this nation," the ad stated. Of course, these "Latin neighbors" encountered the United States through many other means—movies, consumer goods, and, not least, the presence of troops on the ground. Between 1898 and 1941, the U.S. government had dispatched military forces to Mexico, Cuba, Haiti, the Dominican Republic, Puerto Rico, Nicaragua, Guatemala, Honduras, Panama, and El Salvador. But ads such as "The Good Neighbor Who Calls Every Day" conveniently obscured this history of military interventions, suggesting to U.S. audiences that aviation represented a wholly new and wholly benign means of international influence. Like Lindbergh, Pan Am's airborne Uncle Sam promised to win Latin American hearts and minds, promoting U.S. interests through attraction rather than force: "In the swift, clean lines of the Clippers, in the cut of the men who fly them, in the snap and efficiency of the service, [Latin Americans] see something that they can understand and admire," the ad's text continued. Identifying aviation with progress and benevolence, such statements served to reassure readers of Latin Americans' loyalty to the United States and therefore to justify U.S. economic expansion in the region.[47]

Indeed, while FDR's Good Neighbor policy eschewed military intervention, good neighbor discourse could nonetheless validate U.S. hegemony in the neighborhood. "The Good Neighbor Who Calls Every Day" represented the neighbors as friends but not equals. The ad's rhetoric and visual grammar coded Latin America as close to, yet behind and beneath, the United States, both literally and figuratively—stuck in a premodern past, unable to advance into the aerial age (or even to exploit its own "virgin markets") without U.S. tutelage in "snap and efficiency." Aviation, then, brought Latin America within closer reach of the United States, but cultural representations of the airplane often confirmed notions of U.S. superiority and Latin American backwardness. Pan Am's airplanes, moreover, offered a means of influence that presumably would be greeted with deferential doffed hats—like that of the gaucho—rather than the protests and anti-Americanism that resulted from military invasions. Commercial aviation reinvented power as progress, hegemony as neighborly friendship.[48]

The discursive dynamic that effected this reinvention—the cultural construction of Latin America as both close to and beneath the United States, similar to and different from it—can be seen most clearly in the cultural history of that quintessential good neighbor industry: tourism. Because of the high cost of flying, business travelers continued to far outnumber tourist travelers during the 1930s, accounting for nearly 80 percent of Pan Am's passenger revenues during its first decade. One could fly the cocktail circuit from Miami to Havana for just $25 each way, but fares to more distant destinations—$351 to Lima, $438 to Rio de Janeiro, and $500 to Buenos Aires or Montevideo—well exceeded most Americans' travel budgets. Still, Pan Am aggressively courted tourists in its advertising and publicity. And after World War II prevented Americans from traveling to Europe, the Latin American tourism market expanded rapidly. In the fall of 1939, traffic doubled on every inter-American air route, and that winter tourists comprised half of all Pan Am passengers during the Southern Hemisphere's peak travel season.[49]

To appeal to tourists, Pan Am's advertisements, like its newsletter *Pan American Air Ways*, depicted Latin America as alluringly exotic yet safely accessible. And at a time when many people considered flying to be dangerous, Pan Am promoted the airplane as a form of "home," in the sense of both the domestic and the national. "Imagine a cruise by air—with all the comforts of home!" exclaimed a Panagra "Skyway Cruise" ad from 1941, published in the *New Yorker*. While enticing readers to imagine "romance in Rio," "springtime in Chile," or "the picturesque antiquity of Peru," the ad promised that they could experience such marvels without giving up the "comforts of home": efficient transportation, modern plumbing, familiar foods, and English-speaking staff. The Skyway Cruise combined "wondrous sight-seeing in an ancient empire" with "the restful ease of a great airliner"—as suggested by ads that juxtaposed images of traditionally attired peasants and lush natural landscapes against gleaming, modern aircraft. The airplane itself, then, domesticated tourists' encounters with the foreign. Aviation made Latin America more accessible not only by reducing travel times but also by carrying the comforts of home to distant places.[50]

Those who could not afford to book passage on a Skyway Cruise could imagine the experience through reading about it. In the United States, popular magazines and newspapers played a key role in introducing both aviation and Latin America to a broader public. To complement its print advertising, Pan Am sponsored all-expenses-paid trips for reporters in order to "prepare the ground for favorable newspaper publicity" (as John MacGregor, now Panagra's publicity director, acknowledged). Reporters' accounts of these junkets, often written in the style of travel diaries, filled the pages of local papers across the country. Like Pan Am's own ads, these articles depicted Latin America as primitive, mysterious, and romantic, yet also unthreatening and inviting. Thus, even in small towns far from the cosmopolitan travel hubs of Miami and New York, aviation culture taught U.S. citizens to imagine Latin America as both entrancing and accessible.[51]

In the *Wenatchee World*, for example—the local newspaper of tiny Wenatchee, Washington—a series of articles by editor Rufus Woods chronicled a Pan Am–sponsored trip through Central and South America. The series also appeared in the local papers of Seattle, Tacoma, and other

Washington cities and was later published as a 112-page illustrated booklet, *Riding the Wings over South America*. In vivid, breathless prose, Woods
conjured the strange and thrilling people, places, and things that he had
encountered during his journey: "grotesque designs . . . found in the
sculptural work adorning the centuries-old Mayan temples near Mexico
City"; "two old black mammies standing in front of their thatch-roofed
cabins" in Natal, Brazil; "the real Shangri-La" discovered in Colombia
amid a "high chain of the Andes"; "the land of the buccaneers" in the
West Indies. But in addition to chronicling such primitive or natural
wonders, Woods emphasized that Latin America was rapidly modernizing,
and therefore becoming more attractive to U.S. tourists and investors.
"THIS WHOLE COUNTRY [*sic*] FROM MEXICO TO CENTRAL
AMERICA, THEN FROM COLOMBIA TO CHILE, IS IN PRO-
CESS OF CHANGE," he proclaimed (capitalization in original).[52] Tour-
ists could expect "roads, telephones, telegraphs, and better service from
the railroads . . . ready-made dresses for women . . . malted milk stands . . .
Coca-Cola . . . a fine big new hotel—modern style." Meanwhile, Latin
America had much to offer the United States in return. Woods described
Guatemala as "the Land of Bananas and Coffee"; Brazil was a "Land of
Vast Resources" holding "great deposits of tin, chromium, manganese,
molybdenum, and tungsten," key components of industrial materials
and weapons. Reprising the discursive strategies of Pan Am's own advertisements, this type of reportage made Latin America seem accessible and
desirable, marvelously exotic yet safely familiar.[53]

Discourse on tourism typically construed Latin America as both in
need of U.S. influence and receptive to it. In 1941, Basil Brewer, publisher
of the *Standard-Times Mercury* in New Bedford, Massachusetts, took a
Pan Am–sponsored flight through the region and subsequently wrote a
series of twenty-seven articles that were syndicated in at least twenty
other local newspapers and distributed internationally (including to Latin
American countries) by United Press International. Pan Am, Brewer
wrote, was "our best purveyor of the Good Neighbor policy." Its employees, who "know their Latin America as well as their flying," were "wearing down whatever antipathy exists toward United States business in
Latin America." Brewer acknowledged that his fellow citizens had often
been "cast, psychologically, with those Spaniards who came 400 years

ago." But his travels had convinced him that "if we show courtesy . . . we will win a permanent place in the business of Latin America." In the era of the Good Neighbor policy, these words could have appeared on State Department letterhead—and, in fact, a letter from Undersecretary of State Sumner Welles prefaced the booklet edition of Brewer's series. "I have been very favorably impressed by . . . these articles which have been read with care by the officers of the State Department specializing in our relations with the American republics," Welles testified.[54]

Other accounts of commercial air travel in Latin America justified the United States' regional hegemony more explicitly. In her book *Wings over the Americas* (1940), Alice Rogers Hager, a *New York Times* aviation reporter and consultant for the Civil Aeronautics Board, argued that only the United States could redeem its southern neighbors from centuries of backwardness resulting from the Spanish conquest. Looking down on Mexico from the comfort of a Pan Am Clipper, Hager saw a country desperately in need of help from the north. "I rested in my comfortable, familiar, American-built plane seat, watching the gashed arroyos and ragged peaks of this foreign cordillera, the backbone of a passionately loved country, mystical and immense, and wondered a little how different its history might have been if these fair gods of the North, riding their armored charges whose feet are set upon the invisible air, had been projected back into the time of the Spanish conquest," she wrote. Recalling the *New York Times'* rendering of Lindbergh as a "Northern conqueror" who could pacify "fiery Nicaragua," Hager's in-flight fantasy gendered and racialized inter-American relations. She imagined the "fair gods" of North America, aboard their airborne "armored charges," swooping down and rescuing a "mystical," "gashed," and "passionately loved" Mexico from its brutal Spanish conquerors. Whereas the Spanish, in Hager's scenario, embodied the savage masculinity of more primitive times, the United States embodied the modern, civilized manliness of Lindbergh. North America's "fair gods" would seduce Mexico by saving it, using technology as an instrument of rescue and uplift. Like the film *Flying Down to Rio*, this rendering of U.S.-Latin American relations transformed power into love, framing the Good Neighbor policy as a romance facilitated by the airplane.[55]

During the late 1920s and 1930s, then, aviation played a key role in the cultural reinvention of U.S. hegemony in Latin America. Discourse on

Lindbergh's goodwill tour, Pan American Airways' advertisements and public relations strategies, and published accounts of air travel all served to rebrand the United States as a neighbor, friend, or lover rather than a Yankee imperialist. Yet imperial power never really left the neighborhood, even though it now wore the navy blue uniform of Pan Am's "Masters of Ocean Flying Boats" instead of military camouflage. World War II, moreover, both revealed and strengthened the historical links between military and commercial aviation. As Nazi Germany advanced across Europe, and President Roosevelt looked for ways to bolster the military defenses of the Western Hemisphere, the U.S. government once more turned to its chosen instrument in aviation.

"A Polite Fiction"

The resumption of war in Europe raised the stakes of U.S.-Latin American relations, making the Good Neighbor policy a matter of national and hemispheric security. By 1941, the Roosevelt administration had grown increasingly concerned about a German attack on South America. According to the scenario envisioned by economist Eugene Staley in an influential 1941 *Foreign Affairs* essay, German ships could launch an invasion of South America's east coast and from there advance toward the Panama Canal—"which is vital to the naval power on which defense of our own coast depends"—and ultimately to the United States itself. The title of Staley's essay, "The Myth of the Continents," ominously suggested that no true geographical barriers stood between North and South America. An Axis attack anywhere in the Americas could thus imperil the entire hemisphere.[56]

Aviation, and commercial aviation specifically, was key to the Roosevelt administration's hemispheric defense strategy. In August 1939, an interdepartmental committee of State, War, Navy, and Civil Aeronautics Authority officials devised a "Plan for Aeronautical Development in the Western Hemisphere," which concluded that the United States should take a much more active role in the development of Latin American aviation, both military and commercial. Specifically, the report advocated the expansion of Pan American Airways and its local subsidiaries, encouraging the company to provide technical assistance to Latin American airlines and, if possible, to purchase controlling shares of their stock. The report also urged the U.S. government to fund the construction of airfields and navigational facilities throughout the Americas, as well as to fund the training of pilots, mechanics, and radio operators. Finally, it requested that customs procedures be modified so as to increase export sales of U.S.-manufactured aircraft and equipment. Well before Pearl Harbor, then, diplomatic and military officials within the Roosevelt administration had outlined a strategy for maintaining U.S. control over the Western Hemisphere's skyways. Indeed, beyond its immediate objective of securing the Americas against Axis invasion, the plan aimed to ensure the United States' long-term regional preponderance in commercial aviation. And it called for the U.S. government to work collaboratively with private enterprise toward the twin goals of military preparedness and economic supremacy.[57]

As the United States' chosen instrument in commercial aviation, Pan American Airways facilitated the Good Neighbor policy's transformation into a wartime security measure. In 1940, Secretary of War Henry Stimson began discussions with Juan Trippe regarding the development of airfields in Central and South America. Military planners felt that they could not rely on Latin American governments to construct adequate facilities. Yet there was little that the U.S. government could do to improve the situation without appearing to violate its neutrality in the war or its professed commitment to respecting the sovereignty of its southern "neighbors." As a private corporation, however, Pan Am could accomplish what Washington could not. Its operating contracts already granted rights to build or improve airport facilities throughout Latin America, and Roosevelt administration leaders realized that the airline could therefore function as a kind of fig leaf—Pan Am could bolster Latin America's aerial defenses under the guise of its own ongoing expansion. Though ostensibly to be used for commercial purposes, its airports could also, with certain modifications, accommodate heavy bombers for use in the event of an Axis attack.[58]

Trippe's discussions with Roosevelt administration leaders resulted in War Department Contract No. W-1097-eng-2123, signed on November 2, 1940. With an initial appropriation of $12 million from the President's Emergency Fund, the contract authorized Pan Am to develop

"airfields and other facilities for the defense of the Western Hemisphere."
Pan Am, meanwhile, created a special subsidiary, the Airport Development Program (ADP). Under the ADP's aegis, the airline would build or
renovate fifty-one airports and six seaplane bases—along with fueling
facilities, radio towers and transmitter stations, power plants, machine
shops, water and sewage disposal systems, warehouses, barracks, infirma-
ries, and other related installations—in fifteen Latin American countries.
Total War Department appropriations for the ADP exceeded $37 million
by November 1943, when the project concluded. The ADP also required
manpower on a massive scale, employing 1,500 engineers, a New York–
based administrative staff that grew from four to nearly 300, and between
55,000 and 125,000 local laborers (company estimates varied), whose work
involved everything from skilled masonry and carpentry to moving dirt
by hand in five-gallon gasoline tins.[59]

The ADP, as Hap Arnold later described it, was "a polite fiction"—a
quasi-military operation in commercial camouflage, undertaken by civilian engineers instead of soldiers. The purposes of the ADP airfields remained couched in secrecy until long after the United States entered the
war. Latin American governments were informed that the facilities would
assist Pan Am's planned transition from seaplanes to landplanes; U.S. diplomats deliberately concealed the fact that the War Department was
funding the construction. It would be "inadvisable and inexpedient" to
reveal such information, wrote Undersecretary of State Sumner Welles
in June 1942, since "Axis agents" seeking to foment anti-U.S. sentiment
in Latin America could use the airfields as evidence of "the imperialistic
ambitions of the United States." Furthermore, noted Assistant Secretary
of State Adolf Berle, "undue publicity" of the U.S. government's role in
the project could "complicate negotiations for postwar rights to these
airfields." The secrecy even extended to Pan Am's own employees, who
were told that the ADP was nothing more than "a company project in
every respect." In the airline's magazine *New Horizons*, articles reporting
"extensive international construction operations on behalf of services in
the Western Hemisphere" revealed nothing of the airfields' military significance or the U.S. government's involvement in their construction.
Journalists, meanwhile, reported the "polite fiction" that they heard from
Pan Am and the Roosevelt administration. Basil Brewer, for example,

mentioned in his 1941 series that Pan Am was building airfields in Brazil that were capable of accommodating "bombers of the largest size," but he claimed that the airfields had "no military significance."[60]

Pan Am's public relations office, meanwhile, launched a vigorous effort to market the ADP, in its commercial guise, as "a definite part of the government's 'good neighbor policy'" that would improve transportation and communications throughout Latin America. A steady stream of press releases, sent to newspapers in both North and South America, heralded the ADP as "a heroic demonstration of unity between the Americas." One typical press release stated that "North American engineers and Brazilian artisans" were working together to build airfields for the "planes of the two nations." ADP management guidelines, too, emphasized the importance of good neighborly comportment. Supervisory engineers were instructed to "conduct yourself at all times in a gentlemanly manner" so as to maintain the "confidence and good will which now exists between all Latin American countries and Pan American." They were told to hold regular meetings with local employees and to respect their cultural customs and work habits: "if necessary, instruct them how to do the work more efficiently, but don't belittle them." When supervisors did act as good neighbors—as when, for example, technicians in Brazil opened a free night school for local laborers—their efforts received favorable publicity in press releases and *New Horizons*.[61]

Although Pan Am's extensive use of local labor sustained its claims that the ADP was a pan-American endeavor, difficult work conditions also raised questions about the company's good neighborliness. Many of the airfield sites lay in jungles inaccessible to ground transport, meaning that equipment and supplies had to be flown in or transported via interior waterways. During the rainy season, much of the work had to be performed by hand. As a supervisor in French Guiana described it: "Rains, fever, material wants. The monotony is eloquent." Official company statements claimed that the ADP offered local employees "steady work," "excellent pay," and "freshly grown food." But according to reports written by on-the-ground supervisors, construction crews suffered from chronic shortages of food, clothing, tools, and medical supplies. Pan Am headquarters, they reported, often failed to wire payments for local workers, whose daily wages ranged from 38 cents (for an unskilled laborer) to $1.20 (for a mechanic or carpenter). Performing hard labor amid torrential rains with inadequate shelter and nutrition, many developed malaria and other debilitating tropical diseases.[62]

Reports written in the spring of 1942 by Mario de la Torre, the Ecuadorian-born construction supervisor at Amapa (the most remote of the ADP airfields, located in the jungles of northern Brazil), starkly conveyed a litany of hardships. "Again we are short of dynamite for blasting . . . and the hand labor of obtaining rock from river bottom has ceased because of malaria casualties." "Men that came from Belem asked the office there to pay certain amounts of their wages to their respective families. . . . This has not been done. Some of the men concerned have already quit." "Last week we ran out of food for the laborers." "We have had several planes come in . . . but not a single item of fresh supplies came on these planes. . . . We have not had any bread, eggs, or a single item of fresh vegetables or fruit in the last three weeks. . . . Some of the men are breaking out with open sores." "Shortage of bread provoked threat of a strike by the carpenters on June 5. On following day they did strike for half a day." At other sites, supervisors reported that thievery and sabotage were "a constant recurrence." At an airfield in Bolivia, for example, "Lights were broken. Generators were short-circuited. Pieces of scrap iron were tossed into the gravel crusher. Greaselines were cut and crankcases drained. Dirt was put into the gas tanks of various pieces of equipment."[63]

Back at Pan Am's New York headquarters, executives worried that public exposure of such incidents "could create serious embarrassment for the Company," and they accordingly urged supervisors to "combat effectively any conditions which are destroying Pan American's goodwill in Latin America." Meanwhile, Van Dusen's publicity office continued to market the ADP as a pan-American endeavor. No mention of sabotage, strikes, or shortages could be found in official accounts of the project—such as the *New Horizons* article, from November 1942, which described the Amapa airfield as a "Tropical Triumph."[64]

From Pan Am's perspective, the Airport Development Program was indeed a triumph. After the war, Trippe claimed to have undertaken the project out of patriotic motives; forgoing the opportunity to profit from the deal, he had charged the U.S. government only a symbolic fee of one dollar for services rendered. But such claims were somewhat disingenuous, for Pan Am clearly derived long-term benefits from the ADP. On the War Department's dime, the airline was able to create a network of airfields, communications relays, and related installations that promised to vastly increase international air traffic after the war. "Much of future commercial aviation is expected to follow the trail blazed by these new and modern facilities," noted a State Department observer in 1943.[65]

The ADP also fulfilled the objectives of the Roosevelt administration. By creating airfields that could accommodate the latest U.S. bombers, the project played a key role in securing the Western Hemisphere against potential Nazi invasion. Furthermore, as Chapter 4 will explain in greater detail, the Latin American airfields significantly assisted the Allied war effort by facilitating the transportation of Lend-Lease supplies. To circumvent Germany's stranglehold over shipping in the Mediterranean, U.S. transport planes used the Central and South American airfields as refueling bases as they ferried Lend-Lease supplies across the southern Atlantic to Allied positions in Europe, North Africa, and the Middle East.[66]

Of course, the ADP was only the latest instance in which the U.S. government used its chosen instrument in aviation as an informal extension of its military forces. From Pan Am's very inception in early 1927, its Army and Navy founders intended for the airline to serve as a bulwark against further German incursions into the Latin American aviation field, particularly in the vicinity of the Panama Canal. Anticipating postwar developments in aviation, Pan Am's commercial operations fulfilled, and were made possible by, military and strategic objectives. Though the airline rebranded Uncle Sam as a friendly civilian—the "good neighbor who calls every day"—the ADP revealed that his armed compatriots had never completely left the scene.

The history of U.S. aviation in Latin America between the 1920s and 1940s shows pan-Americanism to be entirely compatible with nationalism. Indeed, the Good Neighbor policy served national interests quite well. By the late 1920s, the military interventions and economic imperialism undertaken in the name of the Monroe Doctrine had begun to jeopardize that policy's stated objective of maintaining U.S. preponderance in the Western Hemisphere. Eschewing such overt forms of dominance, the Hoover and Roosevelt administrations, in collaboration with their allies in private enterprise, devised more subtle ways of shaping the region's politics, culture, and economy. Commercial aviation and discourse about aviation rebranded the Yankee imperialist as a good neighbor while simultaneously extending the Monroe Doctrine to the skies. Lindbergh's goodwill flight, Pan Am's expansion of commercial air routes, and the Airport Development Program all contributed to inter-American integration. In each case, however, the airplane also extended the reach of Wall Street and Washington. The good neighborly skyways did not replace U.S. hegemony but became conduits for it.

Even in the days of the cocktail circuit, however, the United States' aerial ambitions—and those of Juan Trippe personally—were never limited to the Western Hemisphere alone. As William Van Dusen stated at the end of 1937, Pan Am was already "looking ahead . . . to the point where transport aircraft could conquer the great ocean trade routes" to Europe and Asia. The airline's experience in Latin America had prepared it well for such conquest. The region, Van Dusen continued, had served as "a practical working laboratory in which could be tested, under the pressure of actual operating experience, advanced equipment, materials, and methods." And so, too, did the Western Hemisphere serve as a "practical working laboratory" for U.S. foreign policy more generally. As Secretary of State Cordell Hull recalled, "In carrying out our policies toward Latin America, it was never my wish to make them exclusively Pan American. . . . We should be more than delighted to share them with the nations of the rest of the world."[67]

Just as Pan American Airways would soon span the Pacific and the Atlantic, the strategies of influence first tested in Latin America—commercial expansion, cultural diplomacy, and covert militarization—would be exported worldwide. And aviation, again, would enable this global Good Neighbor policy to take flight.

3

Global Visions,
National Interests

In his widely influential editorial "The American Century," originally published in the February 1941 issue of *Life*, publisher Henry R. Luce argued that the realities of the so-called air age required the United States to expand its global commitments. Thanks to technologies like aviation, "our world of 2,000,000,000 human beings is for the first time in history one world, fundamentally indivisible." Americans must therefore reject isolationism, Luce argued, and embrace a "truly *American* internationalism," based not on narrow interests but on the universal principles signified by "big words like Democracy and Freedom and Justice." This American internationalism, Luce proposed, should become "as natural to us in our time as the airplane and the radio."

By 1941, aviation and the radio already suggested how the United States might wield global power by exporting its culture and values. In Luce's view, these technologies exemplified distinctly American traits of openness, ingenuity, entrepreneurialism, and individualism. At the same time, though, the airplane and the radio were enormously popular beyond U.S. borders. All over the world, people admired the feats of the Wright brothers and Lindbergh, listened to NBC radio programs, and heard about the latest achievements of U.S. aviation over the radio. Luce's seemingly contradictory notion of a "truly *American* internationalism" rested on his belief—shared by Presidents Hoover and Roosevelt, as well as by many other corporate and government leaders—that the United States' culture and technology had worldwide appeal. Its reigning conceptions of "Democracy and Freedom and Justice" could and should be universalized. Indeed, Luce argued, American culture and technology represented the only true form of internationalism: "American jazz, Hollywood movies, American slang, American machines and patented products, are in fact the only things that every community in the world, from Zanzibar to Hamburg, recognizes in common."[1]

Referring to space as well as time, the concept of the American Century envisioned the world, unified by air routes and radio waves, as one. It also, however, envisioned the United States as uniquely entitled to global leadership. A quintessential expression of "nationalist globalism," Luce's essay reconciled visions of global unity with visions of American exceptionalism and greatness. As "the world's most powerful and vital nation," the United States had a "duty and opportunity . . . to exert upon the world the full impact of our influence, for such purposes as we see fit and by such means as we see fit"—including "the right to go with our ships and our ocean-going airplanes where we wish, when we wish, and as we wish." In 1943, Henry Luce's wife, Congresswoman Clare Boothe Luce, would reiterate these arguments in a speech declaring that Americans had earned the right to "fly everywhere."[2]

More than any other source, the Luce publications shaped World War II–era discourse on aviation and its relation to the United States' role in the world. Aviation, meanwhile, implemented the Luces' vision of the American Century. Indeed, the symbiotic relationship between aviation and the Luces' particular brand of nationalist globalism reflected convergences both political and personal. Having known one another since their undergraduate years at Yale, Henry Luce and Juan Trippe both lived in Greenwich, Connecticut, where they golfed together at the exclusive Round Hill Club. In 1935, Trippe played wingman for Luce during his courtship of Clare Boothe, a glamorous playwright, *Vanity Fair* editor, and fellow resident of Greenwich. (A Pan Am station manager reportedly informed Trippe that Boothe had been on a flight to Havana; Trippe conveyed the tip to his friend, and Luce promptly boarded a Pan Am plane to meet her.) Eight years later, Trippe advised Mrs. Luce on her campaign for Congress, and, once elected, she returned the favor by advocating policies that aimed to increase Pan Am's dominance of international air transport. Mr. Luce, for his part, published flattering profiles of Trippe and Pan Am in his magazines, *Life* (which by 1944 had a domestic circulation of four million), *Time* (domestic circulation, 1.6 million), and *Fortune* (whose 170,000 readers included highly influential corporate executives, political leaders, and intellectuals). Aviation, meanwhile, implemented the Luces' vision of the American Century. The Luces themselves were avid world travelers. The son of missionaries, Henry Luce had been born in China in 1898, where he lived until the age of fourteen; he subsequently attended boarding schools in England and Switzerland, and during his adolescence he traveled widely, often on his own, throughout Europe. During their marriage, the Luces made several extensive trips to Europe and Asia and were frequently photographed disembarking from Pan Am planes. Reflecting this passion for travel and international engagement, Luce explained, in the inaugural issue of *Life*, that the magazine's purpose was "To see life; to see the world."[3]

During the Roosevelt era, the worldwide expansion of U.S. aviation offered ample evidence that the United States had become the world's most "powerful and vital nation." In 1932, Franklin D. Roosevelt became the first presidential candidate to fly to his party's nominating convention, a gesture intended to demonstrate the health of the candidate's body and the safety of aviation. Described by historian Alan Dobson as "a decisive actor in the development of both U.S. domestic and international civil aviation," Roosevelt presided over an unprecedented expansion and consolidation of the nation's domestic aviation infrastructure. As part of the New Deal's broader regulatory agenda, FDR ratified the Air Mail Act of 1934, which broke up monopolistic aviation holding companies and empowered the federal government to grant airmail contracts to private operators, set routes and schedules, determine subsidy rates, and establish pilot licensing and safety standards. The Civil Aeronautics Act of 1938 subsequently created a new federal agency, the Civil Aeronautics Authority (CAA), to provide centralized oversight of aviation. After the outbreak of World War II, FDR ordered a massive buildup of the nation's military airpower, from 800 planes in 1939 to over 300,000 in 1945. The number of U.S. citizens who had traveled by air also increased dramatically during the Roosevelt years, from 475,000 in 1932 to four million in 1941. "Travel Transformed by Wings: America's Airlines, in Ten Years, Have Covered the Nation with a Network of Planes Providing Speed, Safety, and Luxury," declared the *New York Times* in 1938—the same year that Airlines Terminal opened across from Grand Central Station, its sleek Art Deco façade a monument to air-age progress. By the end of World War II, it seemed that Americans would, in fact, soon be flying everywhere, thanks to the United States' domestic aviation infrastructure, Pan Am's expansion across the Pacific and Atlantic Oceans, the phenomenal wartime growth of the U.S. aircraft manufacturing industry, and the nation's worldwide network of military bases.[4]

Through the windows of Pan Am's Clippers, or via the colorful maps of air routes published in such magazines as *Life* and *Time*, Americans saw the world as one. Yet as aviation elevated the United States' geopolitical status, the aerial vantage point simultaneously encouraged Americans to "look down" on the world—to grow accustomed to power and at times to see other nations as inferior. Expressed materially in Pan Am's expansion and discursively in aviation culture, this distinctive form of nationalist globalism rendered the United States' ascent to global power—and the very idea of the American Century—both legible and legitimate.

Across the Oceans by Flying Clipper

Meeting with four of his top executives in the summer of 1934, Juan Trippe declared that within the next year, Pan Am would establish airmail and passenger service from California to China. The other men "stared at him in disbelief," recalled communications chief Hugo Leuteritz. The route that Trippe was proposing spanned 8,700 miles, mostly over open water. At the time, the world's longest transoceanic air route—between Dakar, on the coast of west Africa, and Natal, on the eastern coast of Brazil—comprised just 1,865 miles and was flown by French mail planes carrying no passengers. Although U.S. Navy pilots had recently flown from San Francisco to Honolulu, thirteen pilots had died during previous attempts to cross the Pacific. Given such challenges, Trippe's proposal to establish transpacific passenger flights struck his executives as audacious to the point of absurdity.[5]

A former Navy reservist and the scion of a seafaring family, Trippe had long dreamed of ocean-spanning airways that would connect the United States to the lucrative markets of Europe and Asia. In 1929, just two years after Pan Am's first flight from Miami to Havana, he instructed company engineers to begin researching the meteorological and navigational requirements of transoceanic air transport. The following year, he held discussions with leaders of Air France and Britain's Imperial Airways regarding an exchange of operating rights on transatlantic routes. Yet these meetings proved fruitless after Washington, consistent with its policy of excluding foreign airlines from U.S. skies, refused to allow Air France and Imperial Airways to fly to the United States. Subsequently, Trippe looked toward the Pacific.[6]

Like Latin America, the Pacific basin was a historic crucible of American expansionism. The fabled "China market," with its millions of potential customers, had lured the world's merchants and investors since the days of Clipper sailing ships. During the late nineteenth century, economic crisis in the United States—particularly the depressions of 1873 and 1893, which diminished U.S. consumers' purchasing power and raised concerns about "overproduction"—made the China market especially alluring, and the Cleveland and McKinley administrations sought to bolster U.S. power in the region through both formal and informal imperialism. "East Asia is the prize for which all the energetic nations are grasping," wrote Brooks Adams in 1899. That same year, Secretary of State John Hay elaborated what would become a key doctrine of U.S. diplomacy: the "Open Door" policy, codified in a series of notes that Hay dispatched to the major European powers concerning trade with China. Substituting commerce for colonies, the Open Door allowed the United States to project power internationally "without the embarrassment and inefficiency of traditional colonialism," as historian William Appelman Williams has written. At the same time, however, "traditional colonialism" also enabled the United States to expand its power in Asia, where it annexed its first overseas territories (Hawaii, the Philippines, and Guam, along with Cuba and Puerto Rico) after its victory in the Spanish-American War of 1898.[7]

For the United States, Asia thus offered an accessible and potentially lucrative air transport market. By 1931, Pan Am's competitors had already established themselves in the region—Imperial Airways flew to Singapore and Hong Kong, KLM to Batavia, and Air France to Saigon—and Trippe was eager to get his own share of the traffic. This ambition, however, presented formidable logistical challenges. Because of the earth's curvature, the shortest air routes from North America to Asia proceeded north, over Alaska, Siberia, and Japan. But after Charles and Anne Morrow Lindbergh conducted a 1931 survey flight, famously chronicled in Anne's book *North to the Orient* (1935), they advised Pan Am that harsh weather conditions made the northern passage to Asia too hazardous for passenger transport. Furthermore, the Soviet Union and Japan had refused to grant landing rights to U.S. airlines. To reach the China market, then, Pan Am had no choice but to fly directly across the Pacific—to "tackle the world's broadest ocean in a direct frontal attack" (as a company press release put it).[8]

Just as the Monroe Doctrine had assisted Pan Am's expansion in Latin America, the United States' imperial history in the Pacific opened its routes to the Philippines and China. By using Hawaii, Wake Island, Midway Island, and Guam as refueling stops, Pan Am could cross the Pacific without needing to land on foreign territory. And just as concerns over German influence in South American aviation had prompted Washington to support Pan Am's initial endeavors, concerns about imperial Japan led the Roosevelt administration to look favorably on Trippe's latest ambition. The Japanese Army had recently conquered the Chinese province of Manchuria, and in December 1933, the U.S. ambassador to Tokyo informed the State Department of rumors that Japan's Navy, seeking an equivalent prize, was planning to invade Guam. Exacerbating concerns about Japanese aggression was the fact that the United States' influence in the Pacific had recently begun to wane. The Tydings-McDuffie Act, ratified in 1934, mandated the removal of U.S. troops from the Philippines and promised to grant independence to Manila's government within ten years. Critics of the legislation, including key military leaders, feared that the United States' withdrawal from the Philippines would produce a regional power vacuum that Japan would eagerly fill.

In discussions with Roosevelt and Secretary of the Navy Claude Swanson, Trippe found a receptive audience. Pan Am could act as the Navy's surrogate in the Pacific, he argued—again functioning as a chosen instrument, the airline would maintain U.S. interests and influence even after troops withdrew from the Philippines. Pan Am's radio equipment could be used to monitor Japanese activities in the Pacific, Trippe noted, while its landing docks, airfields, and personnel barracks could be transferred to the Navy in the event of war. Yet because Pan Am was a private corporation and not a branch of the U.S. military, its presence would presumably appear less provocative to Japan than a fleet of battleships. By highlighting the convergences between Pan Am's commercial objectives and Washington's strategic objectives, Trippe persuaded the Roosevelt administration to support his latest endeavor.[9]

In early 1935, Pan Am received Navy permission to construct landing facilities on Midway, Wake, and Guam and to use the Navy's own facilities at San Francisco's Alameda Harbor and Hawaii's Pearl Harbor. In March, the *SS North Haven* departed for the South Pacific carrying 119 men (construction workers, aviation technicians, and a doctor) and over one million items of cargo, from channel buoys, generators, and motor launches to prefabricated plywood housing, toothpicks, and a movie projector. On previously barren Midway and Wake Islands, there would soon be hotels replete with screened verandas, wicker lounge furniture, gardens, and guestroom amenities such as Simmons mattresses and hot showers. Back in the United States, Pan Am commissioned aircraft manufacturers Igor Sikorsky and Glen Martin to build flying boats capable of carrying mail and passengers from California to China.[10]

On November 22, 1935, the *China Clipper*—a Martin M-130 with a cruising speed of 163 miles per hour, a passenger capacity of thirty to forty, and a price tag of $417,000—prepared to lift off from Alameda Harbor on its inaugural airmail flight to Manila. (In spite of its name, this Clipper would not be flying all the way to China, which had not yet given Pan Am landing rights.) Requiring six days, the flight would be divided into five segments: San Francisco to Honolulu (2,410 miles), Honolulu to Midway Island (1,380 miles), Midway to Wake Island (1,260 miles), Wake to Guam (1,450 miles), and Guam to Manila (1,550 miles). As with Pan

Am's previous inaugural flights, the event inspired massive public celebrations. The governor of California proclaimed November 22 Pan American Airways Day, and over 100,000 people thronged the shorelines of San Francisco and Marin County to witness the giant, silver-hulled Clipper ascend into the sky with nearly two tons of mail (for which the Post Office hired a hundred extra clerks to process). Inaugural ceremonies, broadcast nationwide and relayed to radio stations throughout Europe and Asia, featured speeches by Postmaster General James Farley, who read a statement from President Roosevelt; Philippine president Manuel Quezon, speaking over shortwave radio from Manila; and Trippe. "It is particularly significant that America's air service, with American aircraft, manned by American personnel, be the first . . . to remove forever the great barrier of distance between the peoples of the East and the West," he stated, mixing appeals to national pride with visions of global unity.[11]

Newspaper coverage also construed the flight as a great patriotic endeavor, the conquest of yet another western frontier. "The *China Clipper* and her two $400,000 sisters typify, more eloquently than a whole library, the American spirit of adventure, inventive genius, resourcefulness, and tenacity of purpose," wrote Waldo Drake in the *Los Angeles Times*. According to the Associated Press, the *China Clipper* put the Pacific Rim "on a Yankee air line tempo." And in an editorial typical of many, the *Los Angeles Times* described Pan Am's latest accomplishment as "one of the greatest tributes to American skill, American pluck, and American invincibility ever written into the records of our national achievements." The editorial also noted what was at stake in the transpacific air service: "Half of the world's population . . . is opened up to us. . . . $5,000,000,000 a year is the trade stake in the Orient thus brought to the front door of California." America's nineteenth-century romance with the "China market" had been updated for the twentieth-century air age.[12]

When the *China Clipper* returned to San Francisco on December 6, it had traveled 16,420 miles during 123 hours and twelve minutes of flight time (less than the estimated time of 130 hours), setting nineteen world records along the way. In the fall of 1936, Pan Am began scheduled passenger flights to Manila; the following spring, the route was extended to Hong Kong, making it possible to circumnavigate the globe entirely via commercial airlines. Appropriately, Juan and Betty Trippe were among the first passengers to do so, completing a chartered worldwide flight (via Pan Am, Imperial Airways, and the German dirigible *Hindenburg*) in December 1936. The first paying passenger to fly around the world, in April 1937, was S. Davis Winship, an American businessman based in Manila. For those who did not earn executive salaries, fares were prohibitively expensive: $799 for a one-way ticket from San Francisco to Manila and $950 to Hong Kong, well over $10,000 in 2010 dollars.[13]

Even so, passenger revenues proved unable to sustain Pan Am's operating expenses. In 1938, the cost of flying the Pacific consumed all but $75,000 of the $1.2 million in profits earned by the airline's Latin American division. But profit was hardly the sole purpose of the route. Military and diplomatic leaders had supported the enterprise for strategic reasons, believing that Pan Am would play a crucial role in maintaining U.S. influence in the region—particularly in the context of Japan's expanding imperial ambitions—and the Navy continued to render assistance to the airline, transporting supplies and helping to repair and maintain its equipment. As Lieutenant Colonel Melvin J. Maas, the Marine Corps' ranking reserve flier and a member of the House Naval Appropriations Committee, told the *Los Angeles Times* in 1935, Pan Am's presence in the Pacific "is invaluable to the future of American trade in the East, but it is absolutely vital to our national defense. Consider the value to our Navy . . . of the splendid bases, or 'stepping stones,' which Pan American has built at Honolulu, Midway, Wake, Guam, and Manila."[14]

When the newly created CAA held hearings on airmail subsidies in 1938, top Navy officials secretly testified on Pan Am's behalf, stating that its operations were vital to the national interest. Subsequently, the CAA increased Pan Am's airmail subsidies from $2 to $3.35 per mile between San Francisco and Manila and to $7.12 per mile between Manila and Hong Kong—despite dramatically reducing domestic carriers' subsidies. Although Roosevelt had entrusted the CAA to eliminate economic waste and corruption in the airline industry, in this case the agency placed military security above fiscal considerations. The increased subsidies enabled Pan Am's Pacific Division to reduce its annual losses from over $1 million to $200,000, making 1939 the most profitable year in the company's twelve-year history. As in Latin America, Washington's short-term strategic objectives served Pan Am's long-term commercial objectives.[15]

As Pan Am expanded its Pacific operations (adding flights to New Zealand in December 1937), Juan Trippe simultaneously pursued plans to begin flying across the Atlantic. Connecting the world's leading industrial powers and trade markets, the skyway between New York and London was viewed, by both European and American aviation interests, as the world's most prestigious and lucrative route. And Trippe was determined not to allow European companies to dominate transatlantic aviation as they had long dominated transatlantic shipping. Pan Am, stated public relations director William Van Dusen, intended to "assure for the United States a leadership on the Transatlantic trade route which it lost, and has never been able to recover, on the sea." Surveys of possible routes and refueling stops began in 1933, when Pan Am sent the Lindberghs on a flight through Newfoundland, Labrador, Greenland, and Iceland. In 1936, Trippe obtained the necessary governmental permissions to operate a twice-weekly airmail and passenger service between the United States and Britain, with Pan Am and Imperial Airways each providing one weekly round-trip crossing. (In order to reach agreement with the British, Trippe consented to the so-called Clause H, which mandated that Pan Am would not begin transatlantic flights until Imperial Airways had the aircraft to do so also—a concession he would later regret, as Pan Am was ready three years before the British.) Agreements with other European governments soon followed. In 1937, Trippe obtained from Portugal an exclusive, twenty-five-year franchise for Pan Am to fly to and from Lisbon and the Azores; the following year, the United States and France signed a temporary air-rights agreement allowing Pan Am to fly to and from Marseille.[16]

On May 20, 1939—twelve years to the day after Charles Lindbergh's historic solo flight across the Atlantic—Pan Am's *Yankee Clipper* lifted off from New York's Port Washington airfield with a cargo of mail bound for Marseille. Like the departure of the *China Clipper* four years earlier, the event was infused with patriotic symbolism and inspired exuberant nationalist rhetoric. Christened by Eleanor Roosevelt, the *Yankee Clipper* had been named after "the famous sailing clippers which brought such great prestige and glory to this country a hundred years ago," as Trippe explained. The comparison evoked visions of past and future U.S. leadership in global transport. Journalist Harrison Forman, for example, hailed

Pan Am as "America's future merchant marine of the air, harking back to another century when America proudly ruled the trade routes of the world with her Yankee Clipper ships."[17]

During the summer of 1939, the transatlantic air service prospered even as war loomed in Europe. Passenger flights between New York and Marseille began on June 28, with flights between New York and the port city of Southampton, England, beginning on July 8. The first transatlantic ticket went to William J. Eck, a railroad executive who had placed his reservation with Pan Am in 1931; the other twenty inaugural passengers included Colonel William Donovan, a New York attorney who would later head the Office of Strategic Services, predecessor of the Central Intelligence Agency; Cornelius Vanderbilt Whitney and his wife; Louis Gimble Jr., of the famed department store; and executives from General Motors, the American Hawaiian Steamship Company, and United States Lines. Fares were set at $375 one-way and $675 round-trip—comparable to the price of a first-class steamer berth or a flight on the Concorde in the 1980s.[18]

After the outbreak of war in August 1939, the Clippers continued to cross the Atlantic, with flights terminating in the nonbelligerent countries of Portugal and Ireland. Meanwhile, the war transformed the experience of international air travel. Previously a glamorous pastime for the wealthy, it was now more serious business, as men such as Bill Eck found themselves sharing cabin space with intelligence agents, uniformed generals, and war refugees. According to a *Life* cover story from October 1941, transatlantic flights were "so packed with Government agents, military observers, official missions and ordinary U.S. spies, traveling under their own or other people's names, that four-star foreign correspondents are regarded practically as stowaways." Pan Am's presence in Europe, moreover, allowed the War Department to monitor German activities just as its stations in the Pacific were helping to keep tabs on the Japanese. After France fell to the Nazis in June 1940, Pan Am's Lisbon base maintained the only radio station on the continent that was not Axis-controlled, and its employees became vital sources of intelligence. The Portuguese capital "teemed with spies, adventurers, and numerous species of displaced humanity," including refugees desperately seeking passage on the westbound Clippers. Lisbon traffic manager Jack Kelly reportedly "flashed his movie-star smile and graciously spurned bribes for the few available seats." Meanwhile, he discreetly watched his Lufthansa counterpart for "readings of Nazi intentions," reporting anything noteworthy to Trippe— who in turn passed the information to G-2, the War Department's military intelligence division.[19]

By the fall of 1941, Pan American Airways had effectively become, to quote the title of *Life*'s October cover story, "a branch of U.S. defense." Even before the United States officially entered the war, Pan Am's routes and airfields in Latin America, the Pacific, Europe, and Africa had all become vitally important to U.S. security, enhancing Washington's ability to assist the British war effort, to protect the Western Hemisphere, and to monitor Axis movements. The Clippers, noted *Life* writer Noel F. Busch, "put Lisbon within 27 hours and Manila within six days of Washington at the precise juncture in history when this improvement was most warranted." Indeed, he added, "it has sometimes seemed that the rest of the defense effort, still largely composed of creaks and groans in Washington, should become an arm of Pan American," whose personnel "amounts to a kind of informal State Department." At the time of Pearl Harbor, Pan Am and its subsidiaries operated 88,478 route miles—twice as many as the United States' ten largest domestic airlines combined— that connected five continents. At three hundred airports around the world, the airline maintained 162 aircraft (with forty more on order) and 192 radio and weather stations. This worldwide commercial aviation network, constructed with the financial and diplomatic support of the U.S. government, would serve after 1941 as the circuitry of the Allied air war.[20]

Meanwhile, Pan Am became a cultural icon whose symbolic potency matched its strategic importance. Epitomizing luxury and glamour, the airline's great flying boats made air travel the ultimate signifier of class status during the 1930s and 1940s. Pan Am enlisted designer Norman Bel Geddes—a leading figure in the Art Deco movement, whose work defined the Streamline Moderne aesthetic—to create passenger cabins as posh as first-class berths on Cunard liners; steamships, after all, were the Clippers' main competition. Passengers sipped Clipper Cocktails (one pony White-Label Puerto Rican rum, one pony dry vermouth, one half-teaspoon grenadine, served on the rocks with a cherry) while relaxing in clubby, wood-paneled lounges. In the dining room, they enjoyed five-course meals served by white-jacketed stewards on tables set with linens, fine china, crystal, and fresh flowers. They then retired to berths with full-sized beds, soundproofed walls, and fresh-air ducts. Upon waking, they would find their shoes freshly shined.[21]

Although only twenty-eight Clippers took to the skies, their cultural significance far eclipsed their numbers. In San Francisco and New York, crowds that numbered in the thousands witnessed Pan Am's daily departures. At international expositions in Chicago (1933–1934), San Francisco (1939–1940), and New York (1939–1940), visitors from around the world saw the Clippers displayed as embodiments of American industrial genius. And in the popular culture of the 1930s and 1940s, they inspired everything from dances to children's games to such Hollywood movies as *China Clipper* (Warner Brothers, 1936), written by former Navy pilot Frank Wead and starring Humphrey Bogart. The story of an upstart airline that defies all odds to be the first to fly between California and China, *China Clipper* presented a thinly fictionalized version of Pan Am's real-life history, and the airline enthusiastically cooperated in the film's production, allowing Warner Brothers to shoot footage of its aircraft and bases. Bogart played a swashbuckling aviator modeled on chief pilot Ed Musick, who helmed the *China Clipper* on its inaugural flight (and in 1938 perished aboard the *Samoan Clipper* on a survey flight to Auckland); the airline's boss, played by Pat O'Brien, was clearly based on Trippe. A Clipper had an important cameo in the finale of an even more famous Bogart movie, *Casablanca* (Warner Brothers, 1943). American expatriate café owner Rick Blaine (Bogart) watches his lost love, Ilsa Lund (Ingrid Bergman), depart Casablanca on a Pan Am flight to Lisbon, where she and Czech resistance leader Victor Lazslo (Paul Henreid) plan to escape from the Gestapo aboard "the Clipper to America." Elmer Rice's Broadway play *Flight to the West* (1940)—hailed by *New York Times* theater critic Brooks Atkinson as "the most absorbing American drama of the season" and staged in community and college theaters across the country—also portrayed the New York–bound Clippers as ships of salvation for European refugees. And in the film *Bombay Clipper* (Universal, 1942), two Americans prevent Nazi agents from hijacking a Clipper carrying $4 million in diamonds from India. In all four productions, as in Luce's

"American Century" essay, the airplane symbolized progress and freedom, serving as a synecdoche for the United States' self-proclaimed role in the world.[22]

Pan Am's iconic aircraft also starred in radio shows (*The Shadow, Don Winslow of the Navy, Doc Savage*), popular songs including Martha Hastings's "Flying on the Clipper Ship" (1941), and advertisements for products unrelated to aviation. An ad for Four Roses gin portrayed a dapper businessman disembarking from the *Dixie Clipper*, presumably on his way to a martini. Gerber's advertised that Pan Am served its baby food aboard flights. Kellogg's quoted Joe Wuller, chief crewman at Pan Am's Miami base, as saying that "a big bowl of Kellogg's Corn Flakes" kept him going in his job as "Nursemaid to a 20-ton Clipper." A type of ale called Clipper Pale Beer, produced by Grace Brothers Brewing Company in the early 1940s, came in cans featuring an image of the *China Clipper*. The beer was discontinued in 1946, and in 2002, eBay sold one of the cans for $19,299.99—in inflation-adjusted dollars about as much as a round-trip flight on the real *China Clipper* seventy-five years earlier.[23]

Like the aircraft's image, the very name "Clipper" became a form of cultural capital in the 1930s, evoking the cosmopolitan glamour of international air travel. In 1936, Cadillac offered its luxury automobiles in the shade of Clipper Blue Metallic; another automaker, Willys, produced a model called the Clipper, whose streamlined aesthetic resembled that of its airborne namesake. By the end of the decade, restaurants and bars with "Clipper" in their names could be found all over the United States, from hole-in-the-wall Chinese eateries such as the China Clipper in midtown Manhattan (owned by two men who claimed to be "Air Veterans of the Chinese War") to the swanky Clipper Ship Lounge at San Francisco's Chancellor Hotel, whose Art Deco dining room overlooked the bay where Pan Am's flights arrived and departed.[24]

Historians have attributed the "Clipper craze" to escapist impulses in Depression-era U.S. culture. "To depression-weary Americans, the new Martin clipper possessed magical qualities," wrote Robert Gandt in a typical version of this argument. "She conjured up visions of the exotic East, of faraway places and mysterious lands. She was a fantasy craft, a magic carpet built and flown by Americans, destined for adventure."[25] But Depression escapism does not fully explain the Clippers' appeal, for cultural representations of the aircraft tapped into much older narratives about frontier conquest and national expansion. The *China Clipper* evoked the United States' history of conquest in the Pacific; the *Yankee Clipper* recalled its former dominance of merchant shipping. If the Clippers conjured globalist visions of "faraway places and mysterious lands," they also conjured nationalist visions of Manifest Destiny, and the Clipper craze expressed fantasies of power as well as escape. The flying boats transported Americans to foreign lands, literally and imaginatively, but they also carried them home, to a nation imbued with expansionist vigor.

As Pan American Airways increased U.S. citizens' access to the world, discourse on aviation simultaneously shaped their ideas about the world. During World War II, this discourse played a key role in teaching Americans to be global citizens. Commentary on aviation invariably incited speculation about the future of a world with "no distant places"—speculation that took on increasing urgency after the bombing of Pearl Harbor in December 1941. Facilitating commerce and tourism but also signifying the terrifying realities of modern warfare, the airplane clarified the stakes of global integration in both war and peace.

Wings over One World

Reprising the situation at the outset of World War I, when World War II began the United States' air forces were ill prepared for a major military conflict. The Army and Navy had demobilized their fledgling air services immediately after the First World War, and for the next two decades—particularly during the Depression—Congress provided little funding for aeronautical research, aircraft manufacturing, and the development of military airpower. The sorry state of U.S. military aviation became tragically clear in the winter of 1934, when Roosevelt canceled the airlines' airmail contracts (due to widespread reports of corruption within the industry) and instructed the Army Air Corps to carry the mail. With pilots who lacked experience in bad weather and night flying, and outdated aircraft that lacked basic navigational equipment such as illuminated cockpit instruments, the Army lost three of its pilots in training exercises; two more men perished during their first week of carrying the mail, and eight planes were destroyed. Eddie Rickenbacker, the famous World War I "ace" and a vice president of Eastern Airlines, described the debacle as "legalized murder." Yet Roosevelt's disastrous experiment with entrusting airmail service to the Army did not prompt Congress to increase appropriations for airpower. As of August 1939, the Army Air Corps employed just 26,000 personnel, while the British Royal Air Force numbered 100,000 men and the German Luftwaffe 500,000. Together, the Army and Navy possessed around 1,600 planes (although many of them were obsolete), compared with Britain's 1,750 and Germany's 3,750. The U.S. aircraft manufacturing industry fared somewhat better, in 1939 producing its highest output (a total of 5,856 planes) since the Wall Street crash ten years prior. However, many of its clients were British and French concerns that had turned to the United States for assistance in rapid wartime mobilization—and with a workforce of just 64,000, the industry lacked the manpower (and plant equipment) to keep up with European demand. In 1939, aircraft manufacturers had a backlog of $680 million, of which $400 million consisted of foreign orders.[26]

President Roosevelt, a Navy man, had not initially been a strong advocate of airpower, but after the Munich Agreement of September 30, 1938—in which British, French, and Italian leaders infamously "appeased" Adolf Hitler's desire to annex Czechoslovakia's Sudetenland—he came to believe that further Nazi advances across Europe could be stopped only by an air force more formidable than the Luftwaffe. As crisis in Europe escalated prior to the Munich Agreement, William Bullitt, U.S. ambassador to France, had tried to warn the president, sending a cable that stated, "The moral is: If you have enough airplanes you don't have to go to Berchtesgaden" (referring to Hitler's mountain resort, where, earlier that September, British prime minister Neville Chamberlain made his first compromises with the Nazi dictator). In hindsight, FDR understood what Bullitt had meant. In November of 1938, he announced his intention to create a U.S. air force of ten thousand planes and the capacity to produce an additional ten thousand per year. Paraphrasing Bullitt, FDR told his advisors that had the United States possessed such airpower in September, "Hitler would not have dared to take the stand he did" at Munich. Although some Army and Navy leaders continued to express reservations about airpower, the public now sided with FDR: in a poll taken shortly after his announcement, 90 percent of respondents supported a larger air force. In April 1939, Congress responded with a $300 million appropria-tion for the Army Air Corps to procure up to six thousand aircraft.[27]

In the spring of 1940, Germany's devastating aerial bombardment of England, the Blitz, eliminated remaining skepticism about airpower among military leaders and "brought about a complete reassessment of what was needed in aircraft production." On May 16, FDR made a public announcement that dramatically increased his previous production quo-tas, calling now for the manufacturing of fifty thousand planes per year. Yet although the U.S. aviation manufacturing industry had expanded significantly since 1938, its productive capacities—manpower, machinery, even square footage of plants—were far from able to meet such demand, especially since Britain and France also dramatically increased their own aircraft orders in 1940. Thus, manufacturing firms could produce only 12,813 aircraft in 1940 and would not meet FDR's 50,000-per-year rate until 1942. Nonetheless, with presidential, military, and congressional backing—both rhetorical and material—the industry mobilized for war at an astonishing pace, especially after automobile manufacturers such as Ford turned to building planes. U.S. production of military aircraft, ac-cording to historian John Rae, "rose from 3,807 in 1940 to 19,433 in 1941, almost as many as were produced in completely mobilized Britain and 8,000 *more* than in Germany."[28]

The techniques of mass production that had created the United States' wealth during the early twentieth century now enabled it to build the world's largest air force. Beginning in 1942, factory assembly lines oper-ated twenty-four hours a day, seven days a week, churning out a total of some 300,000 military planes by the end of the war, an increase of 1,600 percent since 1939. As of 1944, U.S. aircraft manufacturing firms em-ployed 2.5 million workers (a large percentage of whom were women) and produced 110,000 planes per year—thus producing more than twice as many planes *annually* as had been built in the United States in the *thirty-six years* between the Wright brothers' first flight and the beginning of the war. The U.S. Army Air Force grew at an equivalent rate, numbering 2,411,294 by 1944, over ninety times its 1939 size. Much of this growth resulted from the Civilian Pilot Training Program (CPTP), created by Congress in 1939. Taking the New Deal to the skies, the program pro-vided government funding for U.S. citizens to enroll in private flight training schools. By 1944, the CPTP had trained some 400,000 students; half became military pilots.[29]

Commercial airlines also significantly benefited from the war effort. The War Department funded the construction of air bases around the world, including those that Pan Am had built in Latin America under the ADP, which promised to serve U.S. airlines after the war ended. Also with important consequences for the postwar air transport market, the nation's domestic airlines began flying internationally as they lent their equipment and pilots to the Army's Air Transport Command (ATC), which delivered aircraft, supplies, and mail to combat zones worldwide. The domestic airlines would later lobby for international routes by citing their ATC service as evidence of their ability to fly where only Pan Am had flown before. In a 1944 publicity brochure, American Airlines drew attention to its 150 monthly supply missions across the Atlantic, noting that "skilled American Airlines mechanics have been stationed in bases stretching from Greenland's icy mountains to India's coral strand." By the end of the war, the ATC had become, in the words of a *Fortune* headline, "The World's Greatest Airline," operating in six continents and sending a plane across the Atlantic every thirteen minutes and across the Pacific every ninety minutes.[30]

Meanwhile, the world war transformed American men, women, and children into "air-minded" citizens, bringing aviation directly into the daily lives of millions—whether they participated in air raids, worked in factories building bombers, studied aircraft spotting manuals in order to search the skies for enemy planes, or simply read the latest war news in their morning papers. More than any other event, World War II made the airplane real to Americans, even to those who had seen airplanes only in photographs and on movie screens.

Like the growth of the aviation industry, the wartime "air-conditioning" of the population (as *Time* described it) benefited from the support of the federal government. In elementary and high schools, a movement for "air-age education" began in 1942 on the recommendation of Assistant Secretary of Commerce Robert Hinckley. The Aviation Education Research Group, based at Columbia University's Teachers College and endorsed by the Civil Aeronautics Administration and the Department of Education, published a series of eighteen textbooks with aviation themes.

Titles included *The Biology of Flight, Physical Science in the Air Age, Mathematics in Aviation,* and *Globes, Maps, and Skyways;* two volumes, *Wings for You* and *Flying High,* explored depictions of aviation in literature. In late 1942, copies of the series were distributed to five thousand school systems. By 1945, furthermore, half of the nation's 26,000 high schools had aeronautics courses. Complementing such public initiatives, the airlines also made efforts to educate American youth. Pan Am, for example, published a weekly school newspaper, *Classroom Clipper,* which juxtaposed articles on the science and social effects of flight with travelogue-style profiles of the airline's international destinations.[31]

Academic scholars, meanwhile, employed new methods of social science to determine how aviation could benefit the economy, foreign relations, even medicine and psychology. The most influential of these texts was *America Faces the Air Age* (1944), a two-volume study published by the Brookings Institution and authored by economist J. Parker Van Zandt, who had served as Pan Am's station manager in Manila during the 1930s. The airplane, Van Zandt argued, was a "magic wand" that could "put the world on a solid economic foundation." After the war, he predicted, large numbers of U.S. citizens would be able to "fly down to Rio, see the Acropolis, visit Victoria Falls, the Taj Mahal, Peking, Baghdad, and all the other interesting places of the world." As it brought such distant places into Americans' reach, Van Zandt claimed, aviation promised to stimulate the global economy by increasing trade and tourism. *America Faces the Air Age* received multiple reprintings, and the American Historical Association's GI Roundtable program donated copies to soldiers.[32]

Because the airplane seemed to shrink the distance between continents and countries, proponents of air-age education argued that the "air-conditioning" of U.S. citizens must entail educating them about international affairs. "Teach Them the World," urged American Airlines in a 1944 advertisement published in travel industry trade journals. "The travel agent, with his world knowledge, and American Airlines, with its extensive routes, will join in making travel dreams come true," the ad declared. Indeed, during the 1940s, discourse on aviation in such texts as the Air Age Education series, *America Faces the Air Age,* and "Teach Them the World" helped explain international relations and the United States' changing role in the world. The culture of aviation thus gave Americans a "language of international engagement," to borrow Elizabeth Borgwardt's phrase—a language that taught them the world.[33]

No single text shaped this language of international engagement more than Wendell Willkie's best-selling book *One World* (1943), which chronicled a round-the-world flight that Willkie completed in the fall of 1942. As a gesture of political unity on the American home front, President Roosevelt had selected the Republican populist, his opponent in the 1940 presidential contest, to be his personal emissary on a goodwill tour of the world. On August 26, Willkie boarded a converted four-engine bomber at Long Island's Mitchel Field. Appropriately named *The Gulliver*, the plane traveled 31,000 miles in forty-nine days and visited over a dozen cities, including Cairo, Jerusalem, Baghdad, Ankara, Moscow, Yakutsk, Chungking, and Belem. Back in the United States, Willkie told the popular aviation magazine *Skyways* that "the airplane will be one of the greatest factors in developing political and economic internationalism in the post-war world." Marveling that his global voyage had required just 160 hours aloft, Willkie envisioned "a network of airlines connecting all the nations." But aviation was much more than a means of transportation. The idea of flight, he emphasized, "captured the imagination of men in every country I visited." Aviation could "make men think on a global scale" by liberating their movement and communication from national borders. "Aviation represents a highway to new horizons of brotherhood," he concluded.[34]

In a widely heard national radio address broadcast on October 26, 1942, Willkie repeated his globalist message: "Continents and oceans are plainly only parts of a whole, seen, as I have just seen them, from the air." Across the country, commentators described the speech as sensational. "When Wendell Willkie ceased talking Monday night at 10 o'clock, a new phase of world politics began," proclaimed journalist William Allen White; the *Christian Science Monitor* predicted that Willkie's trip "may turn out to be more important than Phineas Fogg's and Marco Polo's put together." In April 1943, Simon and Schuster published *One World*, which reiterated and amplified the themes of the radio address. Within seven weeks, the book had sold one million copies. By 1945, *One World* had become the nation's best-selling work of nonfiction, with 4.5 million copies sold. In the text, Willkie attributed his globalist sentiments to on-the-ground encounters with ordinary people. Clad in a rumpled suit that ill fit his bulky frame, the gregarious Hoosier utilities lawyer presented himself as the ambassador of the American "common man," and he reported what he heard from other common men on farms, in factories, and on the front. But Willkie acknowledged that his globalism had fundamentally been inspired above the ground, by the world he glimpsed through the windows of *The Gulliver*—a world with "no distant places," united by a "navigable ocean of air."[35]

The tremendous popular success of *One World* cemented the coupling of aviation and globalism in American culture. "Our whole conception of One World is bound up with the airplane," stated a *New York Times* editorial written nearly four years after the book's publication. Yet Willkie was far from alone in promoting the airplane as an instrument of international solidarity. Vice President Henry Wallace, the Roosevelt administration's most outspoken proponent of progressive internationalism, made similar arguments. Wallace had embarked on his own goodwill tour in 1943, flying Pan American Airways' subsidiary Panagra throughout South America. In a lengthy article for the *New York Times Sunday Magazine*, he subsequently advocated a "network of globe-girdling airways" owned and operated by a single, multinational airline and peacekeeping air force. The air age, Wallace believed, required innovative new forms of supranational governance—a New Deal on a global scale. He even suggested to FDR that the federal government offer subsidies for U.S. citizens to travel abroad in order to create "eagerness for worldwide cooperation."[36]

Although Wallace's proposal to internationalize the airways was considered radical, American discourse on aviation during the 1940s overwhelmingly affirmed the idea that air travel would enable the "great mixing up of peoples, so vastly accelerated by war, [to] be continued into the peace," as *Fortune* magazine predicted in 1945. Such sentiments were not confined to elite magazines like *Fortune*. In a letter to *New Horizons*, Margaret Saunders, the wife of Pan Am's Guam station manager, testified that "an evening spent in the lobby of a P.A.A. hotel on an Island station is an interesting experience. People from all parts of the world, on varied and important missions, contribute to conversations that could not take place anywhere else in the world. . . . The cross-section of thought and opinion gleaned at these informal chats with Dutch, French, Chinese, Filipino, and American travelers is an educational experience well worth acquiring." Saunders's account corroborated Wallace's hopes that aviation could be the catalyst of a popular globalism.[37]

But one did not need to actually fly to Guam to partake of the "educational experience" that air travel could provide. The magazine *National Geographic*, which had a circulation of 1.1 million by 1940, served as a leading source of Americans' knowledge about the world, allowing them to visit distant places vicariously. Since long-range flight had first become possible in the mid-1920s, *National Geographic* had regularly published photo essays on international aviation, using the airplane, as it had long used the camera, as a device that yielded intimate views of foreign places and peoples. In one such article, J. Parker Van Zandt, author of *America Faces the Air Age*, described how traveling by airplane "does something to you":

It uproots the most ingrained habits and dissolves the stubbornest prejudice. You're never the same again. For the spell of the skyways is in your blood and cannot be shaken. . . . A new earth opens within the old: a new intimacy made possible between diverse peoples. And you share in the modern magic of aircraft that is fashioning a neighborhood of nations, transforming the whole world into an island community.

To illustrate how international air travel could transform the world into a unified "island community," Van Zandt recounted the following anecdote:

Among the dozen or more passengers already aboard, most of whom had been with the plane for several days, one sensed a pleasant air of comradeship. . . . "Do you happen to know a friend of mine in Bangkok, Siam?" I asked [the steward], naming a prominent traffic agent. "But yes!" beamed the steward. "To him I always bring fresh strawberries from Bandoeng; and he gives me that nice yellow fruit—what do you call it?—mangos." A meal aloft on a K.L.M. airliner is a true example of international amity!

Anticipating the global imaginary of neoliberalism, whose dominant spatial units are metropolitan commercial centers rather than nations, Van Zandt suggested that world trade could render political boundaries insignificant. In his vision of the air age, the names of places merely signified different types of consumer goods—strawberries from Bandung, mangos from Bangkok—that could be pleasurably and profitably exchanged by airborne cosmopolitans.[38]

The publication that did the most important cultural work to promote this air-minded cosmopolitanism was Henry Luce's *Life* magazine. Luce had enthusiastically endorsed Willkie in the 1940 presidential race; according to Alan Brinkley, it was "the first time in his life" that Luce "felt truly passionate about a political figure," and the two Republican internationalists forged "a pragmatic friendship" during late-night conversations over drinks in Luce's office. It was widely rumored that Willkie, if victorious, would appoint the publisher as his secretary of state. Meanwhile, *Life*'s frequent, lavishly illustrated coverage of aviation had extolled the idea of one world even before the publication of Willkie's book. Its June 3, 1940, issue, dedicated to the theme of "America and the World," opened with a photo essay, "*Life* Flies the Atlantic: America to Europe in 23 Hours by Clipper." Pan Am's transatlantic Clippers, the article stated, had become "the best international club in the world," an exclusive airborne enclave populated by "headliners": diplomats, military brass, royalty, spies, journalists, and movie stars. "There is excellent conversation in three or four languages, and a striking absence of social ice to break, for the air and the war combine to produce an easy good fellowship of the Atlantic traveling elite." The airplane thus represented a microcosm of good international relations. World affairs, *Life* implied, could be safely entrusted to the "traveling elite." Echoing *Life*'s arguments, Pan Am itself claimed to be the "diplomats' choice." In its own version of the society pages, *New Horizons* featured photo spreads of "the men and women by whose minds the outlines for a post-war peace, security, and freedom will be formulated"—including Henry and Clare Boothe Luce, who were frequent Pan Am passengers.[39]

In spite of its inclusive rhetoric, however, *Life*'s vision of globalism remained highly *ex*clusive. Its very language—"island community," "international club"—implied that the airplane was not a microcosm of the world at all but an elite enclave populated only by those of high occupational or class status. As *Life* acknowledged, most Americans could not fly across the Atlantic, due to expense and to wartime State Department restrictions that granted travel privileges only to diplomats and other priority passengers. Thus, although Willkie and Wallace cherished both the airplane and the "common man," rarely did these two symbols of midcentury globalism actually meet, except in the arena of combat.

The global color line, too, extended into the air. Pan Am, notably, did not discriminate on the basis of skin color—after all, racial discrimination would have been unprofitable for an airline with routes throughout the world—and nonwhite dignitaries from Latin America, Asia, and Africa regularly appeared in the photo pages of *New Horizons.* Domestic airlines, however, devised subtle ways of keeping the skies white. Reservation agents screened telephone calls by residential address, and when a caller requesting tickets "sounded black," agents would often report that all seats were booked. Throughout the U.S. South, many airport terminals remained segregated until the late 1960s. In short, aviation liberated people only from the pull of gravity, not from forces of social inequality. Though the barriers of geography had become less absolute in the air-age world, the boundaries posed by divisions of class, race, nation, and empire continued to restrict human movement.[40]

In spite of countervailing evidence, however, the idea of the airplane as an instrument of international engagement continued to be extremely compelling. Images of "one world" pervaded the visual landscape of World War II–era aviation culture. Perhaps even more than written commentary, such imagery allowed the public to see how airplanes were unifying the world. The U.S. aviation industry promoted global thinking not only by transporting some Americans to other countries but also by advertising the world to a much larger public. In 1943, for example, readers of *Newsweek, National Geographic,* the *Saturday Evening Post,* or any of the Luce magazines would have encountered numerous advertisements with images of shrinking globes. An Eastern Airlines ad, "It's a Smaller World Now!," featured two images of the earth, one labeled "The World War 1917," the other—much smaller and encircled by an orbiting fleet of airplanes—labeled "The Global War 1943." In an ad for Wright Engines, "Modern Atlas," a man placed a tiny globe into a child's open palm. Consolidated Vultee Aircraft's ad "No Spot on Earth Is More Than 60 Hours From Your Local Airport" showed the globe encircled by four airplanes whose routes linked an unidentified point in the American Midwest to Tokyo, Singapore, Moscow, and Zanzibar.[41]

Aviation advertisements, then, did not simply market products; they sold the very idea of globalism, giving "one world" a compelling cultural iconography. American Airlines' "Air Map" ad, from 1943, reproduced (in two-dimensional form) the "air globe" originally conceived by the Aviation Education Research Group at Teachers College. Entirely white, this image of the earth depicted no continents, nations, oceans, or topographical or national boundaries; its only markings were black dots representing cities. In using this image, the company aimed to "introduce to the American people a new concept of the geographical relationship of all cities and all peoples throughout the world," wrote American Airlines' vice president O. M. Mosier in a letter to Roosevelt's secretary. The ad's visual simplicity expressed "the simplicity of air itself." Yet the abstraction of "Air Map" could also function as a technique of exclusion. Like representations of the airplane as an international club, the image rendered invisible the political, social, and economic realities that differentiated one place from another, as if the air could dissolve geopolitical power altogether. Similar visual grammar and ideology would characterize Pan Am's iconic postwar logo, a stylized sky-blue globe marked only by latitudes and longitudes.[42]

But aviation did not simply inspire cartographic representation; it transformed cartography itself. "Air-age geography," conceived during the 1940s, advanced the idea that the airplane had wrought a fundamental change in the meaning of spatial relationships, diminishing the salience of national and hemispheric divisions. "The airplane is capable of altering the geography of our world—and therefore the history of our world," wrote poet Archibald MacLeish in his eloquent introduction to *The Compass of the World* (1944), an influential anthology whose contributors included renowned geographers Halford Mackinder, Isaiah Bowman, Owen Lattimore, Ellsworth Huntington, and Richard Edes Harrison. MacLeish's essay originally appeared in the *Atlantic*, and air-age geography received further publicity from articles in *Life*, the *Christian Science Monitor*, and *Newsweek*. In *Fortune*'s words, the airplane inspired "a geography unbound from the flat Mercator projection that has dominated human thinking for centuries." The Mercator map, designed during a seafaring era, emphasized latitudinal directions and portrayed the world as a bounded, two-dimensional space divided into Eastern and Western Hemispheres. The map of the "air world," by contrast, adopted an azimuthal equidistant perspective, centering on the poles and representing the earth as an integral sphere. By the 1940s, geographers in the United States had come to favor the azimuthal projection because of its increased representational accuracy. Proponents of aviation, too, preferred this map because it reflected the routes that airplanes actually flew. Because of the earth's curvature, the shortest aerial route between two cities was rarely a straight east-west line, as ships sailed, but an arc over the North Pole—the "great circle" defined by the shortest distance between two points on a sphere. Thus, a flight from Chicago to Calcutta (to take the example used in a 1943 *Newsweek* article) would proceed north rather than west, crossing Canada, Russia, and China. As the article observed: "East is no longer East, or West something off to the left as one faces the North Pole. . . . The United States is no longer a country separated by broad oceans from Europe and Asia. We are a country living next door to the world. Our former vacuum of insulating space has been filled, literally, by air and airplanes."[43]

As suggested by such comments, air-age geography bolstered important shifts in U.S. foreign policy during the Roosevelt era. Now that the United States was "living next door to the world," with air routes traversing Europe and Asia, it was appropriate—indeed imperative—that the nation extend its sphere of influence beyond the Western Hemisphere. This was precisely the line of argument advanced by Secretary of State Cordell Hull, who advocated liberalizing international trade policies as a means of expanding U.S. global influence. Air-age geography depicted the world that Hull envisioned, and prominent geographers echoed his arguments. According to geographer George T. Renner, the economic and cultural initiatives of the Good Neighbor policy should be expanded worldwide, since "there is no Eastern or Western Hemisphere; the world is in one piece." Thus, he proposed, "why not a good neighbor policy for Russia?" On the air routes, after all, Chicago was closer to Moscow than to "half a dozen Latin American capitals." J. Parker Van Zandt, similarly,

advocated replacing the obsolete concept of Eastern and Western Hemispheres with a single "Principal Hemisphere" in the northern half of the
globe, which contained, he estimated, 94 percent of the world's population and 98 percent of its industry. Encompassing Asia and Europe, Van
Zandt's Principal Hemisphere also happened, not coincidentally, to be the
strategic focus of U.S. foreign policy during World War II and the early
Cold War.[44]

Importantly, however, living next door to the world also rendered the
United States newly vulnerable to foreign attack. The Japanese bombing
of Pearl Harbor on December 7, 1941, proved beyond doubt that the
airplane had abolished the nation's geographic insularity. But even
before Pearl Harbor, many Americans had wondered, "Can America Be
Bombed?"—the question posed by a Works Progress Administration–
funded geography exhibition that opened in April 1941 at the Science
Museum of St. Paul, Minnesota. The exhibition consisted of globes, maps,
and pictorial displays showing how the world had "shrunk" between
1840, when a Clipper sailing ship required 150 days to circumnavigate the
world, and 1940, when the same voyage could be made in just eight days
aboard a Pan Am Clipper. While reassuring visitors that the United States
would remain relatively safe as long as no hostile power gained control of
the seas, "Can America Be Bombed?" raised the alarming prospect of German bombers striking Minnesota via a "back door" through the Canadian
wilderness.[45]

Ominous visions of air-age geography even appeared in airline advertisements and aviation industry trade magazines. A 1942 ad for regional
carrier Chicago and Southern Airlines, entitled "Shadow of Death or
Symbol of Peace?," depicted an airplane shadow looming menacingly over
a house. The ad's text aimed to reassure readers that "the airplane was
born to fulfill a dream of peace and understanding among men," but its
title and image suggested otherwise. Bombing, moreover, was not the
only danger; airplanes could also export political, economic, and bio-
logical threats across national borders. In a 1943 *Skyways* editorial, West
Virginia congressman Jennings Randolph warned that the global interde-
pendence of the air age "can make a famine in China the cause of empty
breakfast tables in Kansas City; economic disaster in the Balkans can be
responsible for bread lines in Boston."[46]

However, Pearl Harbor "produced no quick reorientation of American ideas" about aviation, as historian Michael Sherry has argued. While recognizing that the airplane could create "doomsday," Americans in the pre-atomic era continued to see it as a means of "deliverance." Although discourse on aviation during World War II did register increased anxiety about the airplane's destructive capacities—as evidenced in "Can America Be Bombed?" and the Chicago and Southern Airlines advertisement— such expressions of concern were marginal compared with the continued effusion of paeans to airpower, both commercial and military. Even after Pearl Harbor, American ideas about aviation tended, overwhelmingly, to represent the airplane as the guarantor of U.S. national security and the Allies' eventual military triumph in the war, as well as a symbol and instrument of global unity, which would ultimately contribute to world peace. Even in the midst of world war, and even as they publicized the devastation caused by strategic bombing, the Luce publications and other mainstream newspapers and magazines extolled the postwar possibilities of air travel and commerce. To be sure, Americans were *relatively* more fearful of airpower and more concerned about their nation's safety after the attacks of December 7, 1941. Yet aviation continued to be seen as the foundation of national security rather than a threat to it—especially after Roosevelt's call for an unprecedented buildup of U.S. airpower. The apocalyptic scenarios that pervaded British discourse on aviation, epitomized by H. G. Wells's dystopian novel *The War in the Air* (1908), never gained much cultural traction in the United States. Instead, commentary on aviation tended to assure Americans that the "blueprint of international air routes" established during the war would soon become "the blueprint of peace," as *Time* predicted in May 1943.[47]

That summer, New York's Museum of Modern Art (MoMA) opened a landmark exhibition on air-age geography entitled "Airways to Peace: An Exhibition of Geography for the Future." Its purpose, according to MoMA director Monroe Wheeler, was "to assist the layman to orient himself in relation to the air age." Filling the entire second floor of the museum, the exhibition was divided into five parts: "How Man Has Drawn His World," "The Progress of Flight," "War Over the World," "Global Strategy," and "The Nature of the Air," along with a prologue and conclusion. It featured maps and globes—including President Roosevelt's 50-inch,

500-pound globe, on loan for the occasion, as well as a globe 15 feet in circumference into which visitors could walk—along with murals, drawings, photographs, and paintings of flight "from the pterodactyl of fifty million years ago to the latest fighter and transport planes." Wendell Willkie wrote the accompanying text, which paraphrased *One World* in describing the airways as "a web of intimacy, a new scene of mutual advantages, a world-brotherhood." While the exhibition acknowledged that the world war was fundamentally an aerial war, Willkie's words expressed unequivocal confidence in U.S. military airpower, as well as confidence in the airplane's peaceful uses: "We are using this mighty weapon to the utmost to defeat the aggressors. When that job is done, we must determine to dedicate the wings of the world to the purposes of peace." The exhibition, according to MoMA's press release, aimed to convince viewers that "in a world internationalized by the airplane peace can only be built on dynamic idealism." "Airways to Peace" broke MoMA attendance records and won widespread press acclaim, from the *New York Times* ("a unique display of the world's expanding horizons") to the *New York Daily Worker* ("deserves the attention of every victory-minded citizen").[48]

Americans' optimism about aviation during World War II did not simply derive from ignorance or naïveté. Rather, it reflected the nationalism that undergirded dominant American conceptions of globalism. Visions of "airways to peace" reflected implicit assumptions that those airways would be controlled by the United States. Americans could believe they would be safe in the air age because they trusted, as Sherry has written, that "despite all the horrors . . . air power would be preeminently an American weapon." Opinion polls taken after Pearl Harbor consistently showed that the public responded to the attack not by condemning aviation but by demanding that Washington increase the nation's airpower. The leading proponent of this view was Alexander de Seversky, a Russian-born naval aviator and author of the influential 1942 best seller *Victory through Air Power.* Echoing Billy Mitchell, Seversky advocated an independent U.S. air force and an airpower strategy based on the offensive use of long-range bombers. In his view, the United States needed more aviation precisely because aerial warfare was so destructive; no other nation could be entrusted to use it appropriately. Because the United States had "no imperial purposes," Seversky argued, but had entered the war with the benign intention of "remov[ing] a threat to the normal life of the world," it alone had earned the right to deploy the powerful new weapons of the air age. Applying Seversky's arguments to civilian air travel, *Time* illustrated "Skyways to Peace" with a graphic depicting the United States at the center of the globe, with air routes extending to numerous foreign cities. Yet no routes connected these cities to one another; nor were any other nations represented. The image implied that peace would be possible if all skyways led to or from the United States.[49]

Seemingly at odds with one another, globalist and nationalist visions of the air age were, in fact, deeply enmeshed. American representations of "one world" implicitly, if not overtly, positioned the United States as leader and custodian of that world. Even as airplanes unleashed unprecedented destruction, U.S. citizens could imagine "Airways to Peace" because they believed in their ability to control technology and in the benevolence of their own intentions. Indeed, globalist ideas gained credence during the 1940s because of, not in spite of, such undercurrents of nationalism and American exceptionalism. Older nationalist narratives, meanwhile, acquired new globalist inflections that lent cultural legitimacy to the United States' expanded role in the world.

"America's Frontiers Pushed back to Infinity"

In a 1939 interview with NBC radio, Pan Am president Juan Trippe proposed that the air age would have a distinctly American character. Invented on U.S. soil, aviation embodied "the pioneering force" that dated to "our earliest settlements on the New England coast" and continued through the modern technological conquest of time and distance. In describing the United States as "above all a pioneering nation," Trippe echoed Frederick Jackson Turner's famous 1893 address to the American Historical Association, "The Significance of the Frontier in American History." But whereas Turner lamented the passing of the western frontier that, in his view, had molded the American character, Trippe reached a more optimistic conclusion: "We, as a nation, have *not* reached a dead end to development and expansion. . . . We still *do* live in a land of new opportunities and unlimited horizons." Nearly fifty years after Turner's address, the frontier had not disappeared; it had simply moved—into the air. For Trippe, aviation conjured an image of "America's frontiers pushed back to infinity."[50]

Trippe's vision of aviation as an engine of U.S. expansion was a nationalist vision. Yet in contrast to the "America First" isolationism then espoused by Pan Am advisor Charles Lindbergh, this was also a globalist vision, one that viewed the United States as "a country living next door to the world." Like Willkie and Wallace, Trippe believed that the United States had a role to play on the world stage—the starring role. In an address to the University of California's class of 1944, he encouraged the graduates to become "world-minded," since the airplane had "made the world a neighborhood." In other public statements, however, Trippe emphasized the singular accomplishments and virtues of the United States. "America holds undisputed leadership on the airways of the world," he repeatedly said, thanks to the "superiority of American pilot captains" and the "leadership of American aircraft designers." For Trippe, then, the airplane held the key to both world and national destiny. The unification of "one world" was entirely compatible with, and indeed necessary to, the United States' continuing expansion.[51]

Pan American Airways' wartime advertisements expressed Trippe's nationalist globalism in compelling graphic form. "There *Are* No Distant Lands by Flying Clipper!" announced a 1940 ad, which used maps to show how Pan Am's 65,000-mile-route network "brings 55 countries and colonies . . . within hours of your doorstep"—while also bringing "reassurance that America's prestige and traditional leadership will be maintained on the skyways of the world." In "America Meets the Challenge of a Changing World" (1940), familiar patriotic symbols underscored the airline's contributions to the national war effort: a giant eagle stretched its wings over a Clipper's hull, which itself bore the image of a U.S. flag. A series of 1941 ads identified Pan Am as "Uncle Sam's ambassador of good will," which transported "America's traditions of freedom to 55 lands." But the airline also served as "Uncle Sam's strong right arm." Its airports, "America's Outposts of Security and Defense," united "every major sphere of U.S. trade and defense," from the North Atlantic to the South Pacific.[52]

Either as Uncle Sam's ambassador of goodwill or his strong right arm, Pan Am identified itself with, and even as, the nation-state. It purported to carry on the historic mission of U.S. expansion, with aviation as the latest expression of the pioneer spirit. One of Trippe's favorite sayings updated Horace Greeley's appeal, "Go West, young man," to "Go abroad, young man." Newspaper and magazine articles about Pan Am, many of which closely paraphrased company press releases, often likened Trippe to famous explorers and conquerors of frontiers. The magazine *Who* called Trippe a "Modern Magellan"; *Saturday Evening Post* writer Matthew Josephson, in a series of five laudatory articles, crowned him "Columbus of the Airways." In 1944, Josephson expanded his series into the first book-length history of the airline, stating that "the pioneering of Pan American's immense sky routes invites comparison with such industrial epics of the past as the building of the Union Pacific's transcontinental railway across the American prairies."[53]

The Luce publications promoted Pan Am's nationalist globalism most visibly and vigorously. Writer Clare Boothe Luce, like her husband Henry Luce, felt passionately about airplanes, world travel, and the United States. Upon becoming a war correspondent for *Life* in 1940, she joined what that magazine had called "the best international club in the world," the cosmopolitan elite of frequent fliers. Beautiful and stylish, she appeared regularly in the photo pages of *New Horizons*, posing in front of Dinner Key Terminal's signature rotating globe or alighting from the steps of a Clipper with nary a wrinkle on her trim A-line skirt. And fittingly, Mrs. Luce's *Life* stories put a glossy spin on Trippe's arguments about aviation and American national greatness.[54]

In her November 1941 article "Destiny Crosses the Dateline," Luce documented a six-day "high adventure in the skies" aboard Pan Am's *Pacific Clipper* en route from San Francisco to Manila. She shared the cabin with numerous "prominent people," including the American minister to Thailand; an Army engineer traveling to Chungking to build a railroad paralleling the Burma Road; a lawyer acting as counsel to the governments of Manila and Hong Kong; two clergymen assisting missionaries in Asia; "two Manila brewers"; "three or four of the inevitable oilmen"; British socialite Lady Diana Manners and her diplomat husband, Duff Cooper, "en route to Singapore to coordinate the Far East Defenses of the Empire" (along with their secretary, who guarded an enormous suitcase containing "a raft of state documents"); assorted Pan Am and War Department officials; and "a man who will run a beautiful giant shovel [at] Cavite, the U.S.A. naval base, where they are building bomber runways." Architects of global power, the *Pacific Clipper*'s passengers carried leadership and expertise—from religious guidance to legal advice to diplomacy and engineering know-how—across the date lines.[55]

"Destiny Crosses the Dateline" was a story about American empire, a term that Luce freely used and celebrated. Whereas flight had enabled Wendell Willkie to perceive commonalities among Americans and the "common men" of other nations, for Luce the aerial gaze served to demarcate hierarchical distinctions between the United States and its dependencies. Indeed, the *Pacific Clipper*'s itinerary can be read as an imperial performance that reenacted the historical course of U.S. expansion in the Pacific. In Honolulu, the plane's first stop, Luce and her companions enjoyed "Waikiki and surfboarding, trips over the Pali, dinners on moon-drenched lanais with Hawaiians playing exotic quasi-Hawaiian songs." On Midway Island, the "second stepping stone to America's Manifest Destiny," they checked in at the Pan American Inn (built and owned by the airline) and signed a register that "reads like a *Who's Who* of the agents and emissaries of this war." On Guam, "a fine little advertisement for American imperialism," Luce noted that the inhabitants "numbered 12,000 when the Americans came [and] now there are 25,000. Their health is excellent, their death rate low, their birth rate very high." The *Clipper*'s final destination, Manila, was, in 1941, a United States possession. In Luce's rendering of the Philippine capital:

There are American ships, American uniforms, many American faces among the dusky ones all around. . . . The Filipinos who carry your baggage to the waiting car ask, "Everything O.K., Miss?" in an accent as pure as Broadway's. . . . You see Spanish forts not in ruins, but in excellent condition, housing American officials, flying American flags. . . . In the hot marble corridors of the swanky Manila Hotel, "the finest hotel in the East," in groups of three, four, five, always perspiring people of all nationalities are sitting at little wicker tables drinking Cokes, beers, whiskies, "gimlets" (the gin favorite at Oriental bars). . . . They are businessmen, reporters, politicians, pilots from Singapore and the N.E.I. come to pick up the planes American pilots have ferried out this far. . . . Surely this Manila is the Geneva of the Pacific. The clearinghouse for the rumors of all the Orient, the hothouse of many of its intrigues.[56]

Just as Luce's language reproduced stock tropes of imperialist literature—doting "dusky" attendants, hotel bar gimlets, and the hothouse intrigues of the "Orient"—the history of U.S. international aviation derived from, and built on, the legacies of empire. The geography of Pan Am's routes followed the geography of U.S. imperial expansion. From Hawaii to the Philippines, the *Pacific Clipper* retraced the routes of the Navy vessels that had claimed these island "stepping stones" as United States territories at the end of the nineteenth century. Implicitly if not explicitly, Manifest Destiny was the "Destiny" of the article's title. Despite its ostensible novelty, the air age formed a continuum with past instances of U.S. global expansion. Aviation facilitated interventions in and influence over foreign places. Transporting "prominent people" to advise governments, establish military bases, develop natural resources, convert souls, and modernize societies, the airplane pushed the United States' frontiers, if not into infinity, then certainly into distant latitudes and longitudes.

In elaborating such connections between aviation and empire, Clare Boothe Luce was not saying anything particularly new or controversial. The phrase "empire of the air" dated at least to 1881, when French aviation enthusiast Louis Mouillard published his designs for a fixed-wing glider in a book entitled *L'Empire de l'Air*. An English translation appeared in the Smithsonian Institution's 1882 *Annual Report*, and Wilbur Wright later described Mouillard as "one of the greatest missionaries of the flying cause which the nineteenth century produced." Anticipating Willkie, Mouillard imagined that flight would unify the world; the empire of the air would be for all humanity, he claimed, eliminating the need for territorial boundaries.[57]

In reality, however, empires of the air mirrored those of the earth. By World War II, the airplane's utility as an instrument of imperial control was well established. In a 1911 treatise on aviation in colonial settings, the French general Henri-Nicolas Frey advocated the use of airpower to "exercise easy, rapid, and continuous police surveillance on barbarian no-

madic tribes" and to "intervene, with the speed of a bird of prey, in threatened or troubled places." Against modern aerial weapons, Frey predicted,
"the cunning and the ingenious tricks to which the so-called 'inferior'
races resort are reduced to impotence." In 1911—the same year that Frey's
treatise appeared in France—Italy became the first nation to employ aircraft in warfare when it invaded the Ottoman province of Libya. During
the following two decades, Britain, France, Spain, and Italy all used aerial
bombardment to suppress popular rebellions and to enforce colonial rule.
Even as European governments debated outlawing bombing in "civilized"
warfare, they did not hesitate to use airpower in places such as Afghanistan, Iraq, Morocco, and Ethiopia. Commercial aviation, too, became a
handmaiden of empire. "Air transport serves to bring overseas colonies
more closely in touch with the homeland. It aids Great Powers to pene-
trate politically and economically into weaker and more backward countries," wrote Oliver Lissitzyn, a leading theorist of international aviation
law and a captain in the U.S. Army Air Force. By the mid-1930s, Europe's
imperial powers all had established airmail routes to their Asian and African colonies. It was no coincidence that Britain named its first international airline Imperial Airways.[58]

In the United States, the phrase "empire of the air" appeared widely in
mid-twentieth-century commentary on aviation. In 1938, *New York Times*
military correspondent Hanson W. Baldwin described how "America
Builds an Empire of the Air: Her Planes and the Plants to Build Them Set
the Pace for the Whole Air Mad World." Even before the outbreak of
World War II, Baldwin invited his readers to imagine "Curtiss pursuit
planes roar[ing] above China's flooded battlefields; Martin bombers
wing[ing] above Formosa; Consolidated flying boats tak[ing] off from the
icy waters of Russian harbors; Boeing fighters . . . spin[ning] and whirl[ing]
in dog-fights in Spanish skies." This aerial empire, he asserted, "has become an integral and an indispensable part of the greater empire which is
our continent." Another *New York Times* correspondent, travel writer
August Loeb, envisioned "A New Empire of Air Travel" (1940), in which
passenger and transport planes would establish U.S. control over "a far-
flung network of overseas air routes," "blaz[e] ocean trails," bind together
"the New World from the Bering Strait to Buenos Aires," and "carry the
Stars and Stripes into the South Seas." An accompanying photograph of a

Boeing Strato-Clipper bore the caption "America extends its conquest along the skyways of the new world."[59]

As in Europe, the United States' aerial empire derived from—and also extended—its previous history of territorial and commercial expansion. In 1936, James G. Stahlman, publisher of the *Nashville Banner*, was one of fifteen passengers on Pan Am's inaugural passenger flight from the United States to China. His account of the flight ran, in three installments, in the *Banner*'s Sunday magazine; so many readers requested copies that he subsequently published a thirty-page booklet, *Wings to the Orient*. For Stahlman, as for Clare Boothe Luce, the *China Clipper*'s voyage evoked the legacy of American imperialism: "Manila—mention of the name brings back memories of a day in 1898, when as a very small boy, I thrilled at the news of Commodore Dewey's victory." But air travel did not simply remind Stahlman of a distant childhood memory; it allowed him to imagine himself as a participant in the conquest of the Philippines. "I could hear those guns of the Olympia, the Raleigh, the Baltimore and the rest of the American fleet as they pounded away at Admiral Montejo's Asiatic squadron," he wrote, "as if I had been standing beside Dewey, himself, on the Olympia's bridge as he swept past the Spaniards and raked their once mighty armada with a fire that sent them to their doom." Literally and figuratively, the *China Clipper* placed Stahlman at the primal scene of American empire.[60]

Like the vision of one world, the vision of aerial empire reached the American public via cartographic representation. The November 1942 issue of *Life*, for example, featured a centerfold azimuthal map with possible postwar air routes outlined in red. The red lines connected important cities within the United States and stretched, tentacle-like, across South America, Africa, and Asia. Graphically depicting the United States' "conquest" of the world's skyways, such images helped readers to visualize their nation's growing importance in world affairs. Cartographic imagery also pervaded Pan Am's publications. Maps of the world crisscrossed by the airline's ever-expanding routes appeared in timetables, annual reports, advertisements, and posters, becoming as recognizable as its logo, a winged globe. In 1943, Pan Am commissioned famed cartographer Richard Edes Harrison, whose work regularly appeared in *Fortune*, to design a map of its routes for *New Horizons*. Letters to the editor reveal that *New*

Horizons readers greatly appreciated Harrison's map. A professor of aeronautical engineering at the University of Texas requested copies for his students; a woman from Miami desired one suitable for framing, noting that she would place it in a "peachy spot in our living room, where it can be studied by us all." In studying such maps, these citizens were studying the evolution of an American air empire.[61]

Importantly, though, aviation could also signify what made American empire ostensibly different from land- and sea-based empires. Aviation advocates argued that airpower and sea power demarcated two distinct historical eras, such that the United States would dominate the twentieth-century age of airpower just as Britain, with its imperial navy, had dominated the nineteenth-century age of sea power. In 1918, Rear Admiral Robert E. Peary of the U.S. Navy, a record-setting aviator and Arctic explorer, expressed this argument in *National Geographic:* "We are now entering upon an era of air power—a stupendous era—which in the near future will be . . . far superior to the greatest sea power of the present as the unlimited ocean of atmosphere." The dawning age of airpower "presents to the United States, with its unique geographical position, its boundless resources, mechanical and inventive ability, and its splendid reservoir of ideal American manhood, the opportunity to be the *first air power in the world*. This should be the second article in our national creed, the first article in that creed being the Monroe Doctrine," Peary concluded. The Monroe Doctrine had authorized the United States' hegemony in Latin America; now airpower promised to globalize that hegemony.[62]

During World War II, U.S. policymakers revived these arguments in order to explain and justify their vision of the United States' role in the postwar international order. "If the Pax Britannica, such as it was, was kept largely by the British Navy, the Pax Americana will be kept by air power and the United States can and should be the greatest air power in the world," commented one State Department official in a 1942 internal memorandum. The following year, Joseph M. Jones, who would become assistant secretary of state for public affairs in 1946, authored a three-part *Fortune* article contrasting Britain's "order-keeping" of the nineteenth century with America's order-keeping of the twentieth. The former had been unilateral, imperialist, and based on sea power, Jones argued; the latter would be multilateral, inspired by "the principle of freedom," and based on airpower. "At the end of the war the air power of the U.S. will be second to none," Jones maintained—but, he emphasized, the United States would not "play with air power in the twentieth century Britain's role with sea power in the nineteenth." Rather than imposing its will through military conquest, the United States would lead through international organizations, alliances with other great powers, and "the principle of freedom" (recalling Henry Luce's "big words like Democracy and Freedom and Justice"). In the air-age world, international order would thus no longer require the direct control of foreign territory.[63]

Such arguments resonated with American critiques of imperialism during the 1930s and 1940s. To be sure, cultural conceptions of the United States as an anti-imperial power long predated the air age; Puritan leader John Winthrop's seventeenth-century vision of the Massachusetts Bay Colony as a "city upon a hill," whose example the world would follow by choice rather than coercion, foreshadowed how the idiom of American exceptionalism might authorize alternative forms of international influence. During the two world wars, however, Americans' antipathy to territorial empire deepened. After Germany's and Japan's aggressive seizures of foreign territory, the term "empire" came to signify *lebensraum* and militarism. Anti-imperialism also gained credence as a result of the Atlantic Charter and President Roosevelt's rhetorical advocacy of national self-determination (in spite of his reluctance to condemn the imperialist practices of Britain). The Roosevelt administration withdrew the Marines from Haiti, signed a treaty with Cuba nullifying the Platt Amendment, and promised to grant independence to the Philippines upon the conclusion of the war. American historians, too, began to portray 1898 as a "great aberration" (in the words of Samuel Flagg Bemis) in an otherwise anti-imperial history—an interpretation that reigned until the New Left revisionism of the 1960s. And, of course, Henry Luce identified a rejection of territorial imperialism as a defining feature of the American Century: instead of conquering and ruling over some "vastly distant geography," the United States would influence the world by spreading its values of "Democracy and Freedom and Justice."[64]

Discourse on aviation played a crucial role in reconciling anti-imperialism with the United States' ascendance to global power. Instead of seizing colonies, Americans simply wished to "fly everywhere"—as Connecticut's newly elected Republican congresswoman, Clare Boothe Luce, proclaimed in her first speech before the House of Representatives in February 1943. Speaking on the topic of "America's place in the present and post-war civilian air world," she predicted that "the masters of the air will be the masters of the planet . . . for as aviation dominates all military effort today, so will it dominate and influence all peacetime effort tomorrow." Luce left little doubt as to which passport these masters would carry. From the American heartland, she noted, "there is not one important city in the whole world, in Europe, Asia, Africa, South America, that cannot be reached by air within forty-eight hours." Coining a new political buzzword, she described Vice President Henry Wallace's proposal to internationalize the airways as "globaloney." In spite of her uncompromising nationalist sympathies, however, Luce reiterated her husband's "American Century" editorial in insisting that Americans "do not expect and do not want one inch of territory outside our own possessions. We desire neither to grab other people's land nor to dominate any race on earth." Rather, she concluded, "The post-war air policy of these hundreds of thousands of young air-minded Americans is quite simple. It is: 'We want to fly everywhere. Period.' "[65]

Reproduced in syndicated type around the country, the speech catapulted Representative Luce into the national limelight. She received swooning praise from conservatives and equally passionate condemnations from the internationalist left. Vice President Wallace branded her "a new American imperialist," while *The Nation* decried her "mélange of emotionalism and misinformation," which had "stirred up fellow-jingoes" and "cemented a new stretch of road leading to World War III." But Luce's underlying argument—that global power in the postwar air age would derive from cultural and economic influence rather than the conquest of territory—gained purchase across the political spectrum. This argument appealed to conservative nationalists in its justification of an "American Century" based on free trade principles. At the same time, by denying that the United States aimed to use the skies to colonize foreign lands, as

Europe had once used the seas, Luce's argument appealed to the anti-imperialist sympathies of liberal internationalists.[66]

In the last months of the war, a *Fortune* article on "The World's Greatest Airline," the U.S. Army's Air Transport Command, explained the relationship of aviation and empire in a particularly revealing manner. *Fortune*, like the Luces, disdained "the grosser forms of imperialism such as colony grabbing and establishment of spheres of influence." However, the article insisted, "it is natural for the imperialist instinct to be *sublimated* in an urge to extend U.S. airlines around the globe" (emphasis added). Psychoanalytic theory holds that through sublimation, the psyche conceals shameful drives and desires by channeling them into socially sanctioned activities such as work or art. In a parallel dynamic, commercial aviation channeled U.S. global ambitions away from "the grosser forms of imperialism" and into the more benign activities of trade, travel, and tourism. As a cultural representation of global power, the vision of U.S. airlines extending around the globe or of Americans flying everywhere displaced less palatable images of "colony grabbing" and "spheres of influence."[67]

Ultimately, then, aviation bolstered the material infrastructure of American empire while simultaneously sustaining its cultural denial. By seeming to make world leadership contingent on access to markets instead of control over territory, the air age reconciled expansionism with anti-imperialism. The United States' global network of commercial air routes "in a sense tends to delimit what might be called the 'American Empire'—except that we do not propose to make it an empire," wrote Assistant Secretary of State Adolf A. Berle Jr. in a strikingly candid 1944 diary entry. Juan Trippe, likewise, said that Pan Am carried "cargos of goodwill," not "cargos of imperialism and hate." But in an internal company memo, executive Roger Wolin acknowledged that "Pan American is both international and imperialistic. It is a U.S. institution transplanted 3,000–15,000–28,000 miles from home. . . . It considers itself the U.S. industrial ambassador." Such moments of candor corroborated *Fortune*'s diagnosis: commercial aviation sublimated imperial drives.[68]

The polyvalent cultural meanings of aviation—its ability to signify both nationalism and internationalism, the self-proclaimed benevolence of the United States as well as its hegemonic power—proved highly effective in representing the nationalist globalism of the American Century. In the air age, Americans could be globalists and nationalists, proponents of one world but also of an American Century, in which the United States would act like an empire but refuse to call itself by that name. The airplane augmented the United States' military and economic power, while the cultural "logic of the air" sustained the comforting fiction that this power was natural and benevolent, flowing through the skies rather than stamping itself on the disputed terrain of the earth. Indeed, the symbolic connotations of the air supported this fiction. Unlike sea or land, the air seemed to be a universal medium. It was a prized resource, yet seemingly available to all; neutral, yet saturated with geopolitical significance; invisible, yet omnipresent. As American Airlines vice president O. M. Mosier stated in 1944: "There is only one air—and it is everywhere."[69]

American Airlines' 1946 ad "Air Is Everywhere, Impartially" expressed this argument visually. Depicting a U.S. Army transport plane in an African desert, along with an American soldier and two camels, minded by a man in traditional North African dress, it suggested the global reach of U.S. power. The caption, though, proposed that the soldier, the reader of the advertisement, and the African camel herder all inhabited a shared, politically neutral space, just as all living beings breathe the universal and "impartial" air. In the political culture of the 1940s, the air was therefore a fitting channel for U.S. foreign policy. It naturalized American power and universalized American ambitions as aspirations shared by the entire world. It whispered the seductive idea that the United States could be a benign hegemon, unencumbered by the "cargoes of hatred" that burdened other great powers. In short, the air seemed to make possible the global networks of empire without the bloody, costly practices of imperialism.

In reality, of course, the skies were neither politically neutral nor uncontested. The empires of the air grafted onto the empires of the earth: Pan Am, expanding first throughout Latin America and then across the Pacific, retraced the geography of the Monroe Doctrine and 1898, just as European airlines connected metropolitan capitals to their colonial outposts in Asia and Africa. In war and in peace, the airplane offered access to and influence over distant lands. It facilitated trade, dropped bombs, demarcated spheres of influence, and exported culture and ideology.

Moreover, the empires of the air not only reflected but depended on traditional forms of territorial power. Aviation required oil, metals, massive tracts of land for runways and airports, factories in which to build airplanes, schools to train pilots and engineers, and manual laborers to perform tasks that ranged from constructing hangars to riveting bombers. International aviation, furthermore, required diplomatic air-rights treaties between national governments. Thus, rather than transcending earthly boundaries, aviation more often reproduced them.

4

"America's Lifeline to Africa"

On January 11, 1943, President Franklin D. Roosevelt boarded Pan American Airways' *Dixie Clipper* in Miami. He was flying to meet British prime minister Winston Churchill in Casablanca, Morocco, where the two Allied leaders would famously agree to impose terms of unconditional surrender on the Axis powers. Just as FDR had been the first presidential candidate who flew to accept his nomination, he now became the first sitting U.S. president to leave the United States during a time of war and the first to travel by air on a diplomatic mission. Captain Howard M. Cone, a thirty-four-year-old "Master of Ocean Flying Boats" who had logged more than 7,500 flight hours on Pan Am's commercial routes, described the president as "an excellent passenger" who seemed enthralled with flight in spite of the intense physical pain that it caused him due to his paralysis. Roosevelt paid six dollars to be initiated into the elite "short snorters" club of pilots and passengers who had crossed the equator, and on the return flight, he celebrated his sixty-first birthday over Santo Domingo, Haiti, and Cuba, dining on caviar and champagne while the crew sang "Happy Birthday." FDR's aerial passage over the Caribbean islands appeared decidedly less imperialistic than his cousin Theodore Roosevelt's iconic horseback charge up Cuba's San Juan Hill during the

Spanish-American War of 1898. Yet Pan Am's very existence both derived from and intensified U.S. hegemony in Latin America—and by the time of the Casablanca conference, the United States' "empire of the air" extended to Africa as well. On its 15,000-mile round-trip journey between Miami and Casablanca, Roosevelt's plane made refueling stops in Trinidad, Brazil, and the British West African colony of Gambia, along with a side trip to Liberia for a meeting with President Edwin J. Barclay. In each case, the air bases where the president landed had been built or improved by Pan American Airways, with federal funding authorized by FDR himself.[1]

In the fall of 1940, when Pan Am first sent a meteorologist to Bolama to gather data for a potential route across the south Atlantic, Africa was the only populated continent untouched by "America's Merchant Marine of the Air." But by the end of 1942, Pan Am was regularly flying the 4,300 miles between Bathurst and Cairo. Rather than airmail or passengers, however, its planes carried military supplies. In August 1941, the U.S. and British governments had authorized Pan Am to operate a trans-African supply route in order to deliver aircraft, gasoline, and equipment to British forces in North Africa, the Middle East, and beyond. Like the Latin American Airport Development Program (ADP), the African route enabled the Roosevelt administration, through the medium of a private corporation, to provide material assistance to the Allied war effort in spite of the Neutrality Acts passed by Congress during the 1930s.[2]

The project also further enlarged the United States' sphere of influence beyond the Western Hemisphere. Along with 1,200 U.S. employees, Pan Am brought to Africa technology, consumer goods, weapons, popular culture, and an entire apparatus of American modernity, from housing complexes and health clinics to roads and refrigerators. During its sixteen-month stay in Africa, the U.S. airline hired some ten thousand locals for jobs that ranged from bricklaying to bookkeeping. And for American audiences, Pan Am produced a profusion of narratives about Africa and Africans—stories that appeared in company publications, advertisements, and employees' writings, as well as in mainstream newspapers and magazines. These narratives offer a unique window onto American globalism's fraught, complex relationship to race and empire.[3]

The history of Pan Am's wartime "adventure" in Africa shows how the United States collaborated with imperial Britain, both materially and ideologically, while simultaneously moving to supplant Rule Britannia with the American Century. The project originated through Anglo-American cooperation and revealed important cultural affinities between U.S. and European investments in Africa as a site of empire and nation building—and as a site of individual subject formation. But in spite of these affinities, Roosevelt administration leaders and Pan Am employees alike invoked American exceptionalism to distinguish themselves from their British allies. Americans were different from European imperialists, they argued, because rather than colonizing Africa, they merely desired to improve it. Modern technologies such as aviation would bring enlightenment and progress to the "Dark Continent," succeeding where colonization had failed.

These quasi-humanitarian justifications for the United States' expanded presence in Africa were not distinctively American. Indeed, liberal ideologies of social uplift—the French *mission civilisatrice*, the Portuguese *missão civilizadora*, the British notion of the white man's burden—had long sustained European colonial projects.[4] However, the commonly held belief that Americans were different from their European predecessors offered a powerful means of reconciling the United States' global expansion with its traditions of anti-imperialism. The air age itself facilitated this reconciliation. Aviation, construed as a uniquely American contribution to human progress, served to contrast the United States' enlightened modernity to the Dark Continent, on the one hand, and the dark ages of colonialism, on the other. Neither distinction held up in practice. Yet the logic of national and racial difference sustained the coherence of American globalism at the level of ideology.

In her important essay "Left Alone with America" (1993), Amy Kaplan traced the epistemological origins of American Studies to the triangulated histories of the United States, Europe, and Africa. During the 1920s, Perry Miller, a founding father of the discipline, was unloading oil drums off a tanker in the Belgian Congo when he experienced a "sudden epiphany" about the "uniqueness of the American experience." Perceiving Africa as a "barbaric tropic" and Europe in postwar disarray, Miller found

himself "left alone with America." This exceptionalist vantage point—
which inspired *Errand into the Wilderness*, Miller's seminal 1956 study of
the Puritan mind—represented "a coherent America" only by setting it
apart from a primitive Africa and a decaying Europe. However, Kaplan
argues, Miller's epiphany in the Belgian Congo also exposed "an imperial
unconscious of national identity." His very presence in Africa revealed
American investments in empire—economic, cultural, geopolitical, and
emotional—even as he sought to deny any similarities between the United
States and the European imperial powers. Indeed, Miller's very desire to
set the United States apart from the world both reflected and derived
from its increasing global entanglements.[5]

Pan Am's own errand into the wilderness likewise brings to light the
imperial unconscious of the American Century. Like Perry Miller, Pan
Am employees came to understand their nation's identity through their
work in Africa, which seemed to confirm cultural narratives of American
exceptionalism and benevolence. Ironically, though, Pan Am's wartime
activities in Africa problematized those very narratives. Perhaps this was
inevitable. In the air age, after all, no country remained truly alone—not
even America.

Assuming Britain's Burden

Touching down at Bolama in February 1941, Pan Am's *Dixie Clipper* had
flown the first commercial air route linking North America, South America, Europe, and Africa. Spanning the seas once plied by slave ships, the
route created an aerial Atlantic world. Additionally, as Roosevelt's advisor
Robert Sherwood noted, it represented "the first pioneering move toward
establishing one of the most vital strategic lines of communication in the
Second World War." Because air routes "were all two-way streets," FDR
was "determined to control [the south Atlantic] before Hitler could,"
Sherwood recalled. "What I want," the president wrote in a June 1941
memo to Undersecretary of State Sumner Welles, "is the possibility of an
American plane hopping off from Natal, Brazil, with the option of landing at any one of three places in Africa—Bathurst, Freetown, or Liberia. I
regard this as an essential." In fact, Roosevelt and Juan Trippe had been
discussing ways to secure U.S. control of the skyways between South

America and Africa since the summer of 1940. Military leaders concurred that "the inauguration of an American air line across the South Atlantic" would be "most desirable."[6]

After the fall of France in June 1940, the Roosevelt administration had grown increasingly concerned that the Axis could use the French colony of Senegal to launch an attack on South America. The security of the Western Hemisphere thus increasingly depended on the security of western Africa. British prime minister Winston Churchill implored Roosevelt to establish air and naval bases in Liberia—sub-Saharan Africa's only independent nation but one that had been culturally linked to and economically dependent on the United States since its establishment by freed African American slaves in 1847. War Department leaders determined, however, that "the effort and ensuing involvement" in setting up such bases would be "so great as to far outweigh the potential advantages."[7]

Washington's options changed dramatically after Congress passed the Lend-Lease Act in March 1941, which authorized the United States to lend military supplies to the Allies, provided that they repay the U.S. government in kind at the war's end. The legislation prohibited "the entry of any American vessel into a combat area," meaning that U.S. ships and transport planes could not deliver their cargo directly to Britain or other combatant European nations. But they were free to go to Allied territories that were not currently theaters of combat, such as Britain's African colonies. According to the Office of Emergency Management, "the Neutrality Act does not prohibit the United States . . . from authorizing a citizen to operate Government-owned aircraft to British territories in Africa not within any combat area."[8]

This interpretation of the Lend-Lease Act paved the way for Pan American Airways to take over a struggling Royal Air Force (RAF) supply route in Africa. Known as the Takoradi Route, because it originated in the West African port city that is now part of Ghana, the supply line traversed the upper half of the continent, enabling the RAF to airlift supplies to British forces fighting in North Africa. (Axis control of shipping in the Mediterranean prevented the British from using this more direct route.) By early 1941, the Takoradi Route was barely operational, plagued by chronic personnel shortages as RAF pilots were called away for more urgent combat duty. In one month, for example, Air Marshall Arthur

Tedder, the RAF commander in North Africa, received only forty-nine aircraft out of an expected 180.[9]

The ever-enterprising Juan Trippe saw opportunity in the RAF's crisis. Dining with Churchill and Air Ministry leaders during a June 1941 visit to London, he casually proposed that Pan Am take over the Takoradi Route. With its superior resources and experience operating in remote locations, Trippe argued, his airline could restore the supply route's functionality and enable the RAF to concentrate its resources elsewhere. Upon returning to the United States, Trippe was summoned to Washington for further discussions with military and diplomatic officials from both sides of the Atlantic. In early July—before any official contracts had been signed—he created two subsidiary companies and appointed 150 executives and managers to oversee them: Pan American Airways-Ferries (PAA-Ferries), which would fly Lend-Lease supplies from the United States to Africa via the ADP airports in the West Indies and South America, and Pan American Airways-Africa (PAA-Africa), which would operate the route between Takoradi and Khartoum. Shortly thereafter, the Civil Aeronautics Board granted certificates for Pan Am to fly the relevant routes, which, it determined, were "urgently needed in the interest of national defense."[10]

On August 12, 1941, a series of five contracts among Pan Am, the War Department, and the British Air Ministry formally authorized the U.S. airline to begin transporting Lend-Lease supplies across Africa. The endeavor would be funded with $17.8 million in Lend-Lease funds, supplemented by an additional $2.8 million from the President's Emergency Fund. A week after the contracts were signed, Roosevelt made a public announcement that Pan American Airways would be providing "direct and speedy delivery from the 'arsenal of democracy' to a critical point in the front against aggression." While the locations and details of the operation remained classified, the president emphasized that the "importance of this direct line of communication between our country and strategic outposts in Africa cannot be overestimated." Newspaper coverage of FDR's announcement overwhelmingly applauded what the Associated Press described as "a far-reaching step" to defend the Western Hemisphere. In a dissenting opinion, the isolationist-leaning *Chicago Tribune* excoriated Roosevelt's action as an attempt "to enlarge the scope of America's participation in the war"—which, in fact, it was.[11]

African American newspapers, meanwhile, responded to the news with a mixture of caution and enthusiasm. Black journalists hoped, on the one hand, that Africa's new military significance would enhance the continent's international standing, perhaps even inducing European governments to grant independence to their colonies after the war. On the other hand, they realized, the arrangement could intensify foreign domination of Africa. "Poor Africa has only been referred to as geographical points where the Allies and Axis have locked horns," noted the *Chicago Defender.* Moreover, the United States' increased presence in Africa would not necessarily benefit local populations: the *Defender* envisioned "exploitation of the laboring classes" and "clashes between overbearing American officers and resentful native civilians."[12]

The West African press greeted FDR's announcement with a similar mixture of optimism and skepticism. The *West African Pilot*—a prominent nationalist organ published in Lagos by Benjamin Nnamdi "Zik" Azikiwe, a U.S.-educated intellectual who in 1963 would become the first president of independent Nigeria—lauded Pan Am's takeover of the RAF supply route as "one of the first concrete developments in the increase of American aid" to the Allies. Like many African American journalists, Azikiwe expressed hope that this "vital air route" would enhance the region's international visibility and significance. A. M. Wendell Malliet, an influential black intellectual and the foreign news editor of Harlem's *Amsterdam Star-News*, also made this argument. "An accident of history—the conquering armies of Adolf Hitler in Europe—may prove to be the greatest blessing that has come to Africa in a thousand years," he wrote in an editorial for Liberia's *African Nationalist.* Although the United States and Europe had "neglected Liberia for a hundred years," that nation had now become "the most strategic bastion of defense shielding the New World from the wanton destruction of the Old." The war could even result in "a joint program of national reconstruction and New World defense," Malliet believed, that would bring economic progress to Liberia while ensuring the security of the Western Hemisphere.[13]

The Atlantic Charter, declared by Roosevelt and Churchill on August 14, 1941—four days before FDR's announcement about Pan Am's activities in Africa—lent credence to such hopes. Roosevelt and Churchill's joint proclamation affirmed eight principles of global democracy, including "the right of all peoples to choose the form of government under which they will live." Although the Atlantic Charter had no binding power, intellectuals and activists throughout the colonized world viewed it as a watershed development, signifying a potential end to colonialism. Pan Am thus arrived in colonial Africa precisely at the moment when independence seemed newly possible.[14]

In the context of anticolonial politics, aviation acquired important meanings as both a symbol of and a means to national self-determination. During the war, West African newspapers enthusiastically covered aviation developments, publishing regular reports on aerial combat, profiles of famous pilots, and photographs of the latest Allied fighters and bombers. In the British colony of Gold Coast, newspapers coordinated Spitfire Funds, through which ordinary people could donate money for the construction of aircraft. The West African press also publicized the achievements of the Tuskegee Airmen, the famous black flying squadron of the U.S. Army Air Forces. The *West African Pilot*—whose very name conjured images of flight—described African American progress in aviation as a "portent" indicating "the shape of things to come after the war." Urging "European States which have 'possessions' in Africa" to pay heed to "the achievements of the Tuskegee Airmen and others," the paper hinted that the airplane might one day be used against Africa's colonizers. In an editorial entitled "Traveling by Air," the *Pilot* declared: "The African has no intention of allowing others to leave him in the background. . . . In the realm of aviation, persons of African descent have made marks *pari passu* with other races. In other parts of the world, the black folks have taken to aviation as a means of traveling." The editorial did not mention Pan Am by name, but it was published in October 1942, at the height of the airline's African operations.[15]

West African newspapers also excoriated European and American attempts to keep the skies white. The RAF, in particular, had come under fire for its discriminatory practices. The RAF began admitting colonial subjects of all races in 1941, because of manpower shortages, but like the U.S. armed forces, it found ways to maintain the color line. West Africans were prohibited from enlisting, for example, because they supposedly carried malaria. In a series of scathing editorials against this policy, Nnamdi Azikiwe—who had himself attempted unsuccessfully to enlist as a pilot—attacked the RAF as a "closed-union workshop open only to members of the caucasoid race." The malaria prohibition was ultimately lifted, and in the fall of 1942, four men from Nigeria enlisted in the RAF. Yet as both African American and West African newspapers reported, black pilots on both sides of the Atlantic continued to face racist treatment that ranged from job discrimination to physical attacks.[16]

The airplane's mixed meanings for Africans (and for African Americans) also reflected its history as an instrument of empire. The first documented instance of aerial bombing occurred during the Italian invasion of Libya in 1911, and during the 1920s Britain, France, Spain, and Italy all used aviation (employing both bombers and reconnaissance planes) to subdue rebellions in their African colonies. The development of airmail and transport services, too, had strengthened Europe's hold over its colonies. As British colonial secretary Leopold Amery wrote in 1926, "from the point of view of establishing white civilisation as a guiding influence over Africa, it is very important that the region should be in close contact . . . with England." In addition to official carriers such as Imperial Airways and Air France, white settlers in Africa started numerous smaller air services that linked rural estates to colonial capitals. The famed "lady flyer" Beryl Markham, raised by her English father in Kenya, operated such a service in East Africa, as she chronicled in her memoir *West with the Night* (1942). Identifying as African rather than English, Markham nonetheless conflated her love of aviation with the project of empire. "The rest of the world may have grown complacent by then about aeroplanes flying in the night, but our world had barren skies. Ours was a young world, eager for gifts—and this was one," she wrote. The airplane, in this rendering, was a "gift" from white civilization to Africa—a form of modern progress that, for Markham, exemplified the positive effects of Britain's presence in Kenya.[17]

Originating from British imperial air routes, Pan Am's trans-African supply route soon constituted an informal empire in its own right. The inaugural Pan American Air Ferries flight left Miami on October 18, 1941, just two months after the contracts enabling the operation were signed. From the United States, Pan Am transported disassembled aircraft (C-47 cargo transports and B-24 and B-26 bombers), parts and maintenance equipment, and gasoline to Brazil via the ADP airports. Departing from Natal, on Brazil's east coast, they then flew across the

Atlantic to British West Africa and Liberia. The supply line continued for 2,300 miles across Africa's interior, terminating in the cities of Khartoum, Cairo, Asmara, or Aden. From there, Pan Am's operations extended to Tehran, Karachi, New Delhi, and Calcutta—where another subsidiary, the China National Airline Corporation, ferried equipment to China and Burma. The route traversed some of the most inhospitable flying conditions on earth, from the dense jungles of the Congo to the Sahara Desert, where temperatures rose to 140 degrees and winds tested the skills of the most experienced pilots. It also required the maintenance of eighteen bases, located in Bathhurst, Freetown, Fisherman's Lake, Benson Field, Robert's Field, Takoradi, Accra, Lagos, Kano, Maiduguri, Fort Lamy, El Geneina, El Fasher, Khartoum, Luxor, Cairo, Asmara, and Aden.[18]

The African route completed the last link in Pan Am's worldwide chain of air routes: as of February 24, 1942, the U.S. airline encircled the globe. If the sun never set on the British Empire, neither did it set on Pan Am. PAA-Africa personnel manager Voit Gilmore illustrated the point in his final report on the project:

> At times when employees in Teheran were shivering in the bitter cold of Caucasian winter, employees at Accra were sweltering under the heat of Africa's mid-summer sun. While employees at Cairo were night-clubbing, employees at El Geneina were playing checkers under gas lamps in the heart of black man's Africa. While employees were witnessing civil riots in Karachi, employees at Fisherman's Lake were watching preparations for the election of a new President in Africa's only republic. While employees in Lagos crept cautiously through the nightly blackouts, employees at Khartoum drove by company bus to theaters and amusement spots down brightly lighted streets.

Here, then, was another type of empire—an empire of the air, spanning five continents. The African supply route, explained an aviation expert in the *Pittsburgh Press*, forged "the last really difficult link in the Pan American around-the-world-by-air-plan." After the war, it promised to "insure the freedom of American air commerce to all parts of the world." Thus, beyond its immediate contributions to Britain's war effort, Pan Am's presence in Africa could have long-term commercial benefits. Previously the exclusive domain of European airlines, the continent held both "attractive trade possibilities" and "opportunities for vacation travel" (as noted in Pan Am's free school newspaper, *Classroom Clipper*). Many of Pan Am's African bases, moreover, could function as refueling stops along major world air routes.[19]

Yet Pan Am's so-called empire of the air required extensive on-the-ground interactions between U.S. employees and the residents of towns and villages near its eighteen African bases, thousands of whom received temporary jobs with the airline. Operating the supply route involved labor-intensive construction, resource extraction, and equipment maintenance, not to mention clerical and domestic work. During its sixteen-month tenure in Africa, Pan Am thus served as an important local employer and an informal ambassador of the United States. In some locations, noted Voit Gilmore, Pan Am employees "were the only Americans present." And their encounters with Africans—as employers, traders, tourists, and amateur ethnographers—exemplified the diverse and often contradictory means through which the United States wielded global influence in the era of the American Century.[20]

U.S.-African Encounters, in the Air and on the Ground

Modeled on the Airport Development Program, PAA-Africa involved much more than aviation. An epic example of "industrial pioneering" (as Seattle's *Labor News* described it), the project in many ways prefigured the modernization and development projects of the Cold War era. When Pan Am arrived in West Africa in the summer of 1941, its engineers and managers lived, much like locals, in mud and grass huts that were described by one U.S. employee as "quite cool and comfortable." Given the hot and humid climate, he added, "I sometimes think that they may be better than the prefabricated houses that are to be sent out here for our permanent quarters." But reigning perceptions of Africa as "a continent practically devoid of available living facilities" (to quote Voit Gilmore) led Pan Am to import prefabricated housing, food, medical supplies, and "all the accessories for the American standard of living." From the airline's offices high atop the Chrysler Building, a three-person staff organized the herculean task of shipping the following items:

> 400 cases of refrigerators, 500 cases of radio equipment . . . upwards of 1,000,000 board feet of lumber . . . 3,750 cases of beer, pepsi-cola, coca-cola, and soda water; 5,000 bundles of structural steel . . . 10 tractors, 4 road scrapers, 500 cases of automobile parts, 1,500 cases of airplane parts, 15,000 bags of cement, 20 station wagons; 20 sedans, 25 trucks; 15 buses; 15 tank trucks; 24 prefabricated houses, each weighing 88,000 lbs.; 500 barrels of assorted sizes of nails . . . 500 ash cans, and a half-dozen boxes of Christmas presents for the boys on the front; about 2,000 bundles of conduit for electrical equipment, 50 cases of telephone equipment; 50 Lorrimer Diesel engines . . . each weighing 10,000 lbs.; 12 crash trucks; 20 army trucks; 750 reels of copper cable, each reel weighing about 3,000 lbs.; 200 reels of steel cable, each reel weighing about 3,500 lbs.; 500 hand trucks; 35 electrical baggage trucks . . . and about 750 drums of paint.

By December 1943, Pan Am had sent 15,521 tons of material to Africa, which was used to build, at the airline's eighteen bases, an entire infrastructure of American-style modernity:

> acetylene generator buildings, administration buildings, barber shops, battery shops, butcher shops, carpenter shops, cafeteria buildings, chemical laboratories, churches, classrooms, commissary storehouses, dining halls, dormitories, engine overhaul buildings, electric shops, fire equipment buildings, garages, guard houses, hospital, kitchens, lumber storage, link trainer, laundries, mechanical shops, medical inspection buildings, native barracks (with kitchens, laundries, toilets, and showers), oil storage, office buildings, paint shops, pump houses, power houses, pantries, police post, plumbing shops, radio shops and transmission receiving buildings, stockrooms, slaughter houses, shower buildings, staff buildings and quarters, toilet and locker rooms, warehouses, water towers and tanks, wells.[21]

Pan Am's Accra headquarters, home to more than seven hundred personnel, resembled an overseas military base in its attempted replication of a U.S. community. With its own power plant and water tower, four dining rooms, two recreational halls and an outdoor amphitheater, Protestant and Catholic chapels, and a private road to the beach, the Accra complex aimed to make employees feel "at home," enabling them to engage in familiar leisure activities, styles of worship, and eating and grooming habits. According to a construction manager stationed at Accra, "no matter where an American travels, he tries to live American style. And PAA made it possible for all its personnel to maintain this standard."[22]

Each Pan Am base was consciously designed to function as a "self-sufficient, self-contained community," a microcosm of the United States transplanted onto African soil. "Were it not for the African town beyond the camp," observed a reporter for the *New York Herald Tribune*, "the installation might be mistaken for similar ones in the United States." Amenities were simple—employees lived in one-story barracks with screen windows and metal furniture—but each base boasted company-run canteens stocked with U.S. products, a bar to replenish tired workers with beer and Coca-Cola, and a cafeteria serving menus planned by experts at *Good Housekeeping*. "Porterhouse steaks, fresh strawberries, and ice cream are now being served to Pan American Airways pilots right in the heart of African jungles," marveled a *Christian Science Monitor* writer. Indeed, he claimed, "world-wide American military thrusts" now required not only aircraft and weapons but refrigerated warehouses and other such amenities.[23]

Consumer goods also accompanied Pan Am to Africa. Discovering that many local people were as intrigued by American-made knickknacks as Americans were by African goods, U.S. employees eagerly bartered their T-shirts and Mickey Mouse watches for ivory, gold, gems, leather products, jewelry, handicrafts, and live animals. "We are all traders," recalled a pilot, whose purchases included "an ivory bridge carved from a tusk, a pair of camels, a pair of elephants, some beads, bracelets, a pair of candlesticks, and six napkin holders . . . ottoman covers, boots, plate holders, and lizard skins . . . a silk robe, four ladies' purses hand-tooled in Egyptian figures, and nine ounces of perfume." Even when Pan Am's operations were more military than commercial, aviation facilitated

international commerce and the global dissemination of American consumer culture.[24]

Motion pictures, too, played a vital role in introducing American culture to Africans who lived near Pan Am's bases. The airline had long
cultivated a close relationship with Hollywood. Clippers had played starring roles in such pictures as *Flying Down to Rio* (1933) and *China Clipper*
(1936), which were produced with Pan Am's willing cooperation; off the
screen, the Clippers transported Hollywood film reels to other countries.
This collaboration continued during the war, with the studios supplying
their latest releases to Pan Am's foreign outposts. Beyond sustaining U.S.
employees' morale, weekly movie screenings at PAA-Africa bases became
popular local attractions: according to Gilmore, residents of nearby villages "thronged around the windows to catch a glimpse of the screen."
Aware of the popularity of American films, PAA-Africa managers tried to
choose ones that would reflect positively on the United States. After a
screening of *Juke Girl* (Warner Brothers, 1942), a gritty melodrama about
migrant farm workers, PAA-Africa assistant manager John Yeomans sent
a letter of objection to the airline's public relations director. Complaining
that the film "gave foreigners a perverted idea with respect to the life of
poor people" in the United States, Yeomans urged that future films not
only be "of high entertainment value, but also of a type that, when distributed in Cairo and other foreign points, will reflect well on this country."[25]

Public health and sanitation was another arena in which Pan Am tried
to uphold U.S. employees' accustomed standards of living while transforming African communities in ways that would "reflect well" on the
United States. The airline's self-proclaimed "conquest of the white man's
grave" (a phrase that appeared frequently in Pan Am publicity materials)
rated among its most prized accomplishments. The term "white man's
grave" had been coined by early British exploring parties whose numbers
tended to diminish rapidly as men died from insect bites and unfamiliar
diseases. Even the RAF, during its operation of the Takoradi Route, saw
its ranks depleted due to malaria and dysentery. Pan Am, however, arrived
in Africa with its own medical department, headed by Dr. Lowell T.
Coggeshall, an authority on tropical diseases. Coggeshall succeeded in
reducing malaria rates among the airline's U.S. employees from an initial
high of 30 percent to less than 1 percent.[26]

The airline's conquest of the "white man's grave" seemed to offer proof that the United States alone possessed the fortitude and vitality to manage the Dark Continent. Britain, as weakened by war and debt as the RAF men had been by malaria, could no longer maintain control over its empire, this narrative implied. The United States now had to assume that burden. And like previous U.S.-sponsored public health initiatives in Latin America and the Philippines, Pan Am's "medical miracle" suggested how the United States could wield international influence without the formal trappings of empire. Although Dr. Coggeshall's department mostly treated U.S. employees, it also attempted to instruct local communities in malaria prevention, food and water purification, and sanitation. Evidence suggests, however, that Africans adopted these practices selectively: "Although they took our medicine and treatment, they always went later to their witch doctor to finish the job," wrote one employee.[27]

PAA-Africa's bases, then, were never as self-contained as airline officials often suggested. The sites functioned, rather, as contact zones, where Americans and Africans mingled in both work and leisure. In fact, the airlift required substantial African participation. Like the ADP in Latin America, this massive wartime enterprise depended on local labor. According to company figures, Pan Am hired some ten thousand Africans to work as ditchdiggers, cement layers, masons, mechanics' assistants, carpenters, domestic servants, cooks, gardeners, secretaries, bookkeepers, and messengers. The use of local labor, managers argued, compensated for the physical fatigue that even healthy white men suffered in tropical climates. In Gilmore's words, "natives" who were "adapted to climatic conditions" performed "invaluable assistance," whether building runways under the hot sun or working as "room boys," domestic servants who "assist[ed] their masters by saving them energy-taking movements in their rooms" and thereby helping them "conserve their strength for the job of running the airline."[28]

As a source of employment, Pan Am initially received a warm welcome in communities along the air route. In Accra, where PAA-Africa maintained its 700-person headquarters, the *African Morning Post* columnist "Duplex" described Americans as "friends of the Gold Coast people" who were "easy going, trustful, and faithful." American merchant ships had once engaged in a lucrative trade with the Gold Coast, Duplex noted, and the airplane now heralded a return to "the Good old days" of U.S.-African commercial relations. (He did not mention that the Gold Coast had also been the hub of the slave trade.) "We are ready to extend to them the right hand of friendship and a hearty welcome," the column declared. "Their settlement among us shall be the first step against unemployment and the beginning of improvement in the standard of living in post war times. It will afford an opportunity for learning the art of supplying our needs in lieu of imported commodities." Such statements imply that Duplex believed that the U.S. presence would stimulate the Gold Coast's economic development, just as African American journalist A. M. Wendell Malliet had suggested that the airlift might bring about Liberia's "national reconstruction."[29]

The West African press also praised Pan Am employees' efforts to socialize with local people. Referring, apparently, to a performance by a PAA-Africa singing group, *Spectator Daily* columnist "Rubbs" congratulated the airline "on the able manner in which your 'boys' choruses . . . acquitted themselves admiringly at Garrison Theatre last Saturday!" Rubbs also thanked the American "boys" for introducing Accra to their national pastime. In the Gold Coast's first baseball game, the U.S. Army trounced Pan Am's team by thirteen to one. But Rubbs cared less about the score than about the sport itself. "Interesting game. OK Boys! We are all game," he wrote.[30]

In November 1942, the *New York Times* published a letter from Albert Ofori, a mechanic's helper at PAA-Africa's Accra base, under the title "Africans Like Our Men: Natives Find Us Understanding and Ready to Help." Ofori's letter began, "We read in your paper which comes to the camp many stories about troubles between Americans in the United States, black and white." He insisted, though, that "there are no such troubles here in Africa." In Accra, Pan Am's white employees had made many friends and could regularly be spotted mingling with locals in cafés, Ofori claimed. Describing the Americans as friendly and helpful, he contrasted them to the "former foreigners"—the British and other Europeans—who had "left the Africans in darkness" under colonial rule. "A European may love an African but to a limited boundary, beyond which no African can go," Ofori stated. The Americans, however, "seem to be natural as if they were born here. . . . Their simple and straightforward life has enkindled many a young African to adopt it." He concluded, "I wish they might stay here entirely so that they could teach us what others have failed to do."[31]

Ofori's letter suggests that the notion of American exceptionalism could resonate with Africans as well as Americans. The U.S. ideology of liberty and democracy, the anti-imperial rhetoric of the Roosevelt administration, and the United States' own historic revolt against colonial rule all seemed to indicate that the Americans would be an improvement on the "former foreigners." And in many important ways, Pan Am was different from its RAF predecessors. Gilmore and other PAA-Africa employees corroborated Ofori's claim that friendships developed across lines of race and class. Along the air route, Americans introduced local residents to pastimes such as baseball, showed them Hollywood movies, and took them on airplane rides. There were a few isolated reports of trouble: in one instance, the British Secretariat informed the U.S. consul in Accra that "the behaviour of some members of PAA towards Africans when they meet in bars in the town is not all that could be desired . . . creating an unfortunate impression among the African population" that "may affect the good name of the PAA and other U.S. personnel in the Colony." In general, though, PAA-Africa personnel seem to have maintained harmonious relations with local communities.[32]

However, Ofori's comments also suggest that African attitudes toward Pan Am, and toward the United States generally, took shape within the context of their attitudes toward Britain and other colonial powers. As Harvey Neptune argues in reference to the wartime U.S. occupation of Trinidad, "local discourse on America almost always implicated the merits of the British Empire." Thus, "proclamations of pro-American sentiment should not be treated simply as testimonials. Voiced within earshot of British colonial rulers, they were meant as politically charged discourse, composed as rhetorical pulls of Britain's imperial tail." Although we cannot know precisely what Albert Ofori was thinking when he wrote to the *New York Times* in 1942, his letter clearly rebuked the Gold Coast's "former foreigners" as much as it complimented the current ones. Likewise, Accra newspapers' initial expressions of enthusiasm for Pan Am could have reflected frustration with the current colonial order as much as admiration for the United States.[33]

Furthermore, criticisms of Pan Am soon began to surface, particularly in Accra, where PAA-Africa was headquartered. In the local press, the airline came under fire for the wages it paid to its African workers. Pan Am claimed to have rejected the discriminatory hiring practices of the British colonial government; according to the *Africa News Letter*, published by PAA-Africa employees, British protocol "had prevented negroes from being given any work in their native land that could be done by the white man. But since the arrival of Americans, and particularly Pan Americans . . . barriers are considerably lower." Nonetheless, Pan Am adopted the wage scales set by British colonial administrators, which paid Africans a fraction of what white workers received. In an editorial entitled "The African's Poor Lot," the *Spectator Daily* decried Pan Am's capitulation to colonial "authorities" who were "against any high wages being paid above the local standard set by the [British] Government."[34] Appealing to the principles of justice affirmed by the Atlantic Charter, which included "raising the standard of living," the editorial asked: "How is the African's standard of living to be raised if his masters consider he should receive the same old meager wages that in some cases are not more than a pittance?" *Spectator Daily* columnist Rubbs emphasized that the "better world condition" envisioned by the "Roosevelt Churchill Eight-points" included a "high standard of living." If Pan Am, a representative of the United States, could choose to ignore this principle, then perhaps the African had been "left out" of the Atlantic Charter's purview. "Sorry that you have to alter your democratic principles on arrival in this country," Rubbs wrote in a sarcastic piece addressed to Pan American. "Some democracy!"[35]

News of the wage controversy soon reached Pan Am's New York headquarters. PAA-Africa manager Harold Whiteman mailed copies of the critical editorials to public relations director William Van Dusen, informing him that "we have been attacked along with the office of the [British] District Commissioner for our employment policy." Whiteman and PAA-Africa assistant manager John Yeomans arranged a meeting with an Accra editor, Benjamin Muta-Ofei, at which they defended their company's practices by arguing that "as green strangers in a new land we had to seek advice and assistance and naturally turned to the [colonial] government since our sole purpose in being here was to implement the British war effort." If Pan Am offered higher-than-average wages, they claimed, workers would be lured away from other employers. Instead of poaching labor from elsewhere, the airline wanted "to give employment to those now unemployed," Whiteman said. In defending these practices, Whiteman and Yeomans portrayed their company as both innocent ("green strangers in a new land") and benevolent (offering employment to the unemployed). In spite of its collusion with the British colonial government, Pan Am claimed to have Africans' best interests at heart.[36]

In Liberia, too, Pan Am came under criticism but in this case for its connections to a legacy of American hegemony. In 1926, the tire manufacturing company Firestone had loaned $5 million to the Liberian government in order to obtain a ninety-nine-year concession for a million-acre rubber plantation. Although Firestone subsequently became one of the country's largest employers (at one point providing more than half of Liberia's tax revenue), it was widely detested for its labor practices; rubber tappers reportedly performed their grueling work in conditions of virtual slavery. In Firestone, however, Pan Am found a crucial ally. The rubber company's Africa-based managers had advised the airline on where to build airfields and seaplane bases, and PAA-Africa personnel lived on Firestone's plantation until their own housing was constructed.[37]

Because Liberia was an independent nation rather than a colony of Britain, the State Department needed to obtain its government's permission in order for Pan Am to operate the airlift through its territory. President Edwin Barclay had initially refused the request, fearing that the operation would violate his country's neutrality and attract the attention of pro-Axis agents in the neighboring colonies of Vichy France. Barclay did agree, however, to allow Pan Am to establish an airmail and passenger service within Liberia. Pan Am technicians arrived in the country in September 1941, but plans to construct two airfields and a seaplane base stalled after the Liberian government demanded compensation for the airline's use of the land. During negotiations in the winter of 1941 and spring of 1942, Undersecretary of State Sumner Welles instructed U.S. diplomats in Monrovia to emphasize to Barclay that the aviation facilities would benefit Liberia "by knitting the country together . . . facilitating trips of Liberian officials and making medical services quickly available in emergencies." What was good for the United States, in other words, would be good for Liberia. The State Department also pledged $1 million in Lend-Lease aid for a road-building project. Airplanes and roads proved to be excellent bargaining chips. In April 1942, the Liberian government signed a contract granting the United States rights to "construct, control, operate, and defend . . . military and commercial airports" on Liberian territory. In return, Washington agreed to provide a small contingent of troops to protect the facilities from Axis sabotage, an amphibious plane for Pan Am's domestic air service, and the funding for road construction.[38]

When Pan Am inaugurated its first airmail flight in Liberia, a representative of Barclay's government hailed it as "a gesture of goodwill and courtesy" on the part of the United States. Pan Am's reputation quickly tarnished, however. "The Liberian authorities feel that [the air service] has not run sufficiently regularly and has been used too much for the army to the exclusion of Liberian passengers," reported a U.S. diplomat in Monrovia in July 1942. Passengers could book flights only after obtaining approval from U.S. Army headquarters in Accra, a process that entailed "considerable delay and annoyance." A crash destroyed Pan Am's original amphibian plane; its successor lasted less than a year before it, too, was damaged beyond repair. By the end of 1942, the airline's apparent disregard for its contractual obligations had become "a source of increasing friction with the Liberian government," reported the U.S. embassy. A representative of the U.S. Foreign Economic Administration, meanwhile, accused Pan Am of violating its "legal obligations" and "moral responsibility" to provide "at least a minimum of adequate service" in domestic air transport. He further admonished its leaders to remember "Liberia's importance to America as a 'springboard into Africa.'" More damning criticism of Pan Am appeared in the Liberian press. According to an article in *The Whirlwind*, the newspaper of Liberia's ruling political party, "the Pan-American Airways have broken with impunity their share of the agreement, and nothing is being said about it; their thread-hold has become a chain-hold on us just as in the case of Firestone." By comparing Pan Am to the widely detested Firestone, *The Whirlwind* branded the airline as an agent of U.S. imperialism—just as Honduran newspapers had done in the late 1920s by linking Pan Am to the United Fruit Company.[39]

Pan Am's business practices in British colonial Africa and Liberia thus complicated American cultural representations of aviation as an instrument of international goodwill. The airline presented itself as an enlightened alternative to European colonialism, and the Atlantic Charter gave Africans reason to hope that it would be. Yet while the Americans consistently emphasized their good intentions, African leaders and journalists demanded material proof—in the form of higher wages or improvements to local transportation—that the Americans were, in fact, different from their European predecessors. And although Pan Am salvaged the RAF's trans-African supply route, it could not so easily rehabilitate the social and economic legacies of colonialism.

African Adventure Tales

As Pan Am constructed air bases in Africa, its employees constructed narratives about Africa for U.S. audiences. In their personal correspondence and company newsletters, they vividly described jungle hardships, primitive tribes, exotic animals, and local customs both funny and fearsome. American newspapers and magazines, too, delighted in recounting such tales, portraying Pan Am as the air-age successor to Livingstone and Tarzan. The intellectual energies expended on this project of narrative construction rivaled the physical efforts expended on constructing the air route itself. Back home, Americans eagerly consumed Pan Am's African adventure tales: "Do let us have more on life on the west coast of Africa," requested Mrs. J. Stuart Ramsey, a *New Horizons* reader from Roanoke, Virginia. In writing and reading about "life on the west coast of Africa," however, Americans were ultimately writing and reading about themselves—just as Perry Miller discovered "the uniqueness of the American experience" in the Congo.[40]

By the World War II era, Americans had already encountered representations of Africa in a variety of popular cultural texts: Teddy Roosevelt's widely publicized 1909 safari expedition, the lavishly illustrated travelogues and anthropological essays that appeared in *National Geographic*, exhibits of "pygmies" at world's fairs, the activities of missionary organizations, and, perhaps most influentially, Edgar Rice Burrough's best-selling *Tarzan* novels, the first of which appeared in 1912, and their equally popular film adaptations. When writing about their experiences, PAA-Africa employees drew, consciously or not, on these familiar texts. "This is *really* Africa," emphasized *New Horizons* in a typical article on the airlift.

Ironically, though, what seemed most real was often what looked familiar. Reproducing familiar images and tropes, media coverage of PAA-
Africa could very well have simply confirmed what Ramsey and other
U.S. readers already believed about Africa, rather than introducing new
knowledge that would challenge or complicate such beliefs.[41]

American conceptions of Africa were also shaped by European imperialist literature, particularly Joseph Conrad's acclaimed novel *Heart of
Darkness* (1902). The harrowing story of an English steamer captain (Marlow) sent to the Belgian Congo to locate and retrieve an ivory trader
(Kurtz), *Heart of Darkness* construed the jungles of central Africa as sites
of unfathomable natural and moral decay. "The general sense of vague
and oppressive wonder grew upon me. It was like a weary pilgrimage
amongst hints of nightmares," Marlow recalls, in a passage typical of
Conrad's dystopian rendering of the Congo. Americans did not necessarily have to read *Heart of Darkness*, moreover, to be familiar with the story.
In 1938, the Mercury Theater of the Air produced a dramatic radio adaptation of Kurtz and Marlow's saga. Starring Orson Welles, the radio play
aired just one week after the actor's performance of H. G. Wells's novel
War of the Worlds had notoriously incited widespread fears of alien
invasion.[42]

The geography of the air route reinforced associations with texts such
as *Heart of Darkness* and *Tarzan*. Although Pan Am operated bases as far
east as Egypt and the Sudan, its forces were concentrated along Africa's
western coast, where the fictional Tarzan swung from his vines. While
Pan Am employees tended to elide geographical specificity in their writings, representing their subject simply as "Africa," their perceptions reflected the centrality of West Africa and the Congo in the American
cultural imagination. West Africa, moreover, had played a central role in
U.S. history as the epicenter of the slave trade. In and around Accra,
where PAA-Africa maintained its headquarters, the coastal forts built by
European slave traders in the seventeenth century stood as monuments to
the traffic in human beings that brought an estimated twelve million
Africans to the United States.

Reflecting long-standing perceptions of Africa as a "Dark Continent"
in need of (white) enlightenment, Pan Am marketed PAA-Africa as a mission of international goodwill. Just as the airline had previously identified

its Latin American operations with the Good Neighbor policy, company officials again portrayed the airplane as an engine of social uplift that would bring progress and civilization to Africa. According to an internal report, "The improvements in living standards, in medical knowledge, and in business methods that are traceable to PAA will be of long-time benefit to the Africans." Echoing what Sumner Welles proposed during the State Department's negotiations with Liberia, Pan Am leaders contended that aviation would improve Africans' living conditions by uniting far-flung villages, expediting the transit of food, medicine, and commodities, and creating jobs. Such arguments reproduced but also reframed European discourses on Africa, updating the *mission civilisatrice* for the air age.[43]

Like Beryl Markham, PAA-Africa personnel manager Voit Gilmore depicted aviation as white civilization's "gift" to Africa. When Pan Am pilots took local residents for airplane rides, he wrote, they "expanded the geographical vistas of many natives from the two or three miles of Africa in which they had always lived, to the wide expanse of the entire world." In working for the airline, Africans learned "American standards for business and social conduct," he further claimed. The airlift also promised to modernize African societies. In Gilmore's view, "Accra had not changed in 100 years, till Pan Africa came along with innovations like Simmons mattresses, modern plumbing, [and] current movies." Africans' presumed gratitude for such "innovations" would ultimately benefit Pan Am, airline leaders predicted. "The goodwill engendered toward Americans in the hearts of natives" would prove profitable "in later years when Americans go to Africa, often on pleasure, frequently (doubtlessly) on business," concluded one company report. Aviation, then, would not only enable the Allies to win the war; it would also enable the United States to win new allies in Africa.[44]

In conjunction with such expressions of internationalism, Pan Am also marketed its African operations as a manifestation of *national* greatness. Juan Trippe liked to say that the airplane would push the nation's frontiers "to infinity," and by dramatically expanding the United States' presence in Africa, Pan Am helped to push the frontiers of its economic, military, and cultural influence beyond the Western Hemisphere, if not quite to infinity. Pan Am's public relations office depicted the African supply route as an epic, quintessentially American story of pioneering and frontier conquest—and the U.S. news media eagerly reported this version of the story, one of the few positive spins that could be put on the beleaguered Allied war effort in late 1941 to early 1942. "Pioneering is an old story to Pan American," stated the *Saturday Evening Post* in its admiring profile of "America's New Lifeline to Africa" from November 1941. The *New York Herald Tribune* credited Pan Am with imparting an "American tinge" to the African "wilds." The *Pittsburgh Sun-Telegraph* applauded the airline's "Miracle in Africa": "American engineers have modernized entire sections of the dark continent overnight . . . bringing sweeping chances to the old Africa of impassable jungles and uninhabitable deserts." For syndicated columnist Raymond Clapper, PAA-Africa exhibited "the same sort of imagination, the same sort of bold, large-scale planning and efficient execution that went into the building of the American West. Those who thought such qualities dead in our country will know from this that it is not so."[45]

Yet the "building of the American West," after all, had involved brutal warfare between white settlers and Indians who fought against the seizure of their lands. By contrast, Pan Am claimed that Africans welcomed its more modern, benign brand of pioneering. Stories about PAA-Africa in company publications prominently featured local African employees, who were typically portrayed as grateful and content. In a 1942 issue of *New Horizons*, a smiling Accra worker named John posed in a "publicity shot" while reading a copy of the magazine. According to the PAA-Africa manager who had taken the photograph, "It is John's job to lash down everything in the planes that come through our base." His real ambition, though, was "to join the U.S. Navy." But what, exactly, was John publicizing in his "publicity shot"? Beyond giving exposure to *New Horizons*, the image aimed to publicize how Pan Am was fostering pro-American sentiment in Africa, obtaining the loyalty of the locals by offering them employment, education, and even reading material. John's desire to join the Navy further indicated how Pan Am was integrating Africa within a widening U.S. sphere of influence.[46]

Africans' own requests for employment with Pan Am were taken as additional evidence of U.S. benevolence and African loyalty. Calling the U.S. airline "one of the most popular employers in Africa," Voit Gilmore quoted a letter sent by a young resident of Accra, dated July 4, 1942: "I

have got honour most respectfully to ask for a vacancy in your Department. I should be glad if I may be employed either as an office-boy,
pantry-boy, or messenger. I attended the Accra Senior Boys' School. . . . I
was loved by all the teachers and pupils because of my gentility, sports-
manship, and obedience. [Signed] Yr. Mst. Obed. Servant." *New Horizons*
published similar letters. A man named T. Bosere wrote in order to request "work which is suitable for me. Either bootboy or watchnight I
want." Hugh Ellis Jenkins, from Nigeria, desired employment as a clerk.
While writing "with due deference" of his "craving for an appointment,"
he also expressed frustration that, "to my utmost surprise," his previous
inquiries had been "treated with everlasting silence while there had been
an influx of engagement of clerks in your department." As proof of his
qualifications, Jenkins testified that he had previously been an associate
editor for the *Nigerian Eastern Mail* and a proofreader for the *Nigerian
Daily Times.* Another letter, said Gilmore, contained "no less than fourteen bits of evidence to convince us of the sender's worth"; others arrived
with grammar school diplomas and handicrafts. Described by Harold
Whiteman as "tattered letters of testimony . . . curious, touching, pathetic,"
Africans' requests for employment appeared to corroborate Pan Am's public image as an agent of opportunity and progress. The writers' professions
of deference, meanwhile, made Africans themselves seem docile, humble,
and grateful for the "tutelage" that a job with the airline could offer.[47]

Pan Am's posture toward its local workforce can best be characterized
as a form of paternalism, defined by J. Douglas Smith as "a belief in the
social, cultural, intellectual, emotional, and often racial superiority of one
individual or group who presumes to act in the best interest of another
individual or group in exchange for the subordinate party's compliance."
Like a father-child relationship, paternalism mixed care and control,
intertwining expressions of affection with expressions of power. Voit
Gilmore, for example, described his African employees as "unspoiled and
humble" but also noted that they "naturally required patient handling
and supervision." By viewing themselves as father figures whose supervision protected Africans' best interests, PAA-Africa managers naturalized and neutralized social hierarchies. Leading by example rather than
by force, they claimed to have "worked with the natives instead of merely
supervising," wrote construction manager H. E. Baldwin. Consequently,

the workers' "amazement grew to respect, then to emulation." Another PAA-Africa manager, however, described Africans' "attempts to emulate the white man and to place themselves on a mental and social par with him" as, simply, "pathetic." As this statement suggests, paternalism left racial hierarchies intact even as it appeared less virulent than other racial ideologies.[48]

The use of local labor thus served Pan Am's interests in two ways. African workers performed strenuous or tedious tasks that allowed U.S. employees to devote their energies to "running the airline" (as Gilmore put it). But Pan Am also derived symbolic value from its local workforce. Photographs of apparently contented African employees, such as *New Horizons'* "publicity shot" of John, combined with their own requests for employment, seemed to confirm the airline's status as an enlightened representative of American democracy. PAA-Africa managers accurately stated that "the job could not have been done . . . without the help of the native." Yet the careful and conspicuous display of African labor must also be recognized as a public relations strategy designed to identify Pan Am with progress and international goodwill.[49]

An advertisement from December 1943, illustrated by Stevan Dohanos—an artist best known for his *Saturday Evening Post* covers—exemplifies how Pan Am incorporated "the help of the native" into its public image. In the ad's foreground, a white man in safari attire speaks with a uniformed pilot (in spotless dress whites) while taking notes on a clipboard. Centered, but in the background, two shirtless African men assist a delivery of air cargo; one raises his arm, as if saluting the airplane. The image's visual grammar acknowledges the African contribution to the Allied war effort, implying that "America's New Lifeline to Africa" could not function without Africans themselves. More subtly, however, the image reinforces racial and national hierarchies. The dark-skinned workers appear diminutive in comparison to the foregrounded white men and their airplanes, which literally loom over them. Next to the pilot's gleaming dress whites, the men's nakedness is conspicuous, a reminder that their function is merely to perform physical labor. The pilot, meanwhile, is indisputably in charge of the scenario, with machines and other men (the Africans and the safari-attired man, a subordinate who appears to be writing down the pilot's instructions) at his command.[50]

Significantly, though, the pilot executes command with a pleasant smile, for his leadership requires not the exclusion of Africans but rather their willing inclusion—albeit on subordinate terms. Likewise, the United States built its ascent to global leadership on other nations' ostensible willingness to be incorporated into widening spheres of American influence: the dynamic that Geir Lundestad famously dubbed "empire by invitation." The very idea that Africa required an American "lifeline," however, implied that it would be included in this global order on the United States' terms and by the grace of its goodwill. The ad likewise suggested that the United States would lead the world by integrating it, emphasizing connection, affiliation, and the bridging of cultural differences. This integrative global imaginary worked, as Christina Klein has argued, to "legitimate U.S. expansion while denying its coercive or imperial nature."[51]

In emphasizing Africans' affiliations with the United States, such representations did not deny that they were different from Americans. Indeed, the global imaginary of integration could reify notions of racial and national difference even as it envisioned closer relations between Americans and foreign others. In Pan Am's company literature, representations of the laboring black body signified both racial difference and African loyalty to the United States. When Harold Whiteman described, in a letter to public relations director Van Dusen, "a swarm of happy jibbering natives who seem to enjoy their work tremendously," he construed the local workforce as docile, unthreatening, and clearly less civilized than their articulate and individuated supervisors. References to "swarms" and "hordes," classic tropes of racializing discourse, further diminished Africans' humanity by likening them to insects or animals. "There were hundreds of them, their ebony black skins glistening with the warmth. Heads without hair, bare to the sun, with now and then a faded fez to mark a northern Mohammedan influence. Bodies were bare, save for shorts and an occasional tattered shirt. . . . They were strong bodies, with straight backs and thick necks," wrote E. Robin Little, PAA-Africa's industrial and public relations manager, in his portrait of workers building a runway. Stories about PAA-Africa in U.S. newspapers and magazines similarly fixated on the laboring African body as an emblem of racial difference. The *Pittsburgh Sun-Telegraph* described "7,000 ebony-skinned men . . . armies of men . . . scraping down the African earth, carrying off pounds of it in baskets balanced on their heads"; Raymond Clapper described how "hundreds of natives, clad in G-strings" worked "in swarms." These articles were typically accompanied by photographs of semiclothed black laborers. Recalling Pan Am's "Lifeline to Africa" ad, such images reinforced racial ideology that identified Africans with physicality and lack of civilization.[52]

In the context of Pan Am's public relations strategies, these familiar racialized images took on specific meanings. Quite literally, the airline displayed its African workforce as evidence that it was a beneficent employer, an agent of social progress, an enlightened exemplar of American democracy. Its visual and textual documentation of PAA-Africa, however, reflected a mastering gaze—a paternalistic gaze that mixed power with affection. Pan Am included Africans in its wartime global enterprise, yet carefully framed their inclusion so as to maintain the racial and national distinctions on which American globalism, as well as Anglo-American cooperation, depended. "America's New Lifeline to Africa" required African labor but employed Africans on subordinate terms; Pan Am's globalist image required the presence of dark-skinned bodies but placed them under the watchful management of uniformed white authority.

The logic of integration that characterized Pan Am's marketing of PAA-Africa informed broader cultural narratives about the United States' participation in World War II. The story of PAA-Africa, which by 1943 had received ample coverage in the mass media, contributed to an emerging pluralist narrative of the "good war" in which a multiethnic U.S. military battled racist Nazis and Japanese imperialists. This story masked uglier social realities, including Jim Crow segregation in the military itself. Yet representations of the United States as a melting-pot nation, even if not entirely factual, bolstered its claims to global leadership. With a population that encompassed the world, only the United States was fit to lead the world, the argument went. Like Henry Luce's conception of the American Century, the pluralist narrative of the good war supported both global integration and U.S. hegemony. As Pan Am's African adventure tales suggest, pluralist logic could reify racial and national distinctions rather than dismantling them.[53]

Contrary to Voit Gilmore's claim that Americans' "wholesome way of living" represented a "complete departure" from the imperial "mannerisms" of the British, U.S. employees' representations of their experiences revealed affinities among American and European conceptions of race, civilization, and power. Echoing the amateur ethnographies produced by European explorers and colonial administrators, letters and publications by Pan Am employees proffered detailed information about Africans' work habits, leisure activities, religious beliefs, clothing, food, and psychological and physical characteristics. In a typical example, an article in *New Horizons* explained how the "13 different tribes" in Portuguese Guinea could be identified: Mohammedans wore "white 'night gowns' and red hats," Fulas had scars on their temples, Mandingas were the "most stolid," and Bujagos were "the tough gang" known to "steal all male babies." Gender emerged as a particular focus of attention—and a key marker of racial difference. African women were said to perform laborious, masculine work like "paving roads and making bricks"; African men, meanwhile, "paraded about, smoking home-made perfumed cigarettes" while "content[ing] themselves with hand crafts, doing house-boy duties, and driving caravans." This apparent reversal of normative Western gender roles seemed to confirm that African societies were backward and in need of reform.[54]

The trope of "dark Africa," famously elaborated in Conrad's *Heart of Darkness*, pervaded Pan Am employees' accounts of their experiences. By their own admission, many Africans were educated, spoke English, and worked at various skilled trades. Yet employees' writings focused overwhelmingly on "jungle natives" who wore scanty clothing, practiced "ju-ju" and cannibalism, and danced to frenzied drumbeats. "At night the wood and stretched skin drums beat out a crude rhythm which sets them prancing crazily about, shrieking, flinging arms and legs to the wind, eyes and teeth flashing, as happy as the day is long," reported a contributor to *New Horizons*. In two vivid letters to the magazine, PAA-Africa employee Tom McMillan described "wildly-dancing Wogs beating on tom-tom drums," "real, wild-looking native villages," "shells which had been dipped in *human blood*," and a man "putting on a wild dance, swinging a HUMAN THIGH BONE in his right hand" [emphasis in original]. McMillan acknowledged that "the man-eating habit in the eastern region is largely a thing of the past," but he also claimed that "human meat has been found in the local market on occasions." Another letter writer testified, "My most vivid recollection of the place was of the local 'ju-ju' man who wore a cloak covered with shells and blood and who marched through the village brandishing a human thigh bone above his head." Cartoon caricatures of savage natives, meanwhile, adorned the covers of PAA-Africa employees' self-published *Africa News Letter*. One such image, captioned "Who's Cookin', Bud?" showed a group of Africans gathering around a pot with human feet sticking out of it. On another cover, an impish-looking, loin-cloth-wearing pygmy aimed a slingshot at the buttocks of a uniformed white officer.[55]

Discourse on race within the United States and long-standing white stereotypes about African Americans shaped what Pan Am employees saw in Africa. In "a letter to help give the folks back home a better idea of what type of country this really is," William Stempel informed *New Horizons* that Africans "appear much happier here than the colored people in the States. Most of them wear a perpetual smile, if little more." He enclosed a photograph, which the magazine published, of a "local rascal who I doubt ever did a day's work in his life." Reproducing the well-worn racist caricature of the carefree and lazy plantation slave, Stempel's language suggested that color determined character everywhere, in Africa as well as in the United States. As if to reinforce the point, Stempel signed his letter "Yours for more enlightenment of dark Africa."[56]

Aviation itself became an important signifier of racial difference. Africans' lack of familiarity with airplanes, which often manifested as fear or bewilderment, marked them as uncivilized in the eyes of U.S. observers, who enjoyed playing up their reactions for humorous effect. According to the *New York Herald Tribune*, a group of "native chiefs" visiting an airport was "one of the strangest sights to be seen":

[A] chief wouldn't think of making such an important tour without bringing his entire retinue of umbrella bearers, mace bearers, pillow bearers and what not along with him. The result was that a terrific mob showed up, ranging from big chiefs down to black small fry. . . . Several of the chiefs weighed close to 300 pounds and were as black as a black cat hiding in the shadow of a tombstone at midnight. The chiefs were swathed in robes which reached their ankles and were adorned with a variety of colors that would make a kaleidoscope seem drab. . . . They wore flat sandals of carved leather which were all right on level ground but not so good for climbing in and out of airplanes.

The chiefs' purported troubles with airplanes went beyond their ill-suited footwear. They "listened politely but skeptically while it was explained to them that the man with the microphone in his hand was talking to the man in the plane in the air." When the airport's Link trainer (a type of flight simulator) "turned and pointed straight at one chief, the chief leaped quickly from in front of it, amid general laughter. He admitted later that he thought the little airplane might fly out the window and knock him out of the way." The *Herald Tribune* reported that the event "resulted in high merriment on both sides": the chiefs were as amused by the Americans' strange technology as were the Americans by the chiefs' attire and behavior. But the article's lighthearted tone belied a more judgmental subtext. Its humor derived from the reader's presumed ability to recognize that airplanes were not actually strange at all—at least not for civilized people— and therefore to see the chiefs' reactions as the result of irrational, outmoded superstitions.[57]

In a similar depiction of cultural differences on the airfield, the *St. Louis Post-Dispatch* described a "turbaned, Mahometan priest" attempting to sound his call to prayer "above the roar from the near-by motors of big American bombers tearing along the hard-packed runway . . . where only a few months ago there was nothing but wasteland." As the roar of the bombers drowned out the priest's summons, the men working on the runway continued their labors rather than heeding the call to pray. Primitive traditions could not stand up against technological progress, the anecdote implied. Even when Africans attempted to incorporate aviation into indigenous cultural practices, U.S. observers often derided the results. The popular aviation magazine *Skyways*, for example, published a photograph of an African man, described as a "modern witch doctor," wearing a headdress topped with a highly accurate model of a P-40 fighter plane— apparently, *Skyways* claimed, for the purposes of "scaring away the evil spirits of the Nazis in darkest Africa." The photo's caption read, "Voodoo

Today, Vot Tomorrow." Like the *New York Herald Tribune* article, the piece used humor to mock what its writer perceived as African savagery and backwardness.[58]

It is impossible to know what journalists and Pan Am employees "really" thought about race. Perhaps they genuinely believed that black Africans were inferior to whites. More likely, though, they invoked familiar racial stereotypes to weave a colorful story, an air-age African adventure tale for the folks back home. These authors' intentions, however, are less historically significant than the cultural effects of the narratives they produced. In the context of American globalism, such narratives served to underscore Africa's difference from the United States while at the same time affirming its loyalty and need for integration into a U.S. sphere of influence.

Stories about Africa also served to define Americans' own national, racial, and gender identities. Echoing Theodore Roosevelt's paeans to the "strenuous life," the physical hardships PAA-Africa employees endured seemed to confirm that white American men were, in fact, fit to pioneer new global frontiers; modern comforts had not made them soft. In Raymond Clapper's telling, Africa offered a proving ground for white American masculinity, filled with dangers and annoyances that tested men's fortitude: "the malarial mosquito which hovers near the ground"; "sandstorms, tropical heat that slays unless you wear an insulated helmet reaching down over the back of your neck, torrential rains"; "the native labor that insists on doing work in slow, ancient ways that drive Americans to exasperation." Exasperated as they may have been, Pan Am's U.S. employees stood up to such challenges. PAA-Africa appeared to offer evidence that white American men could master the Dark Continent. By extension, the United States could assume the white man's burden, taking the place of European powers that were too preoccupied in their own conflicts to manage Africa properly. Reports on PAA-Africa, in Pan Am's own literature and in the mainstream media, conveyed vivid descriptions of successful American leadership under harsh conditions: "boys from the cool New England states toiling in sunlight that registered 158 degrees, some of them spurring on squads of naked natives who have never heard a locomotive whistle or seen an automobile." These narratives not only reproduced familiar representations of Africa and Africans; they also produced a new type of American subject, a white man capable of taking on the world—the protagonist of an American Century.[59]

In PAA-Africa operations manager George Kraigher, Pan Am found a real-life specimen of this type of white manhood. Ironically, Kraigher was not American by birth; a native of Slovenia and a World War I ace, he had flown with the Serbian Air Force before emigrating to the United States and joining Pan Am as the operations manager of its airport in Brownsville, Texas. In 1941, Kraigher was selected to head Pan Am's operations in Africa. There he became a company legend, the living embodiment of its pioneer mythology. "Physically and mentally, George Kraigher is the epitome of what an Operations Manager in Africa should be," stated *New Horizons*. "His erect and tapered torso, topped by a bush of steely gray hair, his gruff masculine manner, his terrific energy, his self-reliance, his ability to make quick decisions—all these fit him to the job." Kraigher also understood the civilized virtues of moderation: he was "a non-smoker and moderate drinker, a big-but-careful eater (he hates fried foods) [and] a confirmed bachelor." Like Kurtz in *Heart of Darkness*, European men had succumbed to corruption, vice, and diseases in Africa. But Kraigher, a self-made citizen of the United States, exemplified the quintessential Yankee characteristics—temperance, self-reliance, ambition, dedication to work—that enabled American men to survive, and even thrive, in such circumstances.[60]

These same characteristics, company literature claimed, also enabled Kraigher to earn the loyalty and respect of the "natives." Unlike Lord Kitchener—the British imperial field marshal who led the conquest of the Sudan in 1898 and went on to vanquish the Boer rebellion in South Africa—Kraigher did not need to assert his mastery through violence. Rather, like the pilot in Pan Am's "Lifeline to Africa" ad, he ruled with a smile (or, given his "gruff masculine manner," perhaps merely with a commanding gaze), not with a sword.[61]

Led by men such as George Kraigher, Pan Am's errand into the "wilderness" did not appear as a form of conquest. The airline did not seek to colonize territory or to rule over local populations. It sought, rather, to "expound [its] America to the twentieth century"—to quote Perry Miller's description of his own chosen mission following his epiphany in the

Congo. As Pan Am planes carried U.S. technology, consumer goods, business methods, and ways of living to no-longer-distant places, the United States' commercial empire of the air seemed to offer a model of global power that differed fundamentally from the European empires of the earth.

Nonetheless, the history of PAA-Africa reveals important affinities between American and British investments in Africa as a site of empire and nation building. The project originated through Anglo-American cooperation; Pan Am adopted certain British colonial practices, most notoriously wage scales for local workers; and the airline's U.S. employees reproduced the racialized language of imperialist literature in their writings about Africa. Some employees, furthermore, appeared to enjoy privileges typically accorded to rulers. "Airportman Harden was surprised when natives, inaugurating a friendly way of bidding adieu, placed him in a bamboo chair, carried him down to the departing Clipper," reported *New Horizons*. PAA-Africa personnel often wrote about their domestic servants: a "houseboy," according to one, "refers to us as 'Master' or 'Sah' . . . shines [our] shoes, takes care of the laundry, brings [our] tea and biscuits in the afternoon, and is always at [our] beck and call." When Americans drove through African towns, claimed a public relations manager, the local population would "get up and bow low," making them feel like "some East Side New York ward heeler acknowledging the plaudits of his followers." In spite of Pan Am's careful efforts to distinguish itself from its RAF predecessors, these narratives reproduced imperial relationships. Africans appear not as wage-earning employees but as subjects who followed, bowed down to, and even carried their American "masters," whom they referred to as "Sah." Such fantasies of mastery would have resonated with the currents of nostalgia for Southern plantation life in mid-twentieth-century American culture—expressed most famously in Margaret Mitchell's best-selling novel *Gone with the Wind* (1936) and its Oscar-winning film adaptation (1939).[62]

Pan Am and the mainstream U.S. media alike hailed PAA-Africa as a triumph of American exceptionalism—"a complete departure" from British imperialism, as Voit Gilmore put it, and a model for a new kind of global leadership that replaced the sword with the smile, conquest with commerce. Some Africans, too, subscribed to this narrative, as evidenced by Albert Ofori's letter to the *New York Times* and numerous editorials in the West African press. In the era of the Atlantic Charter, they had reason to believe that the United States would fulfill its stated commitments to self-determination, democracy, and high standards of living.

Ultimately, however, the United States distinguished itself from imperial Britain by claiming to be a *better* manager of the "white man's burden." Americans imagined themselves as Kraighers, not Kurtzes or Kitcheners; they would arrive in commercial airplanes instead of armed convoys, integrating Africans into their global enterprise rather than conquering them by force. But there was little doubt, among aviation and government leaders, that PAA-Africa would enhance the United States' power in Africa and globally. Prompted by the immediate emergency of the war but serving long-term U.S. political and economic objectives, Pan Am aimed not only to shore up the British position in Africa but to position the United States' "empire of the air" as the logical successor to Britain's territorial empire. The airline's pluralist discourses and policies, then, were not antithetical to the projection of national power but conducive to it. The global imaginary of integration, which folded Africa into the American Century rather than forcibly demanding its subjection, ultimately proved more efficient and less costly than the global imaginary of imperialism. And for an empire of commerce, efficiency and cost effectiveness were of prime importance.

George Kraigher became the first Pan Am employee to receive a commission in the U.S. Army Air Forces when Air Transport Command (ATC) took over PAA-Africa's supply route on November 1, 1942. With the United States officially at war, it no longer made sense to employ a private company to do the work that the Army, with its much greater manpower and resources, could now legally do.[63]

Turning the operation over to the ATC also allowed Washington to assuage British concerns that Pan Am planned to use its wartime presence in Africa to dominate the continent's postwar air transport market. Churchill's government had never intended to hand Africa's skyways to the United States permanently. On the contrary, London sought to maintain the exclusive rights of the British Overseas Airways Corporation (BOAC) to operate commercial flights in and through British colonial territories. Pan Am, however, hoped to use its wartime airfields to obtain a permanent foothold in Africa. Although Pan Am's contracts stipulated that its wartime services did not guarantee any special rights after the war, British aviation leaders charged that Pan Am was already finding ways to restrict BOAC's usage of the airfields that it had built in Africa. Air Marshal Tedder accused Pan Am of "developing routes quite regardless of our interests, despite the fact that [it was] only able to do so with our active assistance." The "cloven hoof of Pan American," he fumed, had stamped on BOAC's rightful domain. By 1947, the U.S. airline operated regular flights between the United States and several African cities, including Accra, Dakar, Leopoldville, and Cairo, that were once served exclusively by European airlines.[64]

Although short in duration, Pan Am's government-sponsored operations in Africa thus had lasting consequences. After expanding throughout Latin America, Pan Am extended its infrastructure across the Atlantic, where the United States collaborated but also competed for influence and resources with Europe's imperial powers. As Raymond Clapper reminded his many readers, the lands through which PAA-Africa's supply route stretched had been claimed by every historical empire from the Roman to the British—and now, "over these ancient paths sunburned Americans in khaki shorts are already appearing in increasing numbers." These sunburned, khaki-wearing Americans facilitated a gradual transfer of global power from Old World to New, from the European imperial order to an emerging American Century. Flying everywhere, they proposed to replace the unseemly old scramble for colonies with an orderly, nonviolent expansion of global commerce.[65]

Pan Am's wartime errand into the wilderness encapsulated the defining tensions of American globalism: its conflation of universalism and exceptionalism; its integrative logic and its reification of difference; its simultaneous projection and denial of empire. These tensions did not go unnoticed or uncontested. If airplanes carried U.S. influence around the world, they also brought the world to the United States—and in November 1944, Chicago hosted an international conference that would challenge U.S. domination of the commercial skies. An American air empire had indeed taken off by the end of World War II, but its ascent would be neither as smooth nor as rapid as men such as Juan Trippe had hoped.

Figure 1 Seeing the world: The globe at Pan American's Dinner Key Terminal in Miami. Pan American World Airways, Inc. Records, Special Collections, University of Miami Libraries.

Figure 2 A bird's-eye view of the globe at Pan American's Dinner Key Terminal. State Archives of Florida.

Figure 3 Pan American Airways president Juan Trippe plotting world air routes on the globe in his office, a family heirloom dating from the 1840s. Time & Life Pictures/ Getty Images.

(a)

(b)

(c)

Figure 4a–c From the Americas to the world: Pan American Airways logos circa 1927, 1947, and 1957.

Figure 5 Thousands of spectators in Berlin, Germany, watch Orville Wright pilot the Model A Flyer, 1909. Wright Brothers Collection, Special Collections and Archives, Wright State University Libraries.

Figure 6 Charles A. Lindbergh in Paris with U.S. ambassador Myron T. Herrick at the Aero Club de France, after Lindbergh's solo transatlantic flight of May 20–21, 1927. Lindbergh Picture Collection, Manuscripts and Archives, Yale University Library.

Figure 7 Charles Lindbergh and Juan Trippe, 1928. Pan American World Airways, Inc. Records, Special Collections, University of Miami Libraries.

Figure 8 Juan Trippe and Charles Lindbergh in British Guiana during their 1929 Latin American goodwill flight. Lindbergh Picture Collection, Manuscripts and Archives, Yale University Library.

Figure 9 Flying the "cocktail circuit": Pan American Airways' first passenger flight from Miami to Havana, 1927. Pan American World Airways, Inc. Records, Special Collections, University of Miami Libraries.

Figure 10 Crowds gather to witness the inaugural flight of Pan American Airways' *American Clipper*, Miami, Florida, 1930. State Archives of Florida.

Figure 11 A different era of air travel: Passengers playing cards aboard one of Pan Am's Sikorsky S-40 Clippers, 1931. State Archives of Florida.

Figure 12 Encircling the Americas: Pan American Airways poster, 1931. Smithsonian Institution/Corbis.

Figure 13 The Good Neighbor policy takes flight: Pan American Airways advertisement, 1941. Ad*Access Online Project, ad #T2301, John W. Hartman Center for Sales, Advertising, and Marketing History, David M. Rubenstein Rare Book & Manuscript Library, Duke University, library.duke.edu/digitalcollections/adaccess.

Figure 14   Two Clippers: A Pan American Airways flying boat passes over a clipper ship off the Spanish coast, 1938. Getty Images.

Figure 15 Crossing the Pacific: Pan American Airways poster, 1938. Library of Congress, Prints and Photographs Division, LC-USZC4-2315.

Figure 16 Wendell Willkie arriving in Cairo aboard the *Gulliver* during his "one world" goodwill flight, 1942. Time & Life Pictures/ Getty Images.

Figure 17 The walk-in globe at the Museum of Modern Art's "Airways to Peace" exhibition, 1943. Library of Congress, Prints and Photographs Division, LC-G612-T01-43752.

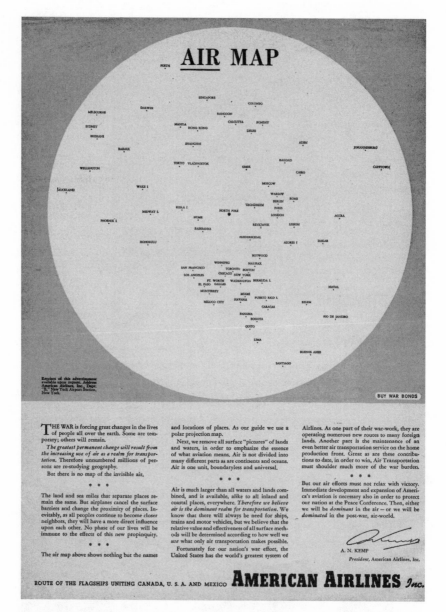

Figure 18 A world with no boundaries: American Airlines advertisement, 1943. Author's collection.

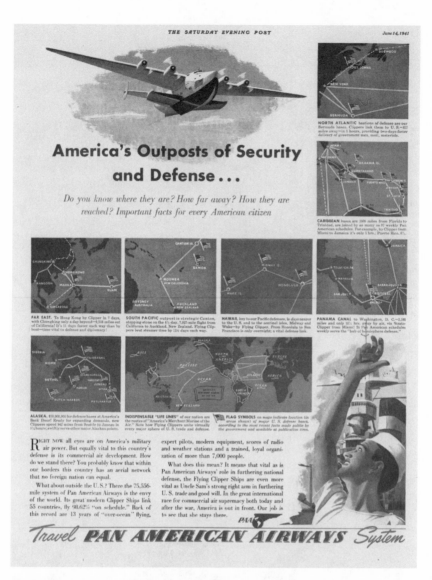

Figure 19 "A branch of U.S. defense": Pan American Airways advertisement, 1941. Ad*Access Online Project, ad #T2300, John W. Hartman Center for Sales, Advertising, and Marketing History, David M. Rubenstein Rare Book & Manuscript Library, Duke University, library.duke.edu/digitalcollections/adaccess.

Figure 20 Workers in West Africa building a runway for Pan American Airways' trans-African supply route, circa early 1941. Pan American World Airways, Inc. Records, Special Collections, University of Miami Libraries.

Figure 21 "Camels to Airliners": Brick-carrying camels and local masons construct installations for Pan American Airways-Africa, 1942. Associated Press.

Figure 22 Passengers of the first commercial around-the-world flight, 1947, including Juan Trippe; Frank Gannett, publisher of Gannett Newspapers; Francis H. Russell, director of the State Department's Office of Public Affairs; M. T. Moore, chairman of the board of *Time-Life-Fortune*; Gardner Cowles, president of the *Des Moines Register and Tribune*; and Barry Faris, editor-in-chief of International News Service. Bettmann/Corbis.

Figure 23 The world on a string: Pan American Airways advertisement, 1948. John W. Hartman Center for Sales, Advertising, and Marketing History, David M. Rubenstein Rare Book & Manuscript Library, Duke University.

He commands your Flagship and your
CONFIDENCE

The pilot of your Flagship truly merits your confidence. He represents a highly selective group, chosen by virtue of intelligence, stability and judgment. He has spent years of continuous training before achieving his 4-stripe grade of Captain.

Your Flagship pilot is a far cry from the barnstorming, stunt-flying "birdman" of a bygone era. Today he is a serious-minded, efficient professional flyer who defines his responsibility in terms of regular and dependable operation. His sole aim is to take you safely and surely to your destination.

More than 400 pilots of the Flagship Fleet have flown over a million miles each for American Airlines. Public confidence in the men who fly the Flagships is matched only by the company's deep sense of pride in their performance.

ALL YEAR 'ROUND, TRAVEL IS *BETTER* BY AIR...*BEST* BY **AMERICAN AIRLINES** INC.

Figure 24   "He Commands Your Flagship and Your Confidence": American Airlines advertisement, 1949. Ad*Access Online Project, ad #T1778, John W. Hartman Center for Sales, Advertising, and Marketing History, David M. Rubenstein Rare Book & Manuscript Library, Duke University, library.duke.edu/digitalcollections/adaccess.

Figure 25 Marketing air travel to women: TWA advertisement, 1953. Ad*Access Online Project, ad #T2098, John W. Hartman Center for Sales, Advertising, and Marketing History, David M. Rubenstein Rare Book & Manuscript Library, Duke University, library.duke.edu/digitalcollections/adaccess.

Figure 26 The jet set: A flight attendant serves cocktails in the lounge of a Pan Am
Boeing 707, 1958. Getty Images.

Figure 27 "Glamor girls of the air": Pan American
Airways flight attendants, circa early 1960s.
State Archives of Florida.

Figure 28 Trans World Flight Center, Idlewild Airport (now Terminal 5, John F. Kennedy International Airport), New York, NY, designed by Eero Saarinen, 1956. Eero Saarinen Collection, Manuscripts and Archives, Yale University Library.

Figure 29 Model of Dulles International Airport, the aerial gateway to the nation's capital, designed by Eero Saarinen, 1958. Eero Saarinen Collection, Manuscripts and Archives, Yale University Library.

Figure 30 A half century of air travel: Pan Am passengers, circa the 1970s (top) and late 1920s (bottom). State Archives of Florida.

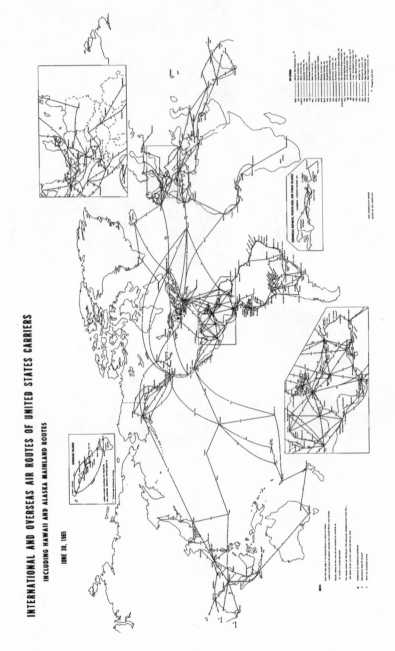

Figure 31 International air routes of U.S. carriers, 1965. Civil Aeronautics Board.

5

From Open Door
to Open Sky

In 1943, a high-level government committee headed by Assistant Secretary of State Adolf A. Berle Jr. painted a grim picture of the future of U.S. international aviation. "Our situation in this hemisphere is relatively good; we hold air entry rights for each of the 20 American Republics to the south except Uruguay," the report began. But, it continued,

> in 42 countries we have no rights of entry at all for peacetime air commerce; at the airports of those countries we cannot disembark or accept passengers, cargo, or mails. Among those countries are most of the nations of Europe, Africa, and the Middle East, as well as Russia, China, India, and Australia. In the United Kingdom, our peace-time landing rights are limited to two per week . . . in France, our landing rights are four per week. In Portugal and the Azores, our landing rights are unlimited as to number, but are restricted to a single United States company [Pan American]. . . . Except for a concession granted in Italy, which has not been used because of the war, we have no other landing rights on the continent of Europe.[1]

Contrary to Clare Boothe Luce's exuberant predictions in her February 1943 speech to Congress, as of that year U.S. airlines could not, in fact, "fly everywhere." Mirroring the borders of the earth, the air was divided into zones of national and imperial sovereignty that corresponded to territorial claims. Flights between two nations could take place only with the permission of both governments, and forty-two governments had thus far refused to grant such permissions to the United States. Most significantly, the vast skyways of the British Empire—including strategically important routes through Asia—remained off-limits to U.S. carriers, due to the system of imperial trade preference under which Britain reserved commercial traffic for its national airline, the British Overseas Airways Corporation (BOAC). Unless this situation was remedied, the report concluded, postwar U.S. commercial aviation would remain largely restricted to the Western Hemisphere.

By 1943, then, the conception of the global that emerged from American air-age discourse had no corresponding diplomatic architecture—and during the last years of World War II, Adolf Berle would dedicate himself to drafting its blueprint. Described by his biographer as "difficult, irascible, prickly, obnoxious, and arrogant," Berle was a lawyer, diplomat, theoretician of capitalism, and original member of FDR's Brains Trust. He saw himself as "the Marx of the shareholding class" but "ultimately wanted to be Marx *and* Machiavelli." In 1913, Berle had graduated from Harvard at age eighteen, with an honor's degree in history; the following year, he earned a master's degree from Harvard and at age twenty-one became the youngest graduate of Harvard Law School. As a corporate lawyer and professor at Columbia Law School, Berle wrote several acclaimed books on economic law, most notably *The Modern Corporation and Private Property* (1932), coauthored with Harvard economist Gardiner Means. Appointed as assistant secretary of state for Latin American affairs in 1938, he became a key player in Roosevelt's postwar planning efforts, responsible for advising the administration on a wide range of economic and diplomatic issues. Like his boss, Secretary of State Cordell Hull, Berle believed that international security fundamentally depended on open markets, and he set about to create such markets in the field of aviation.[2]

Although not himself an "airman," Berle keenly understood aviation's political and economic significance. "I feel that aviation will have a greater influence on American foreign interests and American foreign policy than any other non-political consideration," he wrote in 1942. Along with Henry Luce and Juan Trippe, Berle viewed the airplane as the instrument of a historic transfer of power from the old world to the new, from the European colonial order to an American Century based on trade rather than territorial conquest. "Freedom of the air," he argued, therefore posed "as grave and important a problem" in the mid-twentieth century as did freedom of the seas in the seventeenth century. Berle frequently invoked the name of Hugo Grotius, the great Dutch lawyer whose 1609 treatise *Mare Liberum* (The Free Seas) established the foundations of modern maritime law. Modest in neither talent nor ambition, Berle aspired to become nothing less than a Grotius for the air age.[3]

What was so "grave and important" about freedom of the air? During World War II, the United States developed the world's leading economy, unrivaled in its industrial capital, productive capacities, and financial wealth. Government projections foresaw peacetime employment of 56 million and a $150 billion gross national product. An economy of this magnitude, many economists believed, could continue to thrive only if the United States had liberal access to overseas markets—which included, as Henry Luce had written in "The American Century," "the right to go with our ships and our ocean-going airplanes where we wish, when we wish, and as we wish."[4] American industry desperately needed foreign trade opportunities; otherwise, Roosevelt administration leaders feared, a postwar crisis of surplus production could create a depression perhaps even worse than that of the 1930s. Freedom of the air, like other freedoms of trade, was grave and important indeed.

Toward this end, Berle devised an "open sky" policy, which aimed to dismantle imperial trade preference and other forms of protectionism that restricted U.S. airlines from operating in foreign countries—just as Grotius's free-seas doctrine had dismantled the mercantile cartels of the seventeenth century. The open sky policy's namesake and progenitor was, of course, the Open Door policy codified by Secretary of State John Hay in 1899, which maximized U.S. access to markets in China and elsewhere.

Half a century later, the open sky policy, as Berle envisioned it, would extend the open door to the airways. Part of the Roosevelt administration's broader effort to enshrine free trade principles as the foundation of postwar international security and stability, it built on other New Deal economic initiatives such as the Reciprocal Trade Agreements Act of 1934 and the creation of the Export-Import Bank (both of which Berle also advised).[5] Together with these policies and institutions, the open sky policy aimed to structure the postwar global order, and indeed the very notion of the global, around the principles of American liberal capitalism.

A classic expression of nationalist globalism, the open sky policy placed expansive one-world rhetoric at the service of U.S. national interests. In negotiations with other governments in 1944 and 1945, Berle claimed that open skies would benefit all nations by making the air, like international waters, universally accessible. Yet the political economy of Berle's doctrine implicitly favored planes bearing the Stars and Stripes. By introducing international competition on the airways, the open sky policy promised to reward whichever nation had the most commercial aircraft, the most passengers, and the most goods to trade. And as the war drew to a close, there were few contenders for that title. By 1944, the United States had flown nearly 70 percent of the world's total passenger miles.[6] Well aware of this fact, foreign leaders rightly perceived that the open sky policy accrued maximum benefit to the United States.

Between 1943 and 1946, diplomatic disputes over freedom of the air touched off what Alan P. Dobson has dubbed "the other air battle" of World War II, whose weapons were cables and telegrams instead of fighters and bombers.[7] First within the Roosevelt administration, then in bilateral talks between Washington and London, and finally on a world stage at the International Civil Aviation Conference held in Chicago in November 1944, this diplomatic battle played an important but often unacknowledged part in shaping the postwar global order. For, as Adolf Berle understood, at stake in that conflict was nothing less than the character and leadership of the postwar air age.

Yet if the "other air battle" revealed how the United States attempted to structure the postwar global order in its own image, it also revealed how other nations—particularly Britain and the Soviet Union—resisted the ascendant American Century. These challenges placed limits on the United States' dominance of the postwar skyways. Although a majority of the world's governments ultimately agreed to some elements of Berle's open sky policy, the treaties and policies implemented at the end of World War II were far less expansive, and far less favorable to U.S. interests, than he and other Washington leaders had envisioned.

Nationalizing the Skies

In a 1944 memorandum to President Roosevelt, Adolf Berle described one of the great paradoxes of the air age. Aviation was a globalizing technology par excellence, "the quickest and most immediate method by which our people can come into contact with the peoples of other nations." Although only some nations bordered oceans, "every country has access to the open air." However, international aviation law effectively apportioned the air into closed national zones, corresponding to the borders on the ground—airspace was legally designated as national territory. "You can sail over the high sea without, as a general rule, crossing a national frontier," Berle explained. "But you cannot fly through the open air outside your own country . . . without crossing the boundaries of countries thousands of feet below."[8] International law thus stood at odds with cultural conceptions of the airplane as a harbinger of global unity. Airplanes could soar above the earth's natural boundaries, but they could not so easily transcend its geopolitical divisions.

Late nineteenth-century disputes over the ownership of airspace above buildings had established the legal principle of *cuius est solum eius est usque ad coelum:* he who owns the soil owns up to the sky. But the invention of the airplane raised novel and vexing questions. The *usque ad coelum* doctrine derived from property law, but should the air be considered a form of property or national territory? Flight, noted British aviation law expert J. M. Spaight in 1919, was "inherently and pre-eminently international," problematizing the very notion of national sovereignty. Did an aircraft enter another nation when it landed or simply by entering its airspace? If the latter, exactly how far "up to the sky" did territorial sovereignty extend? Maritime law designated the oceans as international waters beyond specified distances from land; should similar boundaries be drawn in the air, such that at certain altitudes planes would fly in international airspace? As chronicled in Stuart Banner's *Who Owns the Sky?* (2008), lawyers and diplomats on both sides of the Atlantic debated such questions from the earliest days of the air age.[9]

Prior to World War I, proposals to internationalize airspace received serious consideration in both the United States and Europe. An International Conference on Aerial Navigation held in Paris in 1910 endorsed the principle of "free circulation," which would grant pilots liberal rights to fly over foreign territory. The British government, however, threw its support behind the opposing principle of "national air sovereignty," which held that a nation's borders stretched into the atmosphere above its landmass and consequently required aviators to obtain governmental permission before entering foreign airspace. Britain's Air Navigation Acts of 1911 and 1913 subsequently wrote national air sovereignty into law. Other European governments soon followed suit, as concerns about national security—and particularly fears of aerial bombardment—outweighed their initial commitment to free circulation. By 1914, Germany, France, Russia, Italy, and Austria had all passed air sovereignty laws that required foreign aviators to obtain governmental permissions before entering their airspace.[10] Pilots and aviation promoters, meanwhile, objected to such laws, arguing that they imperiled the very future of flight. "If this continues, soon all the frontiers will be closed to aerial navigation," warned the *Bulletin Officiel* of France's La Ligue National Aérienne in 1914.[11]

The use of strategic bombing during World War I, which proved beyond doubt the airplane's destructive capacities, eliminated most remaining support for internationalizing the air. With images of aerial warfare fresh in their minds, delegates to the 1919 peace conference at Versailles ratified a Convention Relating to Air Navigation that enshrined the *usque ad coelum* doctrine as international aviation's governing principle. Signed by twelve nations (Britain, France, Italy, Japan, Cuba, Greece, Romania, Brazil, Serbia, Belgium, Portugal, and the United States), the Convention held that "every State has complete and exclusive sovereignty in the air space above its territory and territorial waters." During the 1920s, both national laws and international treaties further affirmed that the air was an extension of national territory. Although the United States was not subject to the 1919 Convention because of Congress's refusal to ratify the

Treaty of Versailles, the Air Commerce Act of 1926 performed a similar function, proclaiming exclusive sovereignty over the airspace above the continental U.S. and the Panama Canal Zone. And in 1928, the United States and twenty Latin American governments adopted the Havana Convention on Commercial Aviation, whose provisions were nearly identical to the 1919 Convention.[12]

With airspace now defined as sovereign national territory, formal diplomatic permissions became necessary for the establishment of every international air route. "Air rights," as such permissions were called, could be granted in two ways. A government could grant unilateral operating concessions to a foreign airline, as Pan American Airways had managed to obtain throughout Latin America. Alternatively, governments could exchange air rights on a bilateral basis, allowing each other's airlines to operate a certain number of routes per week to specific destinations. The 1936 agreement between the United States and Britain, for example, stipulated that Pan Am and BOAC could each fly two round trips per week between New York and London.[13]

As advances in aircraft design enabled airlines to fly to far-flung destinations, the system of national air sovereignty increasingly seemed to work against the interests of the United States. In response to the explosive growth of the U.S. aviation industry during World War II, foreign governments sought to protect their own national airlines from overwhelming U.S. competition by refusing to sign air rights agreements with Washington. By 1943, Berle noted, there was "a wide and rapidly growing tendency throughout the world to work out arrangements excluding American civil aviation from post-war landing rights and routes"—such that postwar U.S. airlines "could be limited to the continents of North and South America," that is, where Pan Am already operated. Furthermore, the "great circle" shortcuts across the earth's northern reaches traversed Europe and Asia, where the United States mostly lacked air rights. Without such rights, Berle warned FDR, "our flying would be crippled, if not stopped, in great areas of the world. These countries know that; and we know it."[14]

The American public—or, at least, the millions of Americans who read the Luce publications and/or major national newspapers—also knew it, thanks to widespread media coverage of the air rights issue. Even in the midst of war, headlines echoed Berle's concerns about postwar commercial aviation: "Foreign Flying: Nations Fight for Air Space as They Once Sought Colonial Land" *(Wall Street Journal)*; "Air: What's In It for the U.S.?" *(Time)*; "The Future of International Airways: Has the United States a Policy?" *(Harper's)*. Aviation magazine *Skyways* worried that "Americans with heads in the clouds may get kicked in the pants by any one-foot-on-the-ground foreign monopoly"; *Time*, in an article entitled "Aviation: All Dressed Up . . . ," likened U.S. aviation to "a sprightly young man all dressed up in fine clothes but with few place to go outside of his own yard." *Time, Life,* and *Fortune* expressed the strongest criticisms of foreign attempts to curtail Americans' abilities to "fly everywhere." As Clare Boothe Luce had insisted in her famous 1943 congressional speech, the Luce publications maintained that the United States deserved unlimited access to postwar international air routes, due in large part to U.S. aviation's contributions to the Allied war effort. "More than to any other nation, the war has given the world's air to America," wrote Joseph Kastner in *Life*. "The hundreds of thousands of Americans who learned to fly, the thousands of airmen to whom a round trip to India is almost as casual an event as a long weekend drive, the many men who have looked down at the passing continents and felt the round earth shrink in size—none of them will want to give up the air they have taken over."[15]

The Airlines of the United States, an industry consortium, expressed this argument visually in its 1944 advertisement "The Limitless Right of Way," published in national magazines such as *Newsweek*. Portraying Leonardo da Vinci watching a boy release a bird into the air, the image identified "freedom of the air" with liberty and enlightenment. The ad's text, meanwhile, reprised the familiar argument that the United States, more than any other nation, had pioneered advances in aviation: "More than any other one group in the world, the Airlines of the United States have carried the responsibility and the risk of opening the limitless right-of-way of the sky." In all fairness, then, Americans' rights to fly to and from other countries should also be limitless.[16]

Yet air rights had become a growing problem in U.S. foreign policy precisely because the war had "given" the world's air to America (or so it seemed to many observers). In response to the phenomenal wartime expansion of the U.S. aviation industry, other governments sought to protect their own airlines through subsidies, imperial trade preference, and restrictions on U.S. carriers' rights of entry. And when Americans, in turn, demanded to "fly everywhere," they only exacerbated foreign concerns that they sought nothing less than domination of the postwar air. Debates about the meanings and limits of national air sovereignty had thus become a major issue of contention—and tension—between the United States and its allies. By 1943, the Roosevelt administration had grown convinced that the future of U.S. aviation, and indeed the nation's larger economic and strategic interests, would require nothing less than a fundamental restructuring of international aviation policy.

The ICIA and the Making of the Open Sky Policy

During the first week of 1943, Secretary of State Cordell Hull organized the formation of an Interdepartmental Committee on International Aviation (ICIA). Chaired by Adolf Berle, its other members included Civil Aeronautics Board (CAB) chairman L. Welch Pogue, Assistant Secretaries of the Navy Robert A. Lovett and Artemus Gates, Undersecretary of Commerce Wayne Taylor, and the assistant director of the Bureau of the Budget Wayne Coy. Reporting directly to the president, the ICIA was instructed to resolve the air rights issue and to draft a general working blueprint for postwar U.S. policies and objectives in international commercial aviation. Meeting for the first time on January 8, the group convened regularly throughout 1943. Its minutes document robust and, at times, contentious debates that reflected larger institutional and ideological tensions—specifically, tensions between civilian and military policymakers and between the interests of commerce and national security.[17]

Berle's commitment to "open skies" was by no means shared by all Roosevelt administration leaders, particularly those in the military. During the ICIA's initial meetings, Robert Lovett, assistant secretary of the Navy, argued forcefully that the "primary consideration" in U.S. aviation policy "should be that of national security." When commercial interests conflicted with security objectives, he said, they "should be subordinated to [this] more important consideration." In Lovett's view, an open sky was a defenseless sky. Commercial aviation could be used as a cover for espionage or worse: if foreign airlines were granted liberal rights to fly through and into U.S. territory, they could spy on military installations or even drop bombs. Though Lovett's anxieties may now seem far-fetched, it is important to consider the wartime context of the ICIA's deliberations, the rapid and unpredictable development of aviation technologies that could be used for either military or civilian purposes, and the fact that international air travel (particularly across the oceans) was still a relatively new phenomenon. National security concerns thus prompted Lovett to argue, at least initially, against Berle's open sky policy and in favor of sharply restricting foreign carriers' access to U.S. airspace. He also criticized the open sky policy on economic grounds, contending that U.S. airlines could not compete against foreign airlines that received generous government subsidies and employed cheaper labor. International aviation, he argued, should not even involve competition between multiple U.S. airlines but should be delegated to "a single, powerful Government dominated operating company"—the model employed by most European nations. Juan Trippe, unsurprisingly, supported this position, since that company would presumably be Pan Am.[18]

For Berle, by contrast, freedom of commerce was not antithetical but integral to both national security and international stability. The air age could "easily go the bloody way of the history of sea transport," he warned, unless the world's nations could agree to designate the air as "a more or less universal medium of communication and transport service." In addition to maximizing trade, open skies would promote world peace— indeed, Berle believed that peace depended on freedom of commerce. In his 1940 book *New Directions in the New World*, he had argued that restrictive trade policies produced totalitarian governments: "Closed areas are not merely closed in economics. They tend to become closed in politics and in culture, as well," he wrote. If the postwar settlements did not result in an international commitment to lift restrictions on trade, Berle feared a repetition of the circumstances that had brought Hitler to power after World War I: "Great areas in Europe will be in grave physical distress. . . . Populations will be literally starving, naked, and perhaps homeless. . . . It is very nearly a foregone conclusion that there will be, at that time, a great movement of social unrest."[19] Like other liberal internationalists in the Roosevelt administration—including Secretary of State Cordell Hull and

FDR himself—Berle viewed freedom of commerce not only as a means of ensuring national prosperity but as the very foundation of global security and stability. The open sky policy, in turn, was foundational to the liberal capitalist world order that he envisioned.

Ultimately, the ICIA came to favor Berle's position over Lovett's. In April 1943, the committee released a preliminary report stating that "the best interests of the United States are served by the widest generalization of air navigation rights."[20] Its final report, released on August 26, 1943, reaffirmed this principle and outlined a specific plan for the United States to secure favorable air rights treaties before the conclusion of the war. First, the State Department should negotiate a bilateral treaty with Great Britain, which not only controlled "a very large proportion of the strategic points on world airways" but could also "exert substantial diplomatic influence" over its dominions and allies. The report then recommended that the State Department pursue air rights treaties with the Soviet Union, China, France, the Netherlands, and Portugal—countries with strategically important geography, extensive territorial possessions, and/ or potentially competitive aviation industries. Once these initial, most important bilateral agreements were in place, an international conference should be held to settle remaining issues, such as technical and safety standards, on a multilateral basis. Finally, the report affirmed that the United States should secure long-term rights for commercial use of the 228 air bases on Allied territory (including 105 in the British Commonwealth alone) that had been constructed or improved with War Department funds. While the ICIA conceded that it was "not likely that countries with postwar aviation ambitions will freely give us unilateral commercial rights merely because we have been permitted to build airports on their sail in time of war," it also noted "the obvious material benefits resulting from our expenditures may enable us to make reasonable requests with special force"—language that echoed claims that the United States' wartime sacrifices and successes entitled its airlines to a "limitless right of way."[21]

In early November, President Roosevelt met with the ICIA and officially endorsed the recommendations of its final report. U.S. aviation policy, the president affirmed, should promote "a very free interchange" in the commercial skies. Several months later, the War Department issued its own official statement on postwar aviation policy. While maintaining that "national security is of first importance and the national policy in regard to civil aviation must be in accord with the military requirements of national defense," the War Department's report also acknowledged that "maximum expansion and extension of United States air commerce . . . contribute through augmented air power to the national security."[22] By the spring of 1944, then, the Roosevelt administration appeared to have reached a general consensus that U.S. interests would best be served by an open sky policy that minimized restrictions on air rights and maximized access to international markets.

It is important to emphasize, however, that the emerging consensus on U.S. international aviation policy entailed not the defeat of Lovett's explicitly nationalist position but, rather, its incorporation into an internationalist framework. Indeed, Adolf Berle's own arguments revealed how apparently conflicting positions and objectives—nationalism versus internationalism, security versus commerce—were, in fact, entirely complementary. While Berle viewed freedom of commerce as an essential condition of global peace and stability, he also defended the open sky policy in nationalist terms, emphasizing that the United States could retain its world leadership in aviation only if its airlines acquired rights to "fly everywhere." Open skies, he explained, would be the United States' "plainest road to superiority." Notably, Berle's open sky policy did not involve internationalization of the airways, as Vice President Henry Wallace had proposed in a controversial 1943 *New York Times* article advocating a single, multinational air force and airline. Berle explicitly rejected Wallace's proposal as utopian and antithetical to U.S. interests. And in conceptualizing the postwar air age, Berle envisioned not only international integration but also a dramatic expansion of American economic and geopolitical power. The United States would require "preferably unlimited lines of transport and communication throughout the entire world. . . . *No single fraction of the earth's surface can be assumed to be unimportant*," he proclaimed (emphasis added). Berle's globalism, then, was a nationalist globalism that eschewed the conquest of territory, yet authorized the United States, in the pursuit of commerce, to "go, at will, or reasonably so, through and away from practically every country," as he wrote in a 1944 memorandum.[23]

The political economy of the open sky, moreover, decisively worked in the United States' favor. Previous wartime treaties had given the United States a decisive lead over its only true competitor in commercial aviation, Great Britain. In the Arnold-Powers Agreement of 1942, Roosevelt and Churchill had agreed to divide the burdens of wartime aircraft production: the United States would focus on producing transport planes so that Britain could devote its more limited resources and manpower to producing fighters and bombers. This division of labor proved critical to the success of Britain's air war, but it also gave the U.S. aviation industry a flying start in the race for the postwar commercial skies, as its armada of military transports could easily be converted to passenger planes. If postwar air traffic were placed "on a purely economic or commercial basis," as the open sky policy would have it, the United States would, at least initially, face no real challengers for global leadership in commercial aviation. As the chairman of the National Advisory Committee for Aeronautics, J. C. Hunsaker, put it: "For this competition the United States will be well prepared with its vast manufacturing industry, some two million trained air men to select from, and a string of bases with a radio communication system extending from Iceland to the South Pacific."[24]

The United States' "vast manufacturing industry" could also serve as a powerful diplomatic bargaining chip in air-rights negotiations, particularly with Britain and other Allied nations who were receiving large amounts of Lend-Lease aid. In one ICIA meeting, for example, CAB chairman L. Welch Pogue noted that the British currently had four hundred U.S. transport planes on order. "They are very anxious to get those planes, and it is a profitable time to trade when one has something which the other person desires very strongly," Pogue noted. Berle's assistant, Robert G. Hooker, similarly proposed that Washington suspend aircraft shipments until London agreed to a satisfactory air rights agreement; otherwise, he said, the British "will have no incentive . . . to meet us halfway."[25]

As the United States and Britain prepared to begin bilateral aviation policy talks in early 1944, these points were not lost on leaders across the Atlantic. Berle's London counterpart, Lord Max Beaverbrook—an inveterate imperialist and, as Alan P. Dobson has described him, "a man determined that Britain should on no occasion give way more than was necessary to the United States"—denounced the open sky as a euphemism for an American-dominated sky.[26] During the following year, the "other air battle" would strain the Anglo-American alliance and portend a larger conflict over the leadership and character of the postwar international order.

The Anglo-American "Air Battle"

The contest over aviation policy expressed deeper tensions in Anglo-American relations. By the latter years of the war, British leaders had become gravely concerned about their nation's growing economic dependence on its ally. Lend-Lease aid to Britain amounted to $21 billion, and London already owed Washington $3.7 billion from World War I. Under these circumstances, British leaders had limited leverage in negotiating the terms of postwar economic settlements, including those that would shape the future of commercial aviation. The United States, meanwhile, deftly exploited its economic preponderance in its efforts to dismantle the "Ottawa system" of imperial trade preference that restricted U.S. goods from entering British-controlled markets. Article VII of the Lend-Lease Act specified that both the United States and Britain "shall provide against discrimination . . . against the importation of any product originating in either country"—a provision that economist John Maynard Keynes correctly interpreted as a subtle attack on imperial preference. The wording of Article VII was vague, but as the war drew to a close, the Roosevelt administration used Lend-Lease as a bargaining chip to induce the British government to reduce tariffs, to lower its dollar reserves, and indeed to allow U.S. planes into imperial skies.[27]

Such was the context in which the "other air battle" took shape. By 1943, British aviation experts were predicting that their air transport industry would be unable to compete with the United States for at least five years after the war. If the open sky policy prevailed, they argued, U.S. airlines would quickly monopolize key international routes, such as the transatlantic route between New York and London. As Archibald Sinclair, secretary of state for air, stated during a debate on aviation policy in the House of Commons: "After the war we shall be one of two things: either we shall be a small island of 45 million people in a world dominated by the great United States of America, with 130 million . . . or we shall be the center of a great Empire which will be bound together by our air routes."[28] At stake, then, was not simply the matter of whether BOAC would be able to compete with Pan Am but the larger question of whether the postwar global order would be controlled by "the great United States of America" or the "great Empire."

In its broadest implications, the Anglo-American dispute over aviation policy involved a contest between two different visions of global power—geopolitical versus geoeconomic—and two correspondingly different grand strategies. Seeking to preserve hegemony over the skyways of its vast territorial empire, Britain pursued a geopolitical strategy that divided the air, like the land, into sovereign spheres of influence controlled by great powers. The United States, meanwhile, pursued a geoeconomic strategy—the open sky policy—that would allow its planes to go "through and away from every country" (as Berle had written), thus maximizing its access to foreign markets.[29]

Planning for Anglo-American aviation talks began in the fall of 1943, shortly after the ICIA issued its final report that underscored the importance of concluding an air-rights treaty with London. In October, Lord Beaverbrook informed Harry Hopkins that he was ready to negotiate; in Washington, however, delays ensued as Berle awaited President Roosevelt's response to the ICIA's final report and as the State Department dealt with internal disarray following Sumner Welles's unexpected resignation. Further delays resulted from a dispute over whether to invite other nations to join the talks. Beaverbrook wanted to include the Soviet Union and the British dominions, while Berle, not wanting to go up against what he believed would be a dominion united front, insisted on conducting the meetings bilaterally. The parties reached a compromise in February 1944, when FDR approved a preliminary round of negotiations among the United States, Britain, and Canada. Selected other countries—the Soviet Union, China, Brazil, the Netherlands, Norway, Belgium, Portugal, Free France, Iceland, Australia, New Zealand, and South Africa—would be informed of the negotiations but not yet invited to participate.[30]

On April 1, 1944, Berle and Edward Warner—vice chairman of the CAB and an expert on international aviation law—boarded a transatlantic flight to London. During the following week, Berle and Beaverbrook, familiarly known as "The Beaver," developed a friendly rapport, but congeniality did little to resolve their political differences. Berle's last-minute decision to meet with Canadian aviation leaders en route to London angered Beaverbrook, who viewed the strategy, correctly, as an attempt to drive a wedge between Britain and Canada on aviation policy before London could obtain a dominion consensus. The two sides also clashed on the issue of transatlantic passenger traffic quotas. Beaverbrook proposed that the United States and Britain split the traffic evenly, such that Pan Am and BOAC would each operate the same number of flights per week. In response, Berle pointed out that Pan Am had carried a full 80 percent of transatlantic traffic prior to the war; the United States had no intention of accepting a plan that would reduce its share of this market by 30 percent. Beaverbrook also advocated the formation of an international aviation organization that would regulate traffic quotas, flight frequencies, and fares. Consistent with his ideological commitment to freedom of commerce, Berle strongly opposed this plan, which was clearly designed to place restrictions on where and how often U.S. planes could fly.[31]

The London talks ended with the United States and Britain no closer to an air rights agreement. An announcement released to the press on April 7 stated only that the two sides had agreed to convene a multilateral conference in the fall in order to devise international standards on the technical aspects of aviation, such as safety standards and communications procedures. But on the more substantive issues—traffic quotas, international regulation, imperial preference versus open skies—U.S. and British negotiators remained deadlocked. A second round of talks, held in Washington in July, ended as inconclusively as the first. (The only positive outcome of the meeting, it seems, was what Berle described as a "rather fantastic weekend" of hunting, fishing, and drinking at Beaverbrook's family estate in New Brunswick, Canada.) With an imminent Anglo-American air rights agreement looking increasingly unlikely, during the next several months the State Department turned its attention to preliminary bilateral negotiations with France, New Zealand, the Netherlands, and the Soviet Union. London, for its part, focused on obtaining consensus among the dominions.[32]

Meanwhile, the course of the war had turned in the Allies' favor, with a series of major victories in Western Europe during the summer of 1944. The prospect of peace raised the stakes of obtaining international accord on aviation policy, either through a multilateral settlement or bilateral treaties between individual countries. In late August, Berle expressed his concern that in the absence of such a settlement, the war's conclusion would "precipitate at once a competitive air race between Britain and the United States on a violent scale," as the two sides fought to sign exclusive bilateral agreements with third-party countries. By mid-September, as Anglo-American negotiations remained at a deadlock, Berle concluded that a multilateral conference—at which, he believed, the United States could persuade other governments to endorse the open sky policy—would be the best way to settle the issues prior to the war's end. President Roosevelt agreed. Concurring, too, with Berle's suggestion that the meeting be held "in some Midwest City like Chicago, Illinois," he instructed the assistant secretary of state to assemble a delegation.[33]

Chicago was a fitting location for the conference. New York mayor Fiorello LaGuardia, a prominent advocate of commercial aviation and member of the U.S. delegation, grumbled that his city, the nation's flagship metropolis, had not been chosen to host the meeting. Yet Chicago appropriately embodied the possibilities and paradoxes of the air age. On the one hand, the city exemplified aviation's revolutionary effects on space and time: as Clare Boothe Luce had noted in her "fly everywhere" speech, the airplane linked the heartland Midwest, long characterized as the epicenter of American isolationism, to the entire world. On the other hand, Chicago's political culture proved that the air age would not necessarily transform provincial nationalists into cosmopolitan internationalists. The city's leading newspaper, the *Chicago Daily Tribune*, was also the nation's leading organ of anti-internationalist and anti–New Deal opinion, and its editorial pages derided worldly liberals such as Berle as "little men in spats . . . whose craving for social recognition outside their own country is their dominant passion." True to form, the *Tribune* immediately denounced the Roosevelt administration's proposed aviation policies as a "sell-out" of the national interest. Throughout the conference, the paper would lead its editorial pages with sensational (and typically inaccurate) charges that Berle and his fellow "little men in spats" were sacrificing

America's rightful leadership in aviation in order to appease ungrateful foreign governments.[34]

The editors of the *Tribune* need not have worried, however. Far from selling out American interests, Berle aggressively fought for them in Chicago—and far from instantiating the vision of "one world," the conference ironically heralded its demise.

Crafting the "Charter of the Open Sky"

The International Civil Aviation Conference, the most important multilateral meeting to address aviation matters since Versailles, began on November 1, 1944. Representing fifty-two nations, some seven hundred delegates convened at Chicago's Stevens Hotel—which, coincidentally, had opened in 1927, the year of Pan Am's creation and Charles Lindbergh's transatlantic flight. During the opening ceremonies, a message from President Roosevelt described commercial aviation as "the first available means by which we can start to heal the wounds of war." Adolf Berle—conference chairman and head of the ten-member U.S. delegation—expressed similar ideas in his own opening address. Calling the air a "highway given by nature to all men," he argued that the postwar skyways should be made available to all nations, "without prejudice to sovereignty." Freedom of the air would not benefit the United States alone, he emphasized. On the contrary, it would allow "all countries the chance to get into the air," thereby helping to maintain "the conception of one world at peace." "We are thus endeavoring," Berle concluded, "to write the charter of the open sky." Delegates from other nations echoed Berle's effusive globalist language. "Hereafter, national boundaries will be no more significant than the dividing lines between provinces," stated Kia-Ngau of China. The chairman of the Mexican delegation, Pedro A. Chapa, predicted that commercial aviation would "strengthen the bonds that unite the different peoples of the world," enabling "a better and more profound understanding" between those peoples.[35]

Outside the Stevens Hotel, however, the world was quite obviously not united, and the ideological conflicts crystallizing in the late war years—the United States' commitment to open markets versus Soviet communism and Britain's system of imperial trade preference—soon came to dominate the conference's proceedings. Just days before the meeting was scheduled to begin, the Soviet Union abruptly recalled its delegation, purportedly to protest the inclusion of Switzerland, Spain, and Portugal— which, according to a young diplomat named Andrei Gromyko, supported a "pro-fascist policy hostile to the Soviet Union." State Department officials believed, though, that the real reason behind Moscow's withdrawal from the conference was its unwillingness to allow foreign airlines to operate in Soviet skyways.[36] The Soviet withdrawal foreshadowed that the postwar air age would soon be a Cold War air age, with Eastern Europe's airspace enclosed by the iron curtain.

The postwar air age would also begin as an imperial air age. Britain, France, Portugal, and the Netherlands refused to grant "freedom of the air" to their own colonies, and Asian and African nationalist leaders found themselves excluded from aviation policy discussions. In August, editorials in India's nationalist newspaper *Hindustan Times* had attacked the British government for attempting to "barter away [India's] rights" in commercial aviation, and a U.S. diplomat in New Delhi predicted that "nationalist India will not be bound by agreements on post-war civil aviation" made by London without its consent. Although India and several other important British colonies (Egypt, Iraq, Syria, and Lebanon) were represented in Chicago, their delegates had been appointed by the British government and were therefore expected to adhere to the imperial line. Other colonies had no representation at all. According to a report in the *Chicago Defender*, the city's leading African American newspaper, a Detroit physician named Joe T. Thomas arrived at the Stevens Hotel to represent the "Provisional Government of the Congo Free State of West Africa," but he was denied entrance on the grounds that Britain would be representing the Congo. Similarly, the State Department informed Korean nationalist leader Syngman Rhee that Korea could not participate because "it is a conference for sovereign states."[37] Belying Berle's opening remarks, such incidents proved that the sky was not "a highway given by nature to all men" but an extension of geopolitical real estate that reflected balances of power on the earth.

To some extent, however, air-age geography had elevated the status of certain nations that were not otherwise considered great powers. Canada, for example, emerged as a major player at the conference because of the great circle routes that traversed its territory. Ethiopia, similarly, was an important refueling stop along air routes to the Middle East, and as the prominent black journalist George Padmore reported in the *Chicago Defender*, both Britain and the United States attempted to "woo" its delegation by proposing to offer technical assistance in exchange for landing rights. The conference also exposed divisions in great power alliances. In defiance of Britain's desire for a dominion-united front, Australia and New Zealand asserted their independence by putting forth a plan for the international operation of world trunk routes. And during the first week of the conference, nineteen Latin American nations formed a bloc to protest a U.S. plan for governing a proposed international aviation organization. The plan called for a fifteen-member advisory council on which the United States, the Soviet Union, and Britain would each get two seats; Brazil, China, and France would each get one; and the remaining six seats would be divided regionally: three for Europe, two for the Americas, and one each for Asia and Africa. The Latin American bloc protested the division of the council into "'master' powers and 'inferior' powers," with the six regional seats apportioned to "the 'riff-raff' of smaller nations." "We are not riff raff," they asserted.[38]

The Latin American protest succeeded in convincing the U.S. delegation to agree to an alternative proposal, sponsored by Mexico and Cuba, for the aviation organization's entire governing council to be elected by the body of member states. In Latin America, this victory over the United States received widespread and favorable press coverage. On November 6, five of the six principal Mexico City newspapers carried front-page headlines from Chicago: "Spanish America Rebels against the Hegemony of the Great Powers" *(Excélsior)*, "North American Proposal Rejected Because It Involves True Discrimination" *(La Prensa)*, and "A Yankee Plan Rejected" *(El Nacional)*. In Panama, *El Panamá América* carried the headline "19 Latin American Countries Attack the United States' Plan at the Chicago Conference." According to the U.S. ambassador there, "The use of the word 'atacan' (attack) is believed to be of special interest since it conveys the impression to the Spanish-speaking public that a non-conciliatory attitude at the conference exists on the part of the Latin American delegates."[39]

Both the United States and Britain courted allies, moreover, by arguing that their respective proposals would best serve the interests of third-party nations. The British delegation claimed that a strong regulatory authority, with the power to set fares and traffic quotas, would protect other nations from an American "business invasion." During the conference, the London *Daily Express* published an editorial cartoon of an angry eagle menacing two helpless songbirds. "Just exactly what was it that Mr. Adolf Berle said at Chicago about the air being free for all?" queried the caption. Berle might have replied that songbirds could not fly without freedom of the air; the United States' open sky policy, he maintained, would allow all nations to "get into the air" by abolishing BOAC's monopoly on key world air routes. Berle also played anticolonialism to the United States' advantage by emphasizing that the open sky policy was "endeavoring to end the old colonial system in the air." And he argued that what was good for the U.S. aviation industry would be good for the world. As U.S. delegate William A. M. Burden, assistant secretary of commerce, explained to readers of the *Atlantic*, the open sky policy was "desirable for countries with highly developed aircraft industries, which naturally wish to sell their services to as large a market as possible." But, he insisted, a thriving U.S. aviation industry would sell aircraft to other nations, thereby helping them develop their own airlines and air forces. In Chicago, Berle and his team used postwar aircraft sales as an incentive for other nations to cooperate with the United States, raising the issue so often that it practically became a bribe. "Nobody votes against Santa Claus," said one anonymous delegate to the *Chicago Daily Tribune*.[40]

In the end, great-power politics prevailed at Chicago despite the assertions of independence by Latin America and the British dominions. With the Anglo-American dispute taking center stage, at times the conference seemed more bilateral than multilateral, the other delegations mere witnesses to the epic drama unfolding between Washington and London. "For the most part, the great battles were fought, not in the general sessions, but by the traditional 'fifteen men in a smoke-filled room,'" wrote aviation historian Henry Ladd Smith.[41] Like the previous postwar planning conferences held at Bretton Woods and Dumbarton Oaks, the

Chicago conference reaffirmed the existing international balance of power even as it appeared to give all participating nations, regardless of their size and stature, an equal voice in the crafting of postwar aviation policy. In the end, the debate boiled down to which great power would prevail: the United States, playing "Santa Claus," with its leviathan aviation industry; or Great Britain, with its vast territorial empire.

British strategy at Chicago revolved around two objectives: to retain BOAC's privileges on imperial routes and to establish an international aviation organization to regulate competitive practices. A month prior to the conference, London issued a white paper that strongly endorsed these positions, indicating that the months of Anglo-American talks had not yielded any compromises. Lord Beaverbrook, meanwhile, had resigned his aviation portfolio, and his replacement, Viscount Swinton—who had previously served the cause of empire as secretary of state for the colonies—took an even harder line than his predecessor. Upon meeting, Berle and Swinton took an immediate dislike to one another. Berle described his new adversary as "arrogant and inflexible," a man incapable of understanding "the difference between the atmosphere of the coast of the Gulf of Guinea and that of the shores of Lake Michigan."[42] Although Berle exaggerated for effect, his choice of words indicated the degree to which imperial geopolitics shaped British aviation policy.

The United States, equally intransigent and equally motivated by national interests, continued to push for the open sky policy. Specifically, the U.S. delegation sought multilateral ratification of five "freedoms of the air":

1. Freedom of peaceful transit: commercial aircraft from one nation may fly over another.

2. Freedom of technical stop: planes may land in another nation for refueling or in the case of an emergency.

3. Freedom to carry passengers or goods from the home nation to a foreign nation.

4. Freedom to carry passengers or goods from a foreign nation to the home nation.

5. Freedom to pick up and discharge traffic at intermediate points.

The first two freedoms—rights of transit ("innocent passage") and technical stop—were "privileges of flight," more or less essential for international flights to occur at all. The latter three freedoms, conversely, were "privileges to trade," governing the conditions under which airlines could do business in foreign nations. Under the third freedom, Pan Am could fly passengers or cargo from New York to London, for example; under the fourth, it could fly them back to New York; the fifth freedom would allow Pan Am to land in London, discharge and pick up passengers or cargo, and fly on to Paris (or some other third destination).[43]

The fifth freedom was key to the U.S. agenda in Chicago. Because of the United States' geography and lack of overseas possessions, its international flights typically required refueling stops in numerous foreign countries. On Pan Am's New York–Buenos Aires flight, for instance, only 15 percent of passengers traveled all the way to Buenos Aires; the rest disembarked at intermediate stops. In order to make such long-haul routes profitable, Pan Am wanted to be able to pick up additional passengers at these intermediate stops—such that, if one desired to fly from Rio to Buenos Aires, one could choose between flying a U.S. airline (Pan Am) or a Brazilian or Argentinian carrier. Beyond such technicalities of aviation law, the fifth freedom was an expression of the global Open Door policy that the Roosevelt administration sought to implement as the foundation of postwar security and national prosperity. If ratified, it would effectively allow U.S. airlines to do business anywhere in the world, thereby dismantling BOAC's monopoly on air traffic within the British Empire.

Anglo-American discord over the fifth freedom brought the Chicago conference to a virtual stalemate. Britain easily conceded the first two freedoms, as did the other national delegations, and by the second week of November, Berle and Swinton had reached a compromise on the third and fourth freedoms: Berle agreed to traffic quotas on transatlantic trunk routes (Pan Am and BOAC would each operate five flights per week between New York and London, for example), while Swinton agreed to the U.S. demand for a so-called escalator clause, which would allow an airline to schedule additional flights (to "escalate" its frequency quota) if it consistently operated at or above 65 percent passenger capacity. On the fifth freedom, however, the British refused to cede further ground. During the third week of the conference, the two sides seemed "close to agreement" when Swinton provisionally agreed to accept fifth-freedom traffic on the condition that foreign airlines charge higher fares than their local competition (such that the London-to-Paris leg of a Pan Am flight would cost more than the same flight on BOAC). But the tentative compromise fell apart after Berle insisted on a fifth-freedom escalator clause. By the end of November, after nearly four weeks of negotiations, the United States and Britain seemed no closer to an accord. "I do not despair of pulling this thing out yet, though it is going to be close going at best," Berle wrote in his diary.[44]

At this juncture, the United States played its most valuable card: Lend-Lease aid. On November 24, President Roosevelt instructed the U.S. ambassador in London to deliver a message to Churchill in person. "We are doing our best to meet your Lend-Lease needs," FDR wrote, but Congress would likely vote to suspend further aid "if the people feel that the United Kingdom has not agreed to a generally beneficial agreement" on commercial air rights. In his reply, Churchill told Roosevelt that the matter "has caused me much anxiety." He conceded that the United States had already won the race for the postwar skies, with or without the fifth freedom: "You will have the greatest navy in the world," Churchill wrote. "You will have, I hope, the greatest air force. You will have the greatest trade. You have all the gold." The British were not disputing these facts but merely asking that "the American people . . . not give themselves over to vainglorious ambitions, and that justice and fair-play will be the lights that guide them." In response, FDR reiterated Berle's argument that the United States had "no desire to monopolize air traffic anywhere" and was "prepared to make transport aircraft freely available to you on the same terms as our own people can get them." On December 1, Churchill relented, informing Roosevelt that his government had begun to reconsider its stance on the fifth-freedom escalator clause, although it would not reach a final decision on the matter until Swinton returned to London. In the meantime, the British delegation would put forth no further proposals in Chicago. The "other air battle" had reached an armistice, if not exactly a permanent settlement.[45]

As London and Washington wrangled over the fifth freedom, the Chicago conference made progress on the less controversial matter of an international aviation organization. Delegates voted to form a Provisional

International Civil Aviation Organization (PICAO) that would be empowered with establishing and monitoring technical standards and safety guidelines. As the United States had wanted, PICAO would have advisory but not binding authority over competitive practices, including fares, traffic quotas, and route structures. Once ratified by twenty-six signatory governments, the organization would lose its provisional "P" and become the ICAO, governed by a twenty-one-member council whose representatives would be elected by the entire body (as the Latin American bloc had proposed). Headquartered in Montreal, PICAO held its first meeting in August 1945 and became affiliated with the United Nations, as the ICAO, two years later.[46]

The British, meanwhile, found themselves increasingly marginalized in Chicago. On November 27, delegates from a dozen nations—Peru, Venezuela, Panama, Brazil, Ecuador, Nicaragua, Cuba, Mexico, the Netherlands, Sweden, Greece, and China—all spoke in favor of the five freedoms. "When the dust cleared away," Berle wrote, "the British had the support of the French and the Australians and no one else."[47]

But this did not mean that the United States had won the "other air battle." Indeed, the conference's resolution revealed enduring skepticism of the open sky policy. In accordance with a proposal from the Netherlands, the conference ended by producing two separate documents: a Two Freedoms Agreement, which would ratify only the first two rights of transit and technical stop, and a Five Freedoms Agreement, as desired by the United States. On December 7, 1944, all fifty-two participating delegations signed the Two Freedoms Agreement. Only twenty, however, agreed to ratify the Five Freedoms Agreement. In the absence of multilateral consensus on the third through fifth freedoms, national governments would continue to negotiate bilateral air-rights treaties, awarding operating privileges to foreign airlines on a case-by-case basis. Meanwhile, the logistics of the contentious fifth freedom would be referred to PICAO for further study.

Reporting back to Washington, Berle declared the conference a triumph, stating that the creation of PICAO and international consensus on the first two freedoms had "advanced civil air flying by at least twenty years."[48] Such upbeat assessments, however, belied the fact that the Chicago settlements actually fell far short of Berle's aspirations. The Two

Freedoms Agreement, though certainly significant, was hardly the "charter of the open sky" that he had envisioned. The United States continued to lack bilateral air-rights treaties with Britain, France, and most of Europe's other nations. And the Soviets' refusal to participate in the conference portended that a great portion of the world's skyways would remain closed to U.S. aviation in the postwar era. If the conference "advanced civil air flying," it did not do so entirely on the United States' terms. Above all, Chicago signified that the ascent of the American Century would be turbulent. And far from giving rise to "one world," the conference—and the concluding events of the war—augured the demise of that concept.

From One World to Two

When the *Enola Gay* released its nuclear payload over Hiroshima on August 6, 1945, talk of "Airways to Peace" (as New York's Museum of Modern Art had titled its 1943 exhibition on air-age geography) suddenly seemed quaintly naive. "Lady, the atom bomb and the one world idea fit right together, but it isn't the one world you're talking about," said one New Yorker to another in 1946. "With atom bombs the only world is the next world." Divided by ideology and menaced by the threat of nuclear annihilation, the post-1945 international system belied wartime visions of one world. As the air age became the atomic age, FDR's desire to use aviation to "heal the wounds of war" succumbed to darker concerns about *Survival in the Air Age*, as a 1948 report by President Truman's Air Policy Commission would be titled.[49]

The war's conclusion catalyzed major shifts in U.S. aviation policy. In June 1945, the CAB voted to dismantle Pan American Airways' longtime chosen instrument status by awarding international routes to two other U.S. carriers, American Overseas Airlines (AOA) and Transcontinental and Western Airlines (TWA). In the State Department, meanwhile, a change in leadership marked the demise of Adolf Berle's liberal internationalism and the ascendance of a more conservative, overtly nationalistic foreign policy. In November 1944, during the Chicago conference, Secretary of State Cordell Hull resigned due to poor health. He was replaced by Edward R. Stettinius, the former chairman of U.S. Steel, whose back-

slapping, good ol' boy geniality sharply contrasted to Berle's austere intel-
lectualism. The two men reportedly despised one another, and in January
1945, Stettinius sent Berle packing by appointing him as U.S. ambassador
to Brazil. By contrast, Stettinius got along splendidly with Juan Trippe.
The new secretary of state was married to Trippe's sister, Elizabeth, and
his close ties with his in-laws—Juan and Betty named their third son Edward Stettinius Trippe—prompted suspicions that Trippe continued to
pull the State Department's strings.[50]

After Roosevelt died in April 1945, President Harry S. Truman pursued a more conservative course in both domestic and foreign policies.
Aviation policy, too, tacked right, reflecting a growing consensus in the
State Department that Berle's liberal internationalism had failed to
achieve U.S. objectives. In bilateral air rights negotiations with numerous
nations during late 1945 and 1946, the State Department adopted a newly
aggressive posture of overt self-interest. Secretary of State James F.
Byrnes, who replaced Stettinius in July 1945, was, like Truman, a south-
erner who represented the conservative wing of the Democratic Party,
and under his leadership U.S. policymakers used their nation's military
and economic power as leverage in their demands that foreign skies be
opened to U.S. airlines. As pragmatism and power politics came to define
postwar diplomacy, even the rhetoric of U.S. aviation policy changed.
After Berle, State Department leaders no longer spoke of the air as a
"highway given to all men." Now they echoed Henry Luce in arguing
that the air was rightfully America's, earned by its vital contributions to
Allied victory and postwar economic recovery. Gone, too, were Berle's
effusive paeans to one world, open skies, and freedom of the air—visions
that had largely failed to materialize.

Bilateral air rights negotiations between the United States and Britain
resumed in January 1946, amid the more hospitable climate of Bermuda.
Since Chicago, the stakes of obtaining a settlement had become much
more urgent; the Anglo-American rivalry now seemed to pale in significance compared with the escalating Cold War with the Soviet Union.
International commercial flights had also now resumed, and both sides
desired to obtain an accord as a show of Anglo-American unity against
the Soviet Union's closed sky policy. And the United States' relative
economic might continued to be a powerful diplomatic bargaining chip.

Although Britain's recent election of a Labour government prompted fears that London might again embrace protectionism, material necessity ultimately trumped political ideology. Requiring $2–$3 billion to repay its wartime loans from the United States, the British government was in no position to defy "Santa Claus."[51]

During Anglo-American aviation policy negotiations in Bermuda in early 1946, the State Department took full advantage of Britain's economic vulnerability, heeding CAB chairman L. Welch Pogue's advice to "make the satisfaction of certain British needs conditional upon her agreement to cease interfering with our attempts to secure these agreements." The strategy proved successful, allowing the United States to obtain bilaterally in Bermuda all that Adolf Berle had failed to obtain multilaterally in Chicago. On February 11, Britain signed an air rights agreement with the United States that granted all five freedoms and additionally allowed U.S. airlines to use British air bases for a period of ninety-nine years. Washington, by contrast, made few compromises, agreeing only to fare ceilings on flights between the United States and Britain. As President Truman summarized the settlement: "Under the Bermuda Agreement there will be no control of frequencies, and no control of so-called fifth freedom rights on trunk routes operated primarily for through service. It gives to the airline operators the great opportunity of using their initiative and enterprise in developing air transportation over great areas of the world's surface." During the next several years, the Bermuda Agreement would serve as a prototype for other air rights agreements, giving rise to what Anthony J. Sampson described as "a vast cobweb of bilateral international arrangements linking individual pairs of states."[52]

In July 1946, the United States announced its intention to withdraw from the Chicago settlements. Over a year and a half later, only fifteen governments had ratified the Five Freedoms Agreement, and according to a State Department press release, "only two, besides the United States, had developed international air services to any extent." In Washington's view, then, the Chicago accords could not serve "as an effective medium for the establishment of international air routes by United States carriers." The United States would continue to "adhere firmly to the Fifth

Freedom principle," Assistant Secretary of State Dean Acheson affirmed, but it would now seek to achieve that principle through "the bilateral rather than the multilateral approach," following the precedents of the Bermuda Agreement and similar protocols that Washington had signed with seventeen other nations since Chicago.[53]

Why did the State Department so quickly renounce the diplomatic agreements it had fought aggressively to obtain just eighteen months earlier? Berle's biographer, Jordan Schwarz, argues that the fifth freedom had always been a "ploy"—a fig leaf to disguise the United States' true goal at Chicago, the dismantling of British imperial trade preference. "The 'fifth freedom' was a Pyrrhic victory," Schwarz writes, "because it was a totally useless, mischievous ploy that aggravated rather than advanced the American position." He contends, moreover, that Washington never intended to implement the fifth freedom: "the country least likely to approve it, outside of the USSR, was the United States." There is some truth to this assessment. As part of a broader shift toward nationalism in U.S. aviation policy during the late 1940s and 1950s, the Truman and Eisenhower administrations restricted foreign airlines' rights to fly into the United States. The State Department never intended, as Senator Owen Brewster had charged, to "give the world fifty-one franchises to fly into the United States."[54] Adolf Berle's open sky policy, similarly, never intended to give all nations equal access to world air routes; it aimed, rather, to make those routes open to free competition. And a system of free competition served the particular national interests of the United States, with its unrivaled quantities of aircraft, capital, consumers with the money to travel, and cargo to be shipped. However, Schwarz conflates Berle's rhetoric, which proposed "to get all nations into the air," with his objective of ensuring postwar U.S. preponderance in the air. In this instance, as in other examples of American discourse on aviation, globalism was not simply a fig leaf to disguise nationalist objectives but a strategic and necessary means of obtaining them.

The story of the "other air battle" is a story of American hegemony, though not of unalloyed American triumph. Berle's open sky policy elevated the United States' position in the emerging postwar world, facilitating a larger, gradual transformation from Pax Britannica to Pax Americana, from an international order structured in the geopolitical idiom of territorial empire to one structured in the geoeconomic idiom of capital and markets. After 1945, the aviation industry played a vital part in the phenomenal worldwide expansion of the United States' economic power, as well as of the formidable military arsenal that underwrote that power. In American culture, meanwhile, discourse on aviation continued to naturalize U.S. hegemony, universalizing the nation's interests as the world's interests. Still, international disputes over aviation policy at the end of World War II portended that the United States' power in the postwar air age would be contested on multiple fronts. Rival hegemons—first Britain, then the Soviet Union—would assert their own claims to rightful dominance of the skyways. And as the Latin American bloc's "revolt" in Chicago foreshadowed, the world's "small" nations would advance their own aerial interests and aspirations, which often challenged, conflicted with, or differed from those of the United States.

Understood in this context, the history of international aviation policy helps answer a key question in twentieth-century U.S. historiography: why did the "internationalist moment" of the early 1940s wither so quickly? Although the era produced lasting international institutions, laws, and economic arrangements, the ferment of internationalist thought in mainstream American political culture largely did not survive the transitions from world war to Cold War, from air age to atomic age. Historians have cited many causes of internationalism's rapid demise after 1945: fears of nuclear attack; the descent of the iron curtain and the emergence of a bipolar world order; the conservative politics of the Truman administration and the postwar Congress; the consumer prosperity that led many U.S. citizens to turn inward, eschewing engagement with international affairs. An additional explanation, however, might cite the elements of nationalism that had always existed within American globalism, not in opposition to it. The open sky policy epitomized this dynamic. Adolf Berle's desire for international cooperation and unity was absolutely genuine—but so too was his conviction, shared by Henry Luce and other more conservative nationalists, that the postwar global order should be an American Century. Acknowledging the presence of nationalism within the internationalist moment does not minimize the sincerity of wartime cultural visions of one world and airways to peace. It does, however, trouble any easy distinction between the internationalist moment and what followed.

In April 1945, *Fortune* published an article entitled "Americans Fly Everywhere Now," which was both a retrospective analysis of the Chicago conference and a preview of things to come in postwar commercial aviation. "It can be said without boastfulness that no other people on earth know as much about the air as the American does," *Fortune* stated. "He spends more time in it than all the rest of the world. And he designs and builds more airplanes, flies more miles, and altogether does more work in the air than all of his allies and enemies put together. . . . The American airman has been everywhere. He breakfasts in Iceland, lunches in Newfoundland, dines in New York. He buys his shirts in New Delhi and has them laundered at Belem." At the Chicago conference, *Fortune* asserted, "some nations tried to throw a net around the American airman. They wanted him to submit to an international control that would decide where and how often he could fly." Such control, the article suggested, was not only an affront to U.S. aviation's role in attaining Allied victory in the war; it was also unnecessary, since the United States had no imperial ambitions. "An old-fashioned American imperialism aloft would be an evil thing," *Fortune* admitted. "But the American airman does not seek that. His first purpose is commerce, the free movement of goods and people and ideas, at the lowest possible cost, in the largest possible numbers and amounts, between anywhere and everywhere. The world has nothing to fear from that."[55]

The air-age world imagined by *Fortune* was precisely what Adolf Berle's open sky policy had aspired to create: a world in which national and imperial boundaries no longer constrained the free movement of people, planes, and products. Yet *Fortune*'s archetypal airman also carried a U.S. passport and embodied American exceptionalism: no other people on earth knew as much about the air as he did, the magazine claimed. Even the article's title referenced that quintessential expression of air-age nationalism, Clare Boothe Luce's "fly everywhere" speech. Echoing Berle, Trippe, and the Luces, the article condemned "old-fashioned imperialism" as an "evil thing" but meanwhile insisted that the United States be allowed to fly wherever, however, and whenever it wished. The world had nothing to fear from the American airman—but it had better not throw a net around him.

Historian Victoria de Grazia has described the postwar United States as a "market empire"—"a great imperium with the outlook of a great emporium," whose ever-expanding perimeters were marked not by territorial frontiers but by the ambitions of its corporations, the transnational momentum of its capital, the soft power of its civil society norms, and the cultural work of its "ubiquitous brands," which exported consumer desire and an "intimate familiarity with the American way of life."[56] The history of commercial aviation exemplifies, with particular clarity, the origins, character, and effects of this market-driven form of empire. As *Fortune* explicitly stated, the American airman's "first purpose is commerce." He flies and he buys—shirts in New Delhi, breakfasts in Iceland, laundry service in Belem—and he sells, spreading dollars and faith in free enterprise around the world. Benevolent in intention yet limitless in ambition, grounded nowhere while flying everywhere, *Fortune*'s airman embodied and enacted the global imaginary of a market empire—an empire of the air.

The United States' empire of the air would develop, however, not in one world but in a bipolar world divided between capitalism and communism. And in this transformed context, commercial aviation would enable Washington's Cold War to travel—not only by transporting U.S. tourists and dollars abroad but also by remaking "abroad" in the American image.

6

Mass Air Travel and the Routes of the Cold War

As the postwar air age began, Wendell Willkie's vision of "one world" had largely failed to materialize. On the contrary, the world seemed more divided than ever. By early 1946, as the Soviet Union showed no signs of relinquishing its control over Eastern Europe, Winston Churchill famously proclaimed that an "iron curtain" had descended across the continent. Tensions between the United States and the Soviet Union escalated throughout the late 1940s. In 1948, President Truman's Air Policy Commission called for an unprecedented buildup of military airpower; the following year, the Soviets exploded an atomic bomb. The bipolar international order and the ensuing nuclear arms race made the concept of one world seem as obsolete as Pan Am's fleet of Clipper flying boats, which by 1946 had been consigned to the scrap yard and replaced by land planes.[1]

However, the Cold War accelerated rather than stalled the ongoing global expansion of U.S. aviation. Between 1945 and 1960, mass air travel developed in tandem with the national security state—the institutional apparatus of government agencies, private corporations, universities, and think tanks that collectively waged Washington's Cold War. The resurgence of federal defense spending in the late 1940s rescued the aviation industry from its postwar financial crisis, enabling it to fulfill its wartime promises that air travel would give Americans the world, literally and figuratively. The aviation industry, in turn, helped enlist the American public into the Cold War. By promoting tourism as a form of foreign economic aid and cultural diplomacy, the airlines and the State Department worked together to transform Americans' travels into "Cold War holidays" (to borrow Christopher Endy's phrase) that served foreign policy objectives.[2] Pan Am, still the United States' premier international airline, played a leading role in this effort. Its postwar marketing and advertising campaigns construed international travel as a national duty, a patriotic service that would strengthen the United States' economy and export its political culture.

Tourism promotion efforts focused mostly on Western Europe, however, and in the mid-1950s, the geographic epicenter of the Cold War shifted from Europe to the so-called Third World of postcolonial and/or nonaligned nations in Asia, Africa, and Latin America. To increase the United States' influence in these regions deemed especially vulnerable to communism, Washington again turned to the aviation industry. During the 1950s and 1960s, a widespread consensus among U.S. social scientists and policymakers held that modernization—the development of social institutions grounded in advanced technology, capitalist modes of production, and discourses of rationality and efficiency—would, if properly planned and guided by the United States, help prevent communist revolution in "underdeveloped" nations.[3] As part of its larger mission of promoting modernization, the State Department contracted with Pan Am and TWA to develop national airlines in noncommunist Third World nations. These aviation development projects served U.S. airlines' commercial interests by modernizing the world's aviation infrastructure, creating new tourist destinations, and feeding domestic traffic to international flights. Meanwhile, aviation development advanced foreign policy goals by furthering the worldwide dissemination of capitalist business practices and American political and cultural norms.

Commercial aviation thus played a key role in the United States' Cold War strategy—and the Cold War itself ascended to the skyways, as Washington and Moscow competed for preponderance over the world's commercial air routes.[4] Conduits of global power, air routes facilitated transnational flows of money, weapons, technologies, consumer goods, propaganda, and persons of influence. The U.S. government's aviation policies aimed not only to ensure that Americans could fly everywhere but also to prevent the Soviets from flying there first. Airplanes themselves, moreover, acquired symbolic importance in the Cold War ideological contest. With the Stars and Stripes or the Hammer and Sickle painted on their fuselages, airplanes projected superpower influence around the world.

By extending the global reach of U.S. capital and culture, tourism promotion and the development of foreign airlines—both examples of corporate-government cooperation—augmented the United States' power and influence in the world without requiring such overt forms of domination as the Soviets' iron curtain. Yet neither tourism nor foreign aviation development projects entirely succeeded in fulfilling their stated objectives. American tourists did not always act as good ambassadors; U.S. airlines' ongoing international expansion created tensions with NATO allies; and aviation development aid did not necessarily mean that Third World governments would follow Washington's agenda. The routes of the Cold War ultimately proved to be as complex, and as contested, as its roots.

From Postwar Crisis to Cold War Revival

During World War II, the aviation industry had promised the American public that commercial aviation would soon bring the world "right [to] your own back yard!"—as suggested by a 1941 advertisement that depicted a map of air routes superimposed on a suburban lawn. In 1943, *Skyways* had predicted that it would soon be "just a lark to sweep down to an interesting spot somewhere in Brazil for a few days or spend a while during the hot weather on the coast of Newfoundland." More fanciful wartime fantasies envisioned airports in every town (in 1945, *Fortune* foresaw eighteen airports serving the New York metropolitan area), aerial highways with "skytels," and a flying car in every family's garage. Routine air travel would be part of the American Dream.[5]

Consistent with such predictions, the volume of annual traffic on U.S. domestic airlines increased nearly threefold between 1945 and 1950, from six to seventeen million passengers. Although every U.S. town did not yet have an airport, wartime government funding for airport construction had made "international ports of call out of such inland cities as Chicago and Kansas City," as *National Geographic* observed in 1948. Wartime government spending had also revived the national economy, such that Americans had more vacation time and more disposable income to spend on luxuries such as flying. The war itself had proven that flying was safe: thousands of GIs had flown as passengers or pilots, and their families back home witnessed the able performance of U.S. aircraft on newsreels and films.[6]

The war also stimulated advances in technology that benefited commercial as well as military aviation. Boeing's 377 Stratocruiser, the flagship of Pan Am's early postwar fleet, was developed from the B-29 Superfortress bomber; in lieu of bombs, the B-377s carried up to one hundred passengers, who could wile away their flight hours sipping martinis in an upper-deck lounge connected to the main cabin by a spiral staircase. Aboard such planes, stated the *New York Times* in 1946, Americans would soon be traveling "to the four corners of the earth." Pan Am illustrated this idea in a 1948 magazine advertisement entitled "Here Are Wings for Your Dreams," in which miniature globes hung as ornaments on a Christmas tree. Another 1948 ad showed a deck of playing cards depicting continents, implying that aviation placed the whole world in Americans' hands.[7]

The future of air travel thus looked brighter than ever at the end of World War II. For the first time, multiple U.S. airlines offered international service, a result of the Civil Aeronautics Board's 1945 decision to dismantle Pan Am's two-decade monopoly. On flights to and from Europe, Americans could now choose between Pan Am, American Overseas Airlines (AOA, a subsidiary of American Airlines), or Transcontinental and Western Air (TWA, which in 1950 changed its name to Trans World Airlines). Within the Americas, travelers could fly Pan Am, Braniff, United, or American; on flights to Asia, they could choose between Pan Am and Northwestern; and to the Caribbean, Pan Am, National, or Chicago and Southern. By 1946, there were seventy flights per week between the United States and the major capitals of Europe. "The U.S. tourist who wants to go abroad can, for the first time in eight years, go almost anywhere he chooses," observed *U.S. News and World Report* in the spring of 1948.[8]

Of course, international flight remained unaffordable for most Americans. In 1949, a round-trip flight between New York and London cost $630, comparable to first-class steamship accommodations ($450 to $700) but more than twice the cost of tourist-class steamer passage ($300). For Americans in the "middle income bracket"—defined by the Commerce Department as families earning $3,000 to $6,000 per year—a single round-trip transatlantic ticket would require 10 to 20 percent of their annual income. But for business travelers and wealthy vacationers, it was now possible to "le[ave] New York on a Tuesday afternoon, breakfast in Lima, Peru the following morning, ha[ve] one of the world's finest steaks in Buenos Aires Thursday night, le[ave] Buenos Aires Saturday noon and arrive in Oklahoma City Sunday afternoon after that"—as Wayne Parrish, publisher of the influential trade magazine *American Aviation*, did during a typical workweek in 1949.[9]

Such alluring accounts of international air travel filled U.S. magazines and newspapers in the early postwar years. Even if they could not afford to fly overseas, millions of Americans could read about the experience in their hometown newspapers thanks to press flights—airline-sponsored, all-expenses-paid trips for publishers and journalists. When Pan Am inaugurated the first scheduled around-the-world flight in June 1947, its public relations office ensured that the event would make headlines by offering complimentary seats to the publishers of the *Cleveland Plain Dealer*, the *Des Moines Register and Tribune*, the *New Orleans Item*, and the *Nashville Banner*, as well as representatives from syndication services Scripps-Howard and Gannett. Frank Gannett's serialized dispatches from the flight ran in over a dozen of his local newspapers and were later published as a booklet, *Winging 'Round the World*. Meanwhile, national publications such as the *New York Times*, the *Washington Post*, *Newsweek*, *Collier's*, and especially Henry Luce's popular magazines *Life*, *Time*, and *Fortune* reported on everything from the development of jet engines to trends in flight attendant uniforms (a favorite subject of *Life* pictorials). Highbrow magazines *Harper's* and the *New Yorker* offered philosophical perspectives on the air age: *Harper's* columnists Frederick Lewis Allen and Bernard DeVoto penned witty commentaries on the tribulations of the frequent flier, while pilot Wolfgang Langewiesche wrote eloquent musings on the cultural meanings of air travel.[10]

The airlines marketed their services through their own publications, including in-flight magazines, travel guides, and educational newspapers such as *Pan American World Airways Teacher.* Provided to classrooms free of charge, *Pan American World Airways Teacher* publicized the airline's worldwide routes and destinations, explained the science of flight, described the physical sensation of air travel, and even offered tips on what to pack for an overseas trip. Letters from readers suggest how airline publications conveyed ideas about globalism to those who were not personally able to "wing 'round the world." An elementary teacher at a rural school in California wrote that she used *Pan American World Airways Teacher* "as a guide to our imaginary travel scheme to all places we study in social studies," encouraging her students "to draw pictures and make models of planes they like to travel in." Another teacher reported, "I have become a 'rocking chair cosmopolitan.' Really I have traveled very little, but thanks to your publication, my rocker soloing has taken me to many strange and exciting places in the world. . . . I look forward to 'my trip' every two months."[11]

The actual experience of air travel during the early postwar years, however, seldom resembled its depiction in airline publications and advertisements. "The beauty and wonder are about the only compensation for airline travel," griped *Fortune* in 1946. "To travel by plane, a passenger must now sacrifice his comfort, his sleep, and often his baggage. He must endure inconveniences that rise to the level of punishment." Such "inconveniences" had become so commonplace, *Fortune* claimed, "that a person who flew from New York to St. Louis without incident gets more attention than the airline bore with lugubrious tales of his latest mishaps." When flights did proceed without mishap, passengers complained of boredom. "You seem stuck in endlessness and immobility," wrote *Harper's* columnist Bernard DeVoto. In contrast to the "jollity" aboard cruise ships, airplanes encouraged antisocial behavior, DeVoto believed: "nearly everyone reads his two-bit book until he falls asleep."[12]

Airline passengers of the late 1940s had ample reason to be dissatisfied. The nation's air traffic control systems were ill equipped to handle the increased volume of postwar travel; instead of zooming along aerial superhighways, planes often idled in runway queues or circled aloft in holding patterns. Travelers stranded in airports were scarcely better off than those aloft, as the airports of the late 1940s tended to be crowded and spare in amenities. *Fortune* described Chicago's airport as a "slum," its floors strewn with food waste, chewing gum, cigarette butts, newspapers, and piles of misrouted baggage. "In such an atmosphere, the beat-up traveler, interminably waiting for some unexplained reason, has no recourse but to ponder bitterly on the brilliant advertisement that lured him to 'TRAVEL WITH THE EASY SWIFTNESS OF HOMEWARD-WINGING BIRDS,'" *Fortune* wryly observed. International flights involved the added hassles of customs, duty taxes, and sundry varieties of red tape—including, as Langewiesche noted, "the blocked account, the managed currency, the rationed gas, the travel permits, the prohibited areas, the identity cards, the export permits, the forbidden possessions, the visa, the police clearances, the queues."[13]

Safety issues posed more serious hazards. Statistics corroborated the aviation industry's claims that flying was generally safe: passenger fatality rates had steadily declined from 31.8 per million passenger miles in 1930 to 2.07 per million in 1946, below the rate of automobile fatalities (2.9 per million passenger miles). Nonetheless, as more people flew in airplanes, more people died in airplanes—there were 149 fatalities in 1946, compared with thirty-three in 1930—and unlike car wrecks, every commercial aviation disaster made national headlines. In 1947, a series of fatal crashes, including two over Memorial Day weekend that together killed ninety-five people, shook Americans' confidence in air travel.[14] "The average citizen is apparently not convinced that flying is safe; or, if he is convinced, then his wife is not," commented a *Harper's* critic in an article entitled "Our Airsick Airlines."[15]

Exacerbating all of these problems, demobilization had wreaked financial havoc on the U.S. aviation industry. During the war, aircraft manufacturers had feasted on an all-you-can-eat buffet of government funding, and when peace returned, Washington placed them on a crash diet. In 1943, an industry consortium hired blue-chip public relations firm Hill and Knowlton to promote airpower as essential to postwar national security. But the campaign failed to convince the Truman administration to buy into its contention that "Air Power Is Peace Power." By the end of

1945, aircraft manufacturing firms had lost over $21 billion in canceled military contracts. And contrary to wartime predictions, sales of planes to commercial airlines were thus far failing to compensate manufacturers for their lost military revenues. Between 1944 and 1947, aircraft sales dropped off by more than 90 percent, and manufacturers' earnings plummeted from a profit of $67.4 million in 1945 to a net loss of $41.9 million.[16]

The aviation industry's financial crisis developed in tandem with global crisis: the escalating Cold War between the United States and the Soviet Union. In this context, aviation industry promoters amplified their warnings that demobilization gravely threatened national security. "The problem of insuring America's future security is inseparable from the problem of keeping America first in the air," stated John F. Victory, secretary of the National Advisory Committee for Aeronautics (NACA), in 1945. Increased federal funding for aviation research and manufacturing, Victory added, would be nothing less than "an insurance policy on the very life of the nation." According to the editors of *Skyways*, the United States needed a strong air force to serve as "Uncle Sam's Fist"; otherwise, there would be "*nothing* to prevent a large formation of hostile bombers from appearing over vital American targets." Prominent military leaders, including Hap Arnold, Carl Spaatz, and Ira Eaker, issued similar warnings, appealing directly to the American public via mainstream magazines such as *Newsweek*, *Collier's*, and *Look*.[17]

In July 1947, however, increasing concerns about the Soviet threat prompted President Truman to take two major steps that would revive the U.S. aviation industry and conjoin it to the emerging national security state. First, the National Security Act of 1947 designated the Air Force as an independent branch of the U.S. armed forces, fulfilling an objective of airpower advocates since Billy Mitchell. Second, Truman appointed a five-member panel to undertake a comprehensive review of the state of U.S. aviation, both military and commercial. The President's Air Policy Commission (also known as the Finletter Commission, after chairman Thomas K. Finletter—who had previously advised the State Department on foreign economic policy and would serve as secretary of the Air Force between 1950 and 1953) subsequently conducted over two hundred hearings with representatives from all sectors of the aviation industry, summarizing its findings in a report entitled *Survival in the Air Age.* Strongly endorsing airpower as the foundation of national defense, the Finletter Commission called for the Air Force to procure at least 8,100 new planes by January 1, 1953. Because new types of aircraft would take four to seven years to develop, it also urged the federal government to fund aviation research and manufacturing continuously, not just in the event of a national emergency. Preparedness must become permanent, the Commission argued, a way of life in peace as well as war. "Our policy of relative security will compel us to maintain a force in being in peacetime greater than any self-governing people has ever kept," its report concluded.[18]

As the Finletter Commission acknowledged, peacetime military mobilization would be "new in American life." Opposition to standing armies was deeply rooted in American political culture, and concerns that the U.S. government could become a militaristic "garrison state" had intensified after the emergence of totalitarian regimes in Germany, Italy, and the Soviet Union. The Commission maintained, however, that military buildup would not threaten American democracy because "we are not an aggressor nation." Rather, "a strong United States will be a force for peace . . . welcomed by all peace-loving nations." Reiterating ideas that airpower theorist Alexander de Seversky had expressed during the 1940s, the Commission argued that the atomic age would remain peaceful only if the United States, with its ostensibly unique lack of aggressive tendencies, maintained a preponderance of airpower. "Power does not necessarily corrupt if it is accompanied by an equivalent sense of responsibility," wrote Finletter in 1951. "Therefore, it is a prime responsibility on our part to see to it that the Western world maintains a great and unchallengeable superiority in the air."[19]

In spite of Truman's personal wariness of excessive military power, his administration heeded the Finletter Commission's advice, authorizing an unprecedented increase in peacetime defense spending that focused on the development of nuclear weapons and strategic bombing capability. Between 1947 and 1954, government expenditures on military hardware nearly doubled, from 35.5 to 65.7 percent of all federal spending. Once again flush with lucrative defense contracts, the U.S. aviation industry surged out of its postwar slump. By 1955, sales of airframes had quintupled, from half a billion dollars to \$2.5 billion, and the twelve leading aircraft manufacturers, which had reported a collective net loss of \$10.7 million in 1946, posted profits of \$182.6 million.[20]

Commercial aviation likewise benefited from the Cold War revival of federal defense spending. On the Finletter Commission's recommendation, Truman established an Air Coordinating Committee to integrate military and civil aviation policy, composed of representatives from the Departments of State and Commerce, the Air Force, the Navy, the Post Office, and the CAB. During the 1950s and 1960s, the Air Force and the airlines, which had always been closely connected, became even further intertwined as elements of the emerging national security state. Military contracts gave aircraft manufacturers the capital to mass-produce faster and larger airliners along with fighters and bombers, and the airlines invested in new types of equipment, such as jet engines, that were initially developed for military use. The commercial aviation industry also lent support to the armed forces. Through voluntary membership in the Military Air Transport Service, airlines committed to lending their aircraft and equipment to the military in the event of a national emergency. The Air Line Pilots Association, meanwhile, encouraged its members to hold military reserve commissions. And the airlines directly participated in Cold War military activities, just as Pan Am had done in Latin America and Africa during World War II. In 1948 and 1949, commercial airlines airlifted food, fuel, and other necessities to the residents of Berlin after Soviet troops blockaded the city, and during the Korean War, they transported over half of the total air tonnage flown to combat zones.[21]

Beyond supporting overseas military operations, the airlines performed valuable services at home for what President Dwight D. Eisenhower famously described, in his 1961 farewell address, as the "military-industrial complex"—the network of government agencies, defense contractors, and think tanks that formulated Washington's Cold War strategies and built the weapons to implement them. Commercial aviation played an important and often overlooked role in sustaining the military-industrial complex during its formative years. In 1953, for example, Pan Am established a Guided Missiles Range Division (GMRD), which obtained contracts from the U.S. Air Force to manage a 6,000-mile chain of missile testing stations from Cape Canaveral, Florida, through the Caribbean. The

GMRD's employees, numbering seven thousand by 1958, were responsible for tasks that ranged from specialized chemical cleaning and the storage of explosive components to food preparation and medical care. "We're often asked why we got into this," said Pan Am vice president Richard S. Mitchell in an interview with *National Geographic.* "But we have many years experience operating stations in remote areas"—and, he added, the work put Pan Am "in the vanguard of new techniques that eventually will be used by the airlines." In the late 1960s, Pan Am's renamed Aerospace Services Division served the nation's space program as prime contractor to the U.S. Air Force at Cape Kennedy.[22]

Pan Am's 1947 inaugural around-the-world flight epitomized the Cold War revival of the U.S. aviation industry. It also reflected the ever-increasing convergence of military and commercial aviation. At inaugural ceremonies on the morning of the *Clipper America*'s departure from New York, Juan Trippe thanked aircraft manufacturers Lockheed, Douglas, Boeing, Consolidated, Pratt and Whitney, Curtiss-Wright, and Martin for making possible "the peacetime air power which is keeping the American Flag in first place on the world's airways." By sustaining such firms with a steady influx of defense contracts, the national security state not only bolstered military preparedness but also enabled the first around-the-world flight to be accomplished "under the auspices of the American government, by an American company, operating aircraft designed and built in the United States and in charge of an American captain and crew," as Trippe emphasized.[23] Keeping the American flag "in first place on the world's airways" was a Cold War strategy that paralleled, in the civilian arena, the buildup of military airpower recommended by the Finletter Commission.

As with previous Pan Am inaugurals, the around-the-world flight was publicized in the United States and abroad as a national triumph. Recognizing its Cold War propaganda value, the State Department's Office of International Information and Cultural Affairs celebrated the flight in a press release distributed to five hundred journalists and U.S. embassies, consulates, and information bureaus. Indeed, even the meals served aboard the *Clipper America* made a nationalist statement. The thirteen-day flight, as food critic Clementine Paddleford reported, would feature "typically American" dishes such as grapefruit maraschino, filet mignon, deep-fried potato balls, and an ice cream tart. During the next several years, Pan Am continued to publicize the *Clipper America* with a lavish promotional campaign that extended even to fashion: a "round-the-world scarf," available in four colors, could be purchased in department stores throughout the United States. "Hundreds of thousands of women will literally carry our message," proclaimed a sales department newsletter.[24]

The ascent of the *Clipper America*, like the United States' broader postwar ascent as a global superpower, was propelled by the national security state. Passengers aboard Pan Am's around-the-world flights (who numbered 21,240 in 1958 and 45,336 in 1961) did not necessarily connect their travels to Washington's Cold War agenda. However, the fact that the United States remained "in first place on the world's airways" during the 1950s resulted in large part from the Truman and Eisenhower administrations' commitments to peacetime federal spending on airpower. Defense contracts resurrected U.S. aircraft manufacturers from their post-demobilization economic slump, enabling them to produce the aircraft that, during the next decade, would make civilian air travel faster, safer, cheaper, and more comfortable than ever before. Meanwhile, the Cold War itself provided political and economic incentives for Americans to "wing 'round the world." Beginning in 1947, Washington and the aviation industry worked together to promote international travel as an instrument of U.S. foreign policy—a voluntary, low-cost way of exporting American dollars and American dreams. If the Cold War made airpower "the business of every American citizen" (as Hap Arnold had written in 1946), the airline business attempted to make every American citizen a Cold War ambassador.[25]

"Ambassadors of the American Way of Life"

International travel had long served as a cultural crucible in which conceptions of American national identity—and claims to global power—were forged, contested, and refashioned. Beginning in the nineteenth century, as Christopher Endy has argued, overseas travel "formed a cultural or ideological foundation for imperialism and increasing U.S. engagement in world affairs."[26] The State Department began promoting tourism during the Roosevelt administration, when the Office of the Coordinator of Inter-American Affairs (OCIAA) had encouraged travel between the United States and Latin America as part of the Good Neighbor policy. Just as Pan Am's interwar expansion in the Western Hemisphere laid the groundwork for its worldwide expansion, the OCIAA's regional experiments served as the model for the State Department's promotion of tourism during the Cold War. Tourism, in the view of many leading policymakers, could serve Washington's objectives by unifying the "free world," exporting American dollars and values, and undermining the appeal of communism.

Tourism also offered an ideal solution to a major foreign policy issue of the first postwar decade: the balance-of-payments crisis, or "dollar gap." By the end of the 1940s, foreign economic aid, particularly the $13 billion Marshall Plan for the reconstruction of Western Europe, was straining the federal treasury and creating a growing imbalance between imports and exports. Congressional Republicans increasingly attacked foreign aid programs, arguing that taxpayers should not have to foot the bill for other nations' economic recovery. Furthermore, critics maintained, the Marshall Plan and similar foreign aid programs would encourage other governments to become dependent on Washington's coffers in lieu of taking responsibility for their own economic recovery.[27]

Many policymakers and business leaders believed, however, that tourism could serve as a tax-free alternative to federally administered foreign aid. By 1948, U.S. tourists spent a billion dollars annually in other countries. In a 1948 speech entitled "World Prosperity through Foreign Travel," Juan Trippe argued that the dollars spent voluntarily by private individuals had essentially the same effect as government dollars. Simply by traveling abroad, American citizens could help revive foreign economies, reducing the need for expensive, taxpayer-funded aid programs. And because tourists patronized small businesses such as hotels, restaurants, and shops, their activities could encourage entrepreneurial initiative in foreign societies. Finally, Trippe emphasized, tourists were living examples of the American way of life. Arriving with his "cigarette lighter, portable typewriter, luggage, fountain pen, and his wife's compact [and] purse," the tourist advertised the availability of such desirable consumer goods in the United States. Thus, tourism promised to revive capitalist economies and to promote the benefits of capitalism, while simultaneously reducing the expenditures of the federal government.[28]

The Truman administration concurred with Trippe's arguments about the economic and diplomatic benefits of international travel. In 1948 and 1949, the Department of Commerce published studies that showed travel to be "one of the soundest, most significant, and far reaching means at hand for closing the dollar gap." According to its calculations, the thirty million Americans who earned paid vacation time could contribute over $2.5 billion per year to foreign economies. Consequently, the Economic Cooperation Administration (ECA), the agency that administered the Marshall Plan, sought to encourage Americans to go abroad by establishing a Travel Development Section (TDS) in early 1948. With offices in Washington and the sixteen Marshall Plan countries, the TDS surveyed U.S. citizens about their travel habits, publicized European tourist attractions in magazines such as *Life* and *Good Housekeeping*, organized trips for groups of farmers, housewives, and students, lobbied governments to liberalize border-crossing procedures, and instructed European hoteliers and restaurateurs on how to make their businesses conform to American standards of cleanliness and comfort.[29]

TDS director Theodore J. Pozzy, a French-born naturalized U.S. citizen, had been appointed on the recommendation of his friend Senator Owen Brewster (R-Maine), a longtime Washington advocate of Pan Am. Not surprisingly, Pozzy's arguments echoed Trippe's. Tourism, he stressed, would help to close the dollar gap, to stimulate Western Europe's economic revival, and to publicize American capitalism. "American tourists circulating throughout Europe . . . can, if they conduct themselves properly, become personal ambassadors of the American way of life," Pozzy wrote in 1949. "The American tourist can be dramatized to the masses of Europe, not as a Wall Street imperialist bent on global domination, but as a fellow human whose expenditure for pleasure puts bread in the mouths of the railroad switchman, the hotel chambermaid, the waiter, the taxi driver, and, indeed, all the little people susceptible to Communist propaganda." Furthermore, tourism was a potentially unlimited means of influence. The United States could purchase only so many imported goods and provide so much funding for foreign aid programs, Pozzy noted, but tourists could "clamber up Mont. St. Michel or walk through St. Mark's Square or view the Matterhorn over and over again." Sightseeing, moreover, imposed no economic burdens on the U.S. government.[30]

As a Cold War strategy, tourism attracted broad political support in the United States. Conservatives embraced it as a privatized solution to national and global economic problems; for liberals, it was a benign, nonimperialistic means of influencing the development of other nations. Trippe and Pozzy courted both sides of the political spectrum, tailoring their messages accordingly. Notably, though, both conservative and liberal discourses on tourism followed the logic of nationalist globalism expressed in Henry Luce's 1941 "American Century" editorial. While conservatives tilted to the nationalist end of the spectrum and liberals to the globalist, both affirmed that the world's security and prosperity required U.S. global leadership, particularly the exportation of American capital and political culture. Whether conservative or liberal, Cold War proponents of tourism heeded Luce's call for Americans to "exert upon the world the full impact of our influence."[31]

In 1949, however, only 594,000 Americans, less than 1 percent of the population, traveled abroad. An even smaller fraction went by air. A Commerce Department study of Americans traveling from New York to Europe found that over twice as many sailed (129,049) than flew (53,650); a 1950 TDS survey of U.S. travelers leaving France reported that only 26 percent were flying home. Fears of flying and considerations of physical comfort may have influenced some travelers' decisions to book passage on a steamship rather than an airplane. But most Americans eschewed flight because it remained prohibitively expensive. As Trippe told the Senate's Interstate and Foreign Commerce Committee in 1949, international airfares far exceeded the vacation budget of "Mr. Average American."[32]

Trippe's airline, which had initially promoted flight as an exclusive luxury, now spearheaded a campaign to make it affordable for Mr. Average American. In September 1948, Pan Am introduced a "tourist-class" fare on flights between New York City and San Juan, Puerto Rico, reducing the cost of a round-trip ticket from $253 to $150 and a one-way ticket from $133 to $75. By removing the galleys and coatracks on its DC-4s, and thereby increasing seating capacities from fifty-two to sixty-three passengers, Pan Am was able to offer the cheaper fare by packing more people into the cabin. The experiment was an immediate success. In five months, passenger traffic between New York and San Juan tripled, and by 1952, the route accounted for more than a third of the miles flown by Pan Am's entire Latin American Division. As winter descended on the East Coast, sunseekers eagerly snapped up tickets to spend their vacations in Puerto Rico. On the return flights, Pan Am brought thousands of Puerto Ricans to New York, facilitating what one aviation historian has described as "the world's first migration by air." Of the five million Puerto Ricans who emigrated to the United States during the postwar years, approximately two-thirds arrived by airplane.[33]

Trippe next sought to introduce tourist fares on flights between the United States and Europe. Lower fares could permit "six million Americans to go abroad every year," he argued, including such nonelite citizens as "the workman at a lathe receiving three or four hundred dollars a month." Slashing transatlantic fares, however, would not be easy. Whereas Pan Am was the exclusive carrier on the New York–San Juan route, airlines from six other nations (Britain, Belgium, Sweden, Switzerland, Canada, and the Netherlands) operated transatlantic flights—meaning that fare reductions would be subject to approval by the International Air Transport Association (IATA). At the IATA's annual meeting in the winter of 1948, Trippe's proposal incited a fierce debate in which Western European and Canadian delegates expressed broader concerns about the United States' growing domination of the global air transport market. As of 1948, Pan Am, TWA, and AOA each flew some 25 percent of the total transatlantic traffic, with the remaining quarter going to all foreign airlines combined. Pan Am's passenger capacity was 96,724 per year; BOAC and Air France, by comparison, had passenger capacities of 33,540 and 31,356 respectively. In this context, foreign airline leaders viewed tourist fares as an extension of American economic hegemony, dismissing Trippe's populist rhetoric as camouflage for his real objective of shutting his competitors out of the transatlantic market.[34]

The Truman administration and the U.S. news media, meanwhile, came out strongly in favor of tourist fares. The Commerce Department and the ECA's Travel Development Section each published reports corroborating the economic benefits of low-cost air travel. According to Theodore Pozzy, "lower rates on trans-Atlantic travel [are] one of the most practical and the quickest means of increasing European dollar exchange" and thereby closing the dollar gap. Tourist fares also received support on the nation's editorial pages. The *News Sentinel* of Fort Wayne, Indiana, for example, predicted that low-cost international air travel could accomplish "more progress toward peace" than diplomatic conferences. The *New York Enquirer* crowned Juan Trippe the "Henry Ford of the Air," who proposed to make international travel as accessible as automobiles to middle-class Americans. Henry Luce again put Trippe's picture on the cover of *Time*, underneath the caption "Now the world is every man's oyster." Inside, a seven-page article portrayed Trippe himself as Mr. Average American, "a two-Scotches-before-dinner man who likes to hear all about his four children's day in school." Yet with an annual salary of $23,050 (well over $200,000 in today's dollars), Pan Am's president did not have to save war bonds in order to fly around the world—as did the Brooklyn postal worker, living on a salary of $3,000 per year, who was featured in a 1949 article that Trippe wrote for *Reader's Digest*.[35]

The IATA finally approved tourist fares on transatlantic routes at its annual meeting in 1951, after Trippe repeatedly threatened to implement the reductions with or without IATA approval. The new rates, offered by eleven airlines, were slightly higher than Trippe had wanted: one-way fares from New York to London and Paris were set at $270, and round-trip fares from $417 (off-season) to $486 (March through September), about a third less than the regular price. The IATA's decision was widely hailed as a crucial step toward making air travel affordable for the masses. "Instead of the movie star, society leader or businessman traveling on a generous expense account, we find the student, stenographer, GI bride going to visit her family in England," wrote journalist and longtime aviation enthusiast Lowell Thomas. On May 1, 1952, the first transatlantic flight carrying tourist-class passengers departed from New York's Idlewild Airport en route to Paris. Those who had reserved tourist-class tickets, the *New York Times* reported, included a "surprising lack of so-called white collar workers." Many passengers were skilled laborers, salesmen, and farmers in the $5,000-per-year income bracket.[36]

Tourist-class fares proved immensely popular, even if they did not quite yet fulfill Trippe's goal of making international flights affordable for "average" Americans. In 1954, more than 1.7 million Americans traveled abroad by airplane, surpassing steamship travel (1.1 million passengers) for the first time in history. International fliers still comprised a small fraction of the thirty-four million Americans, nearly a fifth of the population, who traveled by air that year. Among those who did fly abroad, however, two-thirds traveled tourist class, and tourist fares generated 54 percent of Pan Am's total revenues. During the next several years, Pan Am introduced additional forms of savings, such as family discounts and the popular "Fly Now, Pay Later" plan, which allowed travelers to pay for their tickets on installment. "Pan Am will fly you around the world for only $135 down," announced a 1955 magazine ad that featured a two-page color photo spread of Spain's picturesque Costa Brava. The airline also offered special incentives for those seeking to emigrate to the United States. Beginning in 1949, U.S. residents could purchase tickets for relatives or friends in other countries, and Pan Am sales agents even helped immigrants to obtain travel funds and documentation.[37]

When Dwight D. Eisenhower assumed the presidency in 1953, the travel industry gained an important new ally in Washington. To an unprecedented degree, the Eisenhower administration committed federal resources to encouraging, monitoring, and managing the travels of U.S. citizens. The fiscally conservative president agreed with Juan Trippe and Theodore Pozzy that tourism offered an ideal solution for closing the dollar gap and stimulating foreign economies without inflating the federal budget. In 1954, Eisenhower's chief of staff, Sherman Adams, instructed all federal agencies to "adopt and support aviation facilitation as an important element in the United States foreign economic program." Furthermore, Eisenhower believed that tourists could play an important role in improving foreign perceptions of the United States and winning the Cold War contest. "If our American ideology is eventually to win out in the great struggle being waged between the two opposing ways of life, it must have the active support of . . . millions of individual Americans acting through person-to-person communication in foreign lands," the president stated in a 1956 press release announcing his People-to-People Partnership, which facilitated international exchanges between business and civic leaders, educators, and other professionals. Meanwhile, the State Department, under the direction of Secretary of State John Foster Dulles, sent technical advisors to reform foreign travel industries, worked to simplify border-crossing procedures, and promoted the benefits of tourism at the United Nations and other international organizations.[38]

As a form of public diplomacy and foreign economic aid, tourism epitomized the "American ideology" that the Eisenhower administration wished to project abroad. In contrast to Moscow's state-directed propaganda initiatives, tourism promised to export U.S. dollars and influence while exemplifying the virtues of private enterprise, limited government, and individual free will. Tourism was therefore a perfect instrument for the diplomacy and political economy of the United States' Cold War.

Yet a crucial question remained to be answered: would Mr. Average American want to go abroad? As the airline industry sought to broaden its market reach during the mid-1950s, it began to sell international travel to new types of consumers—including Mrs. Average American and her family—while Washington worked to ensure that the American flag would prevail on the world's skyways.

Marketing Mass Air Travel

Juan Trippe's vision of mass international air travel ironically coincided with a renewed emphasis on domesticity in American political culture. After fifteen years of depression and warfare, in a Cold War world threatened by nuclear annihilation, many Americans were "homeward bound" (as Elaine Tyler May has argued), desiring to stay in their literal and metaphoric backyards. While they enjoyed more paid vacation time and more disposable income than ever in history, most middle-class Americans chose to take their vacations within the United States, where they could expect familiar conveniences such as refrigerated drinks and private hotel bathrooms. As aviation analyst Wayne Parrish argued in 1948, the travel industry thus faced a difficult challenge in convincing the public "just why one should . . . go to Calcutta or Buenos Aires or some other place where you can't get a nickel cup of coffee, a chocolate malted milk, or a good drive-in hamburger."[39]

Indeed, some experts believed that average Americans could not be convinced to go abroad. In 1950, two ECA labor advisors informed Theodore Pozzy that they doubted "that any great number of U.S. workers are either interested in spending vacations in Europe or are willing and able to spend $800 to $1,000 per person for such vacations." Off-season travel discounts, they noted, did not coincide with most Americans' vacation periods, and even tourist-class airfares remained unaffordable for many. For those who could afford to go abroad, much-publicized foreign annoyances—unsanitary (by U.S. standards) hotels and restaurants, poor road conditions, late trains, anti-American sentiment—offered incentives to vacation in Miami or Manhattan instead of Rio or Paris. In 1949, a TWA survey of passengers returning from Europe revealed a "discouragingly great amount of discontent and resentment on part of American tourists at treatment accorded them in France and Italy." Given such challenges, Juan Trippe's vision of mass air travel would not spontaneously materialize; Americans would have to be persuaded to go abroad.[40]

In 1956, Pan Am launched a massive new advertising campaign, under the direction of renowned agency J. Walter Thompson (JWT), which expressly appealed to the white middle class—the same demographic that purchased Ford automobiles, Kodak cameras, Chesebrough Ponds facial creams, and Parker pens, to name just a few of the other products that the agency touted. Advertising's ability to get Americans into airplanes had already been demonstrated by JWT's previous postwar campaigns for Pan Am. In 1950, tourist-class tickets to Puerto Rico sold out just hours after being publicized on New York radio stations, and the airline received 16,500 inquiries about a package tour advertised just once in *Holiday* and *Time* magazines. Six years later, the air travel market had grown even more competitive as U.S. consumers acquired more disposable income, as aircraft became larger and safer, and as more foreign airlines served U.S. destinations. By 1956, two-thirds of international travelers chose to fly rather than sail abroad. (That same year, the ocean liner *Andrea Doria* sunk off the coast of Nantucket, killing fifty-two passengers and raising public concerns about the safety of ships.) Moreover, in a major shift from the prewar decades, by the mid-1950s nearly 85 percent of airline passengers traveled for pleasure instead of business. Meanwhile, as JWT noted, there had also been "a complete switch in the relative importance from first class to tourist-fare travel." In 1952, two-thirds of Pan Am's transatlantic passengers traveled first class; by 1954, as previously noted, that same percentage flew tourist class. In this changed climate, Pan Am could no longer simply rely on its reputation as "the world's most experienced airline," the slogan that had defined its brand since World War II. JWT's 1956 ad campaign thus aimed to remake the airline of power and luxury into an airline for the masses.[41]

The agency's Review Board records—periodic reports on client accounts compiled for senior-level personnel—offer unusually candid documentation of the logic behind the 1956 campaign. These reports also testify to the changing demographics of the international air travel market and flying's concomitant cultural transformation from elite privilege to normative leisure activity. A Review Board report from December 1956, for example, argued that Pan Am must appeal not only to its traditional customer base of wealthy frequent travelers but to all Americans: "*To whom are we selling?* We are selling to any man or woman in the United States who has any reason to travel abroad or who can be convinced to have a reason to do so." The Review Board emphasized that Pan Am "*cannot afford to rest on the laurels of experience alone;* it must convince its passengers that they are wanted" (emphasis in original).[42]

The themes of Pan Am's 1956 ad campaign consequently emphasized "human" factors over the more impersonal qualities of dependability and experience. A new slogan, "In Every Man's Heart There's a Secret Place He Would Like to Go," explicitly addressed itself to every man, appealing to his heart rather than his head. To heighten the slogan's emotional impact, illustrations showed what JWT described as "warm human and appealing situations." While the new ads continued to combat "Pan Am's most formidable competitor—habit, tradition, laziness, love of the familiar, uneasiness about the unfamiliar," they aimed to "strike far deeper into the reader's heart than if we said, curtly: 'Don't be a slave to habit.' "[43]

To convince Americans that they would be welcomed abroad, the new advertisements also featured "warm, human stories from various countries" that purported to demonstrate foreigners' affection for the United States. An ad from April 1956, which ran in *Life* and the *Saturday Evening Post*, depicted six Europeans dressed in stereotypical national attire. Beneath each individual, a fill-in-the-blank quiz tested readers' knowledge of his or her culture: "He drinks his wine from a *purree*, which is a custom in ____"; "Her blond beauty is typical of a fresh and sparkling land." But the headline underscored what these Europeans had in common: "They all like Americans—they live only an overnight flight away—and they all

have something in common with you." By emphasizing that "Pan American is a builder of good will for America," as a 1956 Review Board report
stated, these ads did not simply aim to persuade U.S. citizens to travel
abroad. They also intended to "make Washington policymakers feel good
about Pan American." As always, Pan Am's advertising and marketing
strategies thus aimed to maximize political capital as well as profits.[44]

JWT's 1956 campaign also aimed to assuage anxieties about international travel by domesticating foreign people and places. During the
1930s and 1940s, Pan Am had marketed its Latin American destinations
by juxtaposing images of airplanes with images of burro-riding peasants,
descriptions of luxurious tourist amenities with references to primitive or
exotic local customs. As the airline attempted to attract new customers
during the Cold War, technology continued to mediate—and indeed to
contain—travelers' encounters with the foreign. Images of Pan Am's
modern fleet of aircraft offered visual evidence that travel abroad did not
have to entail giving up the comforts of home. Aircraft cabins enclosed
the passenger in a space that was recognizably "American," resembling
passengers' own living rooms: windows had upholstered curtains and
seats were modeled on club chairs. The cabin was also domestic in the
sense of being a nationalized space, where passengers could eat American
food, hear English spoken, and expect U.S. standards of service.

Symbolic appropriation of the foreign was another way of neutralizing
its aura of otherness and difference. A series of Pan Am advertisements
from 1957 represented foreign places as objects to be accumulated. Underneath photographs of the airline's destinations, boldfaced type asked
the reader such questions as "Do you collect ancient ruins?"; "Do you collect foreign golf courses?"; "Do you collect exotic islands?" By suggesting
that foreign places could be "collected" as souvenirs, these ads likened
travel to a familiar all-American activity: shopping. As an object of con-
sumption, the foreign looked alluring rather than threatening.[45]

Pan Am's 1956 ad campaign aimed to reach a wide swath of the American public. To court educated and affluent customers, JWT produced a
series of television commercials to run during *Meet the Press*, the esteemed
Sunday morning political talk show that was viewed in fourteen million
households during this election year. These spots pitched Pan Am "to the
cream of our prospect audience," emphasizing "the international impor-

tance, significance, and leadership we can derive from associating Pan Am with the most important people in the world." Print advertisements, meanwhile, targeted a more middlebrow audience. Ads in the local newspapers of twenty-six U.S. cities would be seen by some sixteen million families, JWT estimated. National magazine advertising would reach even more people. In 1956, the average American family received seven magazines per month; as a result, national magazine advertising had increased by 40 percent since 1949, in spite of the advent of television. And because magazine ads could be printed in color, they were best suited to portraying "warm, human, and appealing situations." As JWT's newsletter stated, "They are the mood creators. They offer an unusual opportunity to show real travelers against specific foreign backgrounds and to emphasize specific Pan Am advantages."[46]

JWT's national magazine campaign followed two strategies, each designed to market Pan Am to a different demographic. For publications catering to elites and "constant travelers" (including the *New Yorker*, *Holiday*, and *National Geographic*), the agency showcased exotic destinations, on-board amenities, and Pan Am's record of experience. Meanwhile, in mass-circulation magazines including *Life*, *Time*, and *Newsweek*, ads with "human" themes—such as the "In every man's heart . . ." series—appealed to would-be passengers who presumably needed reassurance that globe-trotting was affordable, safe, and fun. For these ads, JWT commissioned artist Norman Rockwell, described in a sales pitch as "the graphic historian of everyman's daily life," to fly around the world and create a "sketch book" of foreign people and places that would "bring traditions of people throughout the world into the American home." Rendered in Rockwell's recognizably homespun style, the foreign again became familiar. His drawings depicted casually attired tourists (prominently toting Pan Am shoulder bags) in front of such landmarks as London's Buckingham Palace and Rome's Trevi Fountain. Accompanying the illustrations, "short, folksy captions" offered tips for the novice traveler: "For real sukiyaki in Tokyo, you leave your shoes outside, sit on the floor"; "Gay Paree has the best floor shows, and the best menus." The series premiered in March 1956 in *Life* and the *Saturday Evening Post*, the two largest U.S. weekly magazines with a combined circulation of over ten million.[47]

Altogether, JWT estimated that Pan Am's 1956 advertising campaign would deliver "nearly 18 sales talks to every family in the country," reaching 25 percent of television homes through *Meet the Press*, 34 percent of newspaper readers, and the two-thirds of Americans aged ten and over who read national magazines.[48] But the new campaign did not simply aim to attract more passengers; it also aimed to attract different types of passengers. Advertising increasingly divided the air travel market into class segments, as evidenced by JWT's two different magazine campaigns. And for the first time in history, ads began to appeal to women as a distinct category of potential airline passengers. If the Cold War necessitated that airpower become the business of every U.S. citizen, as Hap Arnold had argued, the airlines would need to appeal to "Mrs. and Ms. Average American" as well as to their traditional male client base.

In a 1952 survey of U.S. travel agents, the Research Department of *Ladies' Home Journal* found that 94.8 percent of respondents enthusiastically approved of advertising that targeted women. Their responses included such statements as "We feel that women control better than 80% of vacation travel" (Vanstrum Travel Service, Austin, Minnesota); "without the women, I'd soon be out of business" (Dougherty Travel Service, Modesto, California); and "Are you kidding! Women—God bless them—*are* the travel business" (Sherman Travel Service, Chicago). Enclosed in the survey was a copy of a TWA ad featuring a perky travel agent named Mary Gordon. Published widely in *Ladies' Home Journal* and other women's magazines, TWA's Mary Gordon series was among the first airline ads to appeal directly to women. Still, the series invoked and reaffirmed traditional conceptions of femininity. Mary Gordon advised readers, for example, to "ask your husband to fly TWA so that he can spend more time at home with you." She also offered tips on what kind of clothing to pack and touted in-flight amenities such as pillows, magazines, and "gracious, attentive hostesses."[49]

The airline industry's appeals to women often focused on the family, consistent with travel agents' beliefs (as indicated by the *Ladies' Home Journal* study) that women determined their families' travel plans. As part of its makeover into an airline for the masses, Pan Am offered family discounts such as its "Family Plan to Europe," which promised savings of $1,000 for a family of five. In a 1955 *Life* ad created by JWT, the airplane itself became a catalyst of family unity, allowing wives and children to "be there together" with men traveling abroad on business. The nation's largest domestic carrier, American Airlines (AA), offered half-priced tickets for women and children and, in a series of ads published in *Good Housekeeping*, touted its safety by claiming to be mother-approved: "Columbus may have discovered America, but *Mother* discovered American!" Another AA ad promised mothers that flying would make baby "a cheerful cherub," such that "*you* yourself will . . . feel rested and relaxed" upon landing.[50]

Yet the *Ladies' Home Journal* study also indicated that more women were traveling by themselves, whether on business or pleasure. "Practically all of our vacation cruise and tour business can be attributed to the average working girl taking her annual vacation," wrote Elma C. Robinson, a travel agent from Butler, Pennsylvania. Seeing a lucrative new customer base, the airlines began marketing their services to women as travelers rather than simply as wives and mothers. In magazines read by single "working girls," including *Glamour* and *Charm*, Pan Am explained how its "Fly Now, Pay Later" plan would allow young women to fly to Bermuda, Hawaii, or even Europe "on a secretary's paycheck." Following the success of its "round-the-world scarf" promotion, Pan Am again advertised through fashion: in 1955, dresses by Rosecrest—a brand popular among young working women and sold at some twelve thousand U.S. stores—came with tags bearing a membership card for "air-travel-minded secretaries and business girls" to join the airline's New Horizons Club. In cities such as London, Paris, and Rome, club members received hotel and restaurant discounts, admission to cultural events and nightclubs, special sightseeing tours, and even half-priced Parisian perfume. JWT, which conceived the promotion, noted that according to 1950 census figures, there were seven million single women under the age of thirty-five—out of whom, the agency claimed, "one million can easily afford a five-hundred-dollar, three-country trip to Europe." TWA made its own appeal to "single girls" through sassy ad copy ("Who says it's a man's world?"), eye-catching color photos of fashionably dressed fliers, and reassurances that women could "travel alone without a worry in the world" while "enjoying service that befits a queen." Indeed, airline advertising used images of solo female travelers as evidence that flying was both comfortable and safe. In a 1955 Pan Am ad, published in *Foreign Affairs*, a woman spoke with an airline agent against the backdrop of a tropical setting. "When you are alone in a foreign land, a Pan Am office is as *reassuring* as the American flag," declared the headline.[51]

Even as it appealed to increasingly diverse constituencies, however, airline advertising remained limited in its inclusiveness. Nonwhite faces did not begin to appear in ads until the late 1960s, and even then, ads featuring African Americans ran mostly in black magazines such as *Ebony*. The airline industry, likewise, was deeply stratified by race and gender. Even as airline cabins diversified during the 1950s and 1960s, the cockpit remained an overwhelmingly white male enclave. Continental Airlines hired the first African American commercial pilot, Marlon Green, in 1963. However, the nine-year Air Force veteran had previously been denied employment by ten airlines, and Continental hired him only after the Supreme Court, in a unanimous decision, legally compelled the company to do so. The first female airline pilot, Helen Richey, began flying for regional carrier Central Airlines in 1934, but she resigned within a year due to opposition from the all-male pilots' union. During World War II, Richey flew as a Women Airforce Service Pilot; unable to find aviation employment after the war, she tragically committed suicide in 1947.[52]

Reflecting racial and gender divisions in the airline industry labor force, images of pilots as white father figures became a stock trope of airline advertising. Two visually similar advertisements for different airlines—American Airlines' 1949 ad "He Commands Your Flagship and Your Confidence" and Pan Am's 1956 Norman Rockwell–illustrated ad "Eyes That See Around the World"—featured close-up portraits of white male pilots who appeared to be in their mid- to late fifties, their faces friendly yet wisely serious. The pilots' age, masculinity, and whiteness, along with their sharply tailored, gold-braided uniforms, signified normative conceptions of authority and experience. Pan Am took this symbolism a step further in a series of 1956 ads that likened the pilot to Uncle Sam, the figurative embodiment of the American nation. Depicting a uniformed Clipper captain greeting a mother and her three small children as they boarded a plane, the ad declared, "It's nice to know Uncle Sam's your skipper when you fly to faraway places." The rhetorical nationalization of the pilot and his aircraft again served to domesticate foreign travel, reassuring passengers that the familiar comforts of home, including white paternal authority, would ensure their safe journey to "faraway places."[53]

As the jet age began, only a minority of Americans had traveled by air, because of concerns about the cost and safety of flying. However, as tourist-class fares made air travel more affordable, and as the industry's advertising campaigns appealed to a broader base of consumers, the ranks of airline passengers steadily grew. A 1964 survey conducted by the University of Michigan and sponsored by Boeing, United Airlines, the Federal Aviation Administration (FAA), *Time*, and the Port Authority of New York found that respondents who identified themselves as "experienced air travelers" had increased from 23 percent in 1955 to 36 percent by 1962. And although international flights remained relatively expensive, tourist fares, along with family discounts and "Fly Now, Pay Later" plans, reduced the absolute cost of air travel dramatically. If they carefully saved a portion of each paycheck, secretaries and salesmen could afford to take an international vacation. (In the 1964 study, notably, more than twice as many respondents ranked fear rather than cost as their primary reason for not flying.) Meanwhile, advertising campaigns that featured images of fatherly pilots, carefree single "working girls," happy families, and Norman Rockwell's quaint renderings of foreign places—along with copy that emphasized the safety, convenience, and all-American comforts of air travel—encouraged Mr. and Ms. Average American to view a trip abroad as a desirable, unthreatening consumer experience.[54]

The political implications of mass air travel, however, did not always fulfill the expectations of Cold War policymakers. In fact, as more Americans traveled abroad, many in Washington began to worry that tourist-class tourists would do more harm than good to the United States' international image. Beginning in the late 1940s, the State Department inserted a booklet into U.S. passports that warned, "Tourists who assume an air of arrogance or who transcend the bounds of decency in human conduct can do more in the course of an hour to break down the elements of friendly approach between peoples than the Government can do in the course of a year in trying to stimulate friendly relations." To further disseminate this message, State Department officials and travel industry leaders authored magazine articles, such as "The Knack of Intelligent

Travel" (*Reader's Digest*, September 1952), which instructed readers on
how to behave when traveling abroad. In spite of such efforts, though, by
the early 1960s the term "Ugly American"—originally the title of a book
criticizing the behavior of U.S. diplomats—had come to refer to tourists
who chipped off pieces of landmarks, loudly demanded to be served
American food, or arrived at palaces wearing "dungarees" (to name just a
few of the offenses observed by one travel writer).[55]

The foreign policy implications of tourism faltered, ironically, on the
very success of the mid-1950s travel boom. As more U.S. citizens went
abroad, it became increasingly difficult for the State Department and the
travel industry to monitor or manage them. Precisely because tourism
was voluntary, there was no way to ensure that American travelers would
comport themselves as good ambassadors. Passport inserts, magazine articles, advertising, and other forms of public persuasion could serve to
educate but ultimately had no compulsory power. Moreover, neither policymakers nor airline executives could control the meanings of international travel. Those who went abroad in the 1950s and 1960s did not
necessarily see their travels as opportunities to serve Washington's Cold
War agenda. Rather, as Christopher Endy has argued, many viewed their
vacations as a holiday from the Cold War. Travel offered a welcome escape from everyday realities, including anxieties about communism and
nuclear weapons.[56]

Mass tourism also jeopardized foreign policy goals by causing tensions
within the North Atlantic Treaty Organization (NATO) alliance. The
IATA controversy over tourist fares was not the first time that European
airline leaders objected to what they perceived as unfair competition from
across the Atlantic. In 1950, President Truman overruled a CAB decision
and allowed Pan Am to acquire its competitor AOA, which was also authorized to fly transatlantic routes. The French and the British were out-
raged by Truman's decision. Together, Pan Am and TWA already had
more capital, more passengers, and more aircraft than all of their European rivals combined, and with AOA's resources added to Pan Am's, the
two U.S. carriers stood to nearly monopolize the transatlantic traffic. Air
France would be "crushed between two giants," stated France's civil aviation secretary, who feared its "virtual elimination" from its most lucrative

international route. The Truman administration tried to sweeten the deal by loaning France $50 million for civil aviation, but the money did not silence French criticism. The pro-Gaullist Parisian newspaper *Le Rassemblement* described Pan Am as "une nouvelle menace extrêmement grave" (an extremely grave new menace), noting that its fleet of 152 airplanes dwarfed Air France's forty-seven planes. Another editorial, "Notre Aviation Menacée par un Trust Américain" (Our Aviation Menaced by an American Trust) charged that Truman had jeopardized the future of the entire French aviation industry. The leftist newspaper *Libération* concurred that the consequence of Pan Am's expansion "en serait sans doute la liquidation d'Air France" (would be, undoubtedly, the liquidation of Air France). The French airline, of course, was not liquidated (it has outlasted Pan Am), but by February 1951, Pan Am and TWA were operating more flights out of Paris than all other airlines combined.[57]

European criticisms of U.S. aviation policy offered yet more evidence that mass air travel did not always advance Cold War objectives. France's opposition to the Pan Am–AOA merger—just one example of many such conflicts during the 1950s and early 1960s—indicated that the United States' ongoing economic expansion could conflict with its geopolitical priorities, in this case NATO unity. Just as American tourists did not necessarily see their vacations as opportunities to propagandize for the "free world," commercial aviation created tension between the United States and its allies even as it seemed to be a potent weapon in the Cold War contest for foreign allegiances.

By the mid-1950s, however, NATO allies' objections to U.S. aviation policy appeared less urgent as the focus of the Cold War shifted from Europe, where the bipolar balance of power remained relatively stable, to the so-called Third World of Latin America, Asia, and Africa. In the aftermath of the Korean War, the State Department had grown increasingly concerned about Soviet support of socialist and/or anticolonial movements in these regions. Meanwhile, the Soviet Union had begun to show new interest in commercial aviation, establishing worldwide air routes that linked Moscow to many developing nations deemed vulnerable to communism. In this changing geopolitical context, the skyways assumed even greater importance as a means of containing Soviet ambitions.

Aviation Development and the Global Cold War

During the first decade after World War II, the United States had faced no real competition from the Soviet Union in commercial aviation. As Josef Stalin concentrated on industrialization and the development of military airpower, the state-run Soviet airline Aeroflot languished as "little more than a national barnstorming operation" (to quote one U.S. analyst), whose obsolete aircraft rarely ventured beyond the iron curtain. But Stalin's successor, Nikita Khrushchev, recognized the commercial airplane's political utility. Rejecting the Stalinist doctrine of "socialism in one country," Khrushchev sought to expand the Soviet Union's global presence, and he viewed the airplane as an ideal instrument for such expansion. In the summer of 1953, just months after Stalin's death, the State Department began receiving reports that Aeroflot, in its first attempt to fly to noncommunist nations, was seeking landing rights in Greece and Turkey. By 1955, State Department leaders concluded that the Soviet Union had embarked on "a determined and vigorous program to enter international air routes." Confirming such reports, Aeroflot stunned the world by debuting a turbojet, the Tupolev Tu-104, on a flight from Moscow to London in March 1956—thus becoming the second airline, after BOAC, to offer regular jetliner service. Pan Am would not inaugurate jets until October 1958.[58]

The State Department's growing concerns about Aeroflot also reflected the changing geography of the Cold War itself. In a March 1958 statement on international aviation policy, Secretary of State Dulles warned that the USSR "has great interest in flying over or transiting Western territories in order to reach more easily the sensitive areas of the Middle East, Asia, Africa, and Latin America." Envisioning an aerial domino effect, Dulles believed that if Aeroflot obtained operating rights in key nonaligned nations such as India or Afghanistan, the entire developing world would be vulnerable to Soviet influence. After the Soviet Union and India began negotiating a reciprocal air-rights treaty, which was signed in the summer of 1958, the State Department informed U.S. embassies in the region that they should be "not so concerned about Aeroflot operations [in] New Delhi per se as implications [of] Soviet airline operations [in] intermediate countries and extension beyond India to Burma and possibly points beyond." Aeroflot's passenger traffic was likely to be "Moscow-generated," Dulles noted, as diplomats, propagandists, and technicians would use the airline to gain access to "sensitive areas" in India and elsewhere. Furthermore, both U.S. and Soviet leaders understood that commercial airplanes had important symbolic power. According to a 1957 RAND Corporation report, the Soviets desired global air routes "not because they want to compete for traffic with Pan Am or TWA, but because it will make the flag on the new Soviet jets look brighter and more impressive when they land in Cairo, Delhi, and Rangoon." *Business Week*, likewise, argued that the Soviets would use Aeroflot "as a prestige weapon" to "score a propaganda victory in the Middle East and Asia."[59]

To counteract Aeroflot's growing presence in the world's skyways, Washington not only supported the ongoing international expansion of U.S. airlines but also underwrote the development of national airlines in Asia, Africa, and Latin America. As part of its broader effort to promote modernization and development in these regions, the State Department implemented an ambitious program of aviation technical assistance, which provided nascent airlines with funding, equipment, and advisors. Such efforts to develop aviation in countries of strategic interest to the United States were nothing new; indeed, Pan Am's wartime Airport Development Program (ADP) in Latin America served as a direct precedent for the projects of the Cold War. Just as the Roosevelt administration had viewed (and funded) the ADP as a hemispheric defense measure, the Truman and Eisenhower administrations believed that aviation development projects were essential to the security and prosperity of the "free world." Thriving, U.S.-guided airlines, policymakers believed, would foster entrepreneurial culture in Third World nations and strengthen those nations' economic and political ties to the United States.

The Cold War phase of aviation development assistance began with the 1948 U.S. Information and Educational Exchange Act, also known as the Smith-Mundt Act. The first major U.S. government effort to counteract Soviet propaganda, the act funded the expansion of educational initiatives such as the Fulbright Program, Voice of America radio programming, and similar forms of public diplomacy. It also authorized the Civil Aeronautics Administration (CAA) to send advisors to foreign governments that desired assistance in developing national airlines. The first CAA missions went to Turkey and Pakistan, where they trained pilots and ground crew, supervised airport construction, and provided advice on airline management. Additional funding and support for such projects came via President Truman's Point IV program, so called because Truman announced it in the fourth point of his 1949 inaugural address. Described by the president as "a bold new program for making the benefits of our scientific advances and industrial progress available for the improvement and growth of underdeveloped areas," Point IV identified Third World modernization and development as a major priority of U.S. foreign policy. And it identified technology—"our scientific advances and industrial progress"—as the primary means of ensuring that developing nations aligned with the United States, not with the Soviet Union.[60]

The Eisenhower administration formalized and expanded on these initiatives. After the Marshall Plan's termination in 1951, the developing world became the major beneficiary of U.S. foreign economic assistance. In 1954, Eisenhower created the International Cooperation Administration (ICA), a subsidiary of the State Department, to coordinate Washington's ever-increasing outflow of aid dollars. ICA-funded aviation development projects supplemented those already administered by the CAA, with the Development Loan Fund and the Export-Import Bank providing additional sources of aid. Between 1956 and 1961, the U.S. government devoted nearly $300 million to foreign aviation development. The major recipients of grants were Afghanistan (receiving $45.4 million), Ethiopia ($24.8 million), Chile ($14.6 million), Pakistan ($12.3 million), Vietnam ($8 million), Turkey ($5.9 million), and Tunisia ($5.3 million)—all countries deemed especially vulnerable to communist influence. Aid revenue enabled recipient governments to finance aircraft purchases, to employ U.S. technical advisors, and to construct or improve airports and communications facilities.[61]

Meanwhile, U.S. airlines administered their own foreign aid projects. In 1946, TWA signed a contract with the government of Saudi Arabia to develop a national airline and to train Saudi pilots.[62] U.S. diplomats in the country enthusiastically endorsed the contract, predicting that TWA would establish "great good will at a time when our prestige and Arab confidence in us are seriously threatened." Furthermore, noted Secretary of State Dean Acheson, the Dhahran airfield, "a vital link in world wide

United States commercial air service," happened to be "located in the center of the richest oil field outside the United States for which American firms have the sole concession."[63] TWA's assistance to the Saudi airline thus served both diplomatic and strategic interests.

For its most important aviation development projects, the State Department enlisted Pan Am, its longtime "chosen instrument" in the field of commercial aviation. During the Eisenhower years, Pan Am received ICA contracts to administer technical assistance to airlines in Pakistan, Thailand, Turkey, and Afghanistan. With ICA funding, Pan Am's Technical Assistance Program (TAP) sent teams of twenty-five or more advisors to each of these countries, where they provided one-on-one tutoring to their local counterparts in every aspect of airline operation, from pilot and mechanic training to ticket sales. According to company reports, the initial results of this "Private Enterprise Point IV" (as Pan Am's public relations office called it) were highly encouraging. Within two years, Pakistan International Airways reported a 43 percent increase in passenger traffic and an 82 percent increase in flight hours; Thai Airways, 37 and 44 percent, respectively; and Turkish Airlines, 73 and 84 percent. Just as Pan Am had previously described the ADP as an expression of the Good Neighbor policy, its press releases now construed TAP as a form of altruistic internationalism—a Good Neighbor policy not just for the Americas but for the entire noncommunist world.[64]

Like the ADP, however, the TAP also served Pan Am's commercial interests. The improvement of foreign aviation infrastructure promised to send domestic feeder-line traffic to the U.S. airline's international routes. And according to executive vice president Roger Lewis, who oversaw TAP, the program would "set up new airline[s] in accordance with our tradition"—such that Pan Am's business model would become a worldwide standard. "Demonstrating the same business principles to the developing nations as we applied in our own country is the best form of foreign aid," Lewis wrote in a letter to the *Nashville Tennessean*, which had previously published an editorial calling TAP more successful than the New Deal–era Tennessee Valley Authority (TVA) in promoting global modernization and development.[65]

The newspaper's reference to the TVA was fitting. Aviation technical assistance projects did not simply result in the creation of airlines and airports; they brought to developing countries an entire apparatus of American modernity, from mechanical equipment to consumer goods to management techniques. Aviation development promised, in short, to modernize the Third World in the mold of the United States. When author and pilot Wolfgang Langewiesche described flying from Damascus to Baghdad in a 1949 *Harper's* article, he noted how the United States' growing presence in the Middle East could literally be seen from the air: "the desert suddenly was no longer desert. . . . It was just like some Western American landscapes: a string of high-power civilization stretched with magnificent disregard across the wasteland." But the airplane did not simply make such transformations visible; it propelled them. "What sat there on the naked desert beside a huge airport looked like a piece of Concentrate of U.S.A.," Langewiesche recalled—and that "huge airport" made it possible for "Concentrate of U.S.A." to be flown in and distilled throughout the country.[66]

The history of Pan Am's involvement in Afghanistan, however, underscores that on the ground, U.S. aviation development projects were more complex and more locally determined than Langewiesche's aerial image suggested. Rather than simply unloading "Concentrate of U.S.A." onto inert and passive foreign soil, Pan Am actively collaborated with the Afghan government to develop the nation's commercial carrier, Ariana Afghan Airlines. In the process, Afghan leaders used Ariana to advance their own political, economic, and cultural objectives, deftly exploiting Cold War rivalries in order to obtain U.S. aid on their own terms. In the process, Afghans imparted distinct nationalist meanings to aviation, such that the airplane was never merely an instrument of U.S. power and interests.[67]

In January 1955, the government of King Mohammed Zahir Shah established Aryana Afghan Airways, named after the appellation given to Central Asia by the ancient Greek geographer Strabo. At the time of Aryana's founding, Afghanistan had no airports and only a few World War II–era planes that belonged to its fledgling Royal Air Force—in contrast to nearby Pakistan, India, and Iran, all of which had thriving national airlines. However, the king and his powerful prime minister, Mohammed Daoud Khan, were European-educated technocrats who believed that Afghanistan must modernize in order to regain its rightful status among the world's great powers. They also believed that transportation held the key to Afghanistan's modernization. Once a hub of the Silk Road, Afghanistan now lay astride the great-circle air routes between Western Europe and Asia. For Afghan leaders, then, aviation promised to restore their country's former grandeur as a "crossroads of the world."[68]

Because of its nonaligned status and proximity to the Soviet Union, Afghanistan was also a crossroads of the Cold War, and the royal government's interest in aviation attracted notice in both Washington and Moscow. Aeroflot had been operating flights between Tashkent and Kabul since 1951, and in March 1955—three months after Aryana's establishment—the Soviet and Afghan governments signed a bilateral air-rights treaty. That same day, the Afghan Foreign Office received word that the United States was proposing to extend aid for Aryana's development. As its timing suggests, the offer was motivated not by good neighborly altruism but by Cold War politics. As Dulles warned U.S. embassies in Kabul and Karachi, the "Soviets undoubtedly [are] willing to undertake" a program of technical assistance for Afghan aviation "if [the] US does not."[69]

To administer the program, the State Department turned, once again, to Pan Am. In addition to providing funds for technical assistance, the ICA authorized the U.S. airline to purchase 49 percent of Aryana's stock. "We must strike while iron is hot or forget American airroute via Kandahar to southeast Asia and the strengthening of Afghan ties with free world through mechanism of American managed air line," wrote the U.S. ambassador to Afghanistan, Sheldon Mills, in a telegram emphasizing the diplomatic significance of the arrangement. In Washington, State Department leaders concurred that Pan Am's investment in Aryana would "present Afghanistan with [an] effective link with [the] Free World and would assist in counterbalancing its substantial involvement with [the] Soviets." Although the postwar CAB typically deterred U.S. airlines from acquiring financial interests in foreign companies, in this case it concurred with the State Department that Pan Am's stock purchase offered insurance against further Soviet penetration of Afghan aviation and was therefore "in the national interest."[70]

In April 1957, a team of twenty-five U.S. technicians and four DC-4 airplanes arrived in Kabul. According to Pan Am's contract, the Afghan royal government would continue to own 51 percent of Aryana's stock and appoint its president; officials from Pan Am, however, would select Aryana's other executives, in addition to hiring and training its pilots, flight attendants, and ground crew. The ICA, meanwhile, had given Afghanistan $14 million in aviation assistance, including $5 million to underwrite Pan Am's expenses and another $5 million for aircraft and equipment purchases. The remainder of the money would fund improvements to aviation infrastructure, including the construction of a modern international airport in the southern city of Kandahar. Additionally, the CAA would send its own team of advisors to help the royal government organize a Department of Civil Aviation.[71]

As a result of these efforts, Afghanistan's aviation developed rapidly. Just six months after Pan Am's arrival in Kabul, Ariana Afghan Airlines had a new, more Westernized name and had done the same volume of business that it had reported for the entire previous year. The U.S. embassy reported that "the PAA management group has made friends in high places, and at all levels there appears to be widespread satisfaction with Ariana's recent performance." By 1959, Ariana flew weekly from Kabul to Frankfurt, where it opened a stylish sales office filled with brochures and timetables in both English and German. By 1962, it operated 8,371 miles of routes and employed 569 Afghan citizens, including twenty-six pilots licensed by the U.S. Federal Aviation Administration.[72] By 1967, the year that Washington discontinued aid for Afghan aviation, Ariana's revenues had quintupled. While much of its revenues derived from the annual Muslim pilgrimage to Mecca, Ariana also conducted a thriving cargo business, increasing the nation's exports of fruit, textiles, and karakul (the prized wool of the fat-tailed sheep).[73]

In the United States, Pan Am eagerly took credit for Ariana's success. Press releases described the challenges and rewards of "trying to teach people to operate an airline when they, for the most part, have never seen an airplane before." U.S. newspapers, too, enthusiastically reported how Pan Am had made it possible for Afghanistan to transition from "Camels to Airliners." The project received coverage not only in the *New York Times* and *Washington Post* but also in smaller papers such as the *Manchester Union Leader*, the *Long Island Star-Journal*, and the *Brooklyn Daily*, which all carried articles about locals working in Afghanistan as advisors to Ariana.[74]

Recalling how it had publicized its Latin American and African projects during World War II, Pan Am gave special attention to employees' efforts to uplift Afghanistan's population—particularly, in this case, women. While the airline's managers and technical advisors were busy modernizing Ariana, their wives attempted to modernize Afghan women by reforming how the latter dressed. In 1959, a group of Pan Am wives began voluntarily teaching sewing classes at the Women's Welfare School for Girls in Kabul. Sewing lessons, they believed, would encourage Afghan women to abandon the burqa in favor of more "liberated" Western clothing. Jeanne Beecher, the wife of an Ariana executive, wrote to Pan Am's New York office to request sewing patterns, and a secretary there contacted the *Vogue* Pattern Service, which immediately rushed two hundred patterns to Kabul, along with copies of *Vogue* and *Glamour.* To give Afghan women a live glimpse of the latest American styles, Beecher later organized a fashion show at the city's United States Information Center auditorium.[75]

At a time when the Soviet Union seemed to be winning the contest for hearts and minds in the Third World, not to mention the Cold War air and space race (Sputnik had been launched on October 4, 1957), such stories about Pan Am's partnership with Ariana allowed Americans to feel good about their nation's role in the world. Media coverage of Ariana affirmed popular notions that the United States could and should be a model for developing nations to follow. In this context, the airplane promised to be not simply a form of transportation but an agent of social transformation—an engine of progress that would modernize Afghanistan in the United States' own image.

The full story, however, proved to be more complicated. In Afghanistan, the king, his prime minister, and other technocratic elites embraced aviation as a means of modernization. In contrast to their U.S. counterparts, Afghan leaders believed that aviation would make Afghanistan more independent and less in need of foreign guidance. Tellingly, they rarely credited Pan Am or the United States for Ariana's success, instead attributing it to Afghan greatness. Over Kabul radio—a key medium of communication in a country with a largely illiterate population—officials from the Department of Civil Aviation regularly broadcast updates on airport construction, statistics on how many passengers Ariana had carried, and reports on students enrolled in aeronautical training courses. When speaking of aviation, Afghan leaders emphasized their nation's historical significance as a crossroads of world transportation. "With the completion of the present projects in civil aviation Afghanistan will regain her importance as the crossroad serving the East and the West," predicted government news bulletin *Bakhtar*. In reference to this history, Ariana advertised its flights to Europe as the "Marco Polo Route." Meanwhile, Afghan government officials frequently objected to U.S. aviation assistance policies. Criticizing Afghanistan's initial aid package as insufficient and Ariana's planes (which had been selected by Pan Am) as outdated, they succeeded in obtaining more funding and more advanced aircraft by threatening to approach Moscow if Washington did not up the ante. Thus, rather than simply acceding to the United States, Afghan leaders imparted nationalist meanings to aviation and manipulated Cold War politics in order to modernize on their own terms.[76]

Increasing infusions of dollars, however, did not stop Afghanistan from also accepting rubles. The royal government claimed to be *bi-tarafi* ("without sides" in Farsi), but by the end of the 1950s, it was clear that "the Russian influence is growing very notable," as Pan Am vice president Roger Lewis observed. In 1963, Ariana moved its headquarters from Kandahar, whose airport had been funded and built by the United States, to Kabul, where the Soviets had constructed an airport. Within a decade, the balance tipped decisively in the Soviet Union's favor: in 1968 and 1969, Moscow sent $32.04 million in direct aid to Kabul, nearly twice the $16.9 million sent by Washington.[77]

In late 1979, the Soviet Union invaded Afghanistan, beginning a brutal war that would last for the next decade. Ariana, like millions of Afghan civilians, became collateral damage. During the first year of the Soviet occupation, 250 of the airline's employees, including three-quarters of its pilots, defected to Pakistan, West Germany, and the United States in order to avoid being forced to fly Soviet troops and weaponry to combat areas. Pan Am retained its 49 percent share of Ariana throughout the first four years of the occupation but finally terminated the partnership in 1984 after significant financial losses. Under Soviet control, Ariana served just three foreign destinations: Moscow, Tashkent, and Prague.[78]

Like tourism, then, technical assistance proved to be an inconsistent Cold War strategy. By the end of the 1960s, the U.S. Agency for International Development, the ICA's successor, had largely ceased funding aviation development. The FAA (the renamed and reorganized CAA) continued to send small teams of advisors to governments that requested them. But especially as the costs of the Vietnam War escalated, Washington could no longer afford expensive aid packages for foreign airlines. From the perspective of many experts, moreover, aviation development projects had largely failed to accomplish their objectives. In 1964, RAND Corporation analyst Hans Heymann Jr. published a highly critical report concluding that in most countries, commercial aviation had contributed little to overall modernization and development. Most projects had focused on developing showpiece international airlines, Heymann emphasized, while the populations of recipient countries would have been better served by domestic carriers. Advisors, meanwhile, often did not understand local conditions, resulting in unfortunate decisions—such as the decision to build an international jet airport in the remote Afghan town of Kandahar. Finally, Heymann argued that the ideological significance of commercial aviation had diminished. In the jet age, he wrote, the airplane "is no longer a novelty, not even in remotest Africa where modern jet transports have by now become a familiar sight. . . . The marginal 'impact' value of an added display of jet aircraft with U.S. markings, therefore, may not be an impressive one."[79]

Aviation development also proved unreliable as a Cold War strategy because rather than simply following the United States' lead, developing nations insisted on shaping the course of their own modernization. As exemplified by the case of Ariana Afghan Airlines, foreign governments accepted U.S. aviation assistance for reasons that often had nothing to do with Washington's Cold War agenda. In countries around the world, airports and airlines signified national greatness more than loyalty to the United States. Indeed, aviation was such a potent instrument of modernization not simply because it advanced U.S. interests but because it could also be appropriated in the service of rival nationalisms and competing modernities.

As the jet age began, such complexities would continue to characterize both aviation and American globalism. The 1950s had confirmed the United States' hegemony in the world's commercial skyways: by 1958, the year that Pan Am inaugurated its first jet flights, 82.9 percent of the planes operated by all airlines worldwide had been manufactured in the United States. In 1960, U.S. airlines flew 58 percent of global air traffic; Europe had a 24 percent share and the rest of the world only 18 percent. The United States' ascent as the world's premier aerial power resulted, however, not only from the unrivaled prosperity of American consumers, nor from intrinsic national qualities of ambition and ingenuity (as Juan Trippe and other industry leaders often claimed), but from the strategically crafted corporate-government partnerships, publicity campaigns, and policies of the Cold War national security state. For the Truman and Eisenhower administrations and their allies in private enterprise, "survival in the air age" necessitated continuous federal funding for aerospace research and development, the exportation of American dollars and dreams via mass air travel, and the development of pro-U.S. airlines in foreign countries, particularly in the Third World. As a result of such initiatives, the period between 1945 and 1960 seemed to be one of uninterrupted ascendance for the United States' aerial empire, as aircraft bearing the Stars and Stripes transported U.S. capital, consumer goods, and culture—not to mention military troops and weaponry—around the world.[80]

In the jet age, though, commercial aviation would also reveal unexpected limits and challenges to American globalism. Just as jets themselves did not fly as smoothly as airline advertisements claimed, the U.S. aviation industry experienced increasing turbulence during the 1960s, as did the nation as a whole. While jet-age culture conjured fantasies of unlimited power and prosperity for the United States, the trajectory of the American Century would not inevitably continue onward and upward.

7

The Jet Age
and the Limits of
American Power

On October 26, 1958, Pan American World Airways Flight 114 departed from New York's Idlewild Airport en route to Paris, inaugurating daily jet service between the United States and Europe. The Boeing 707 arrived in Paris after seven hours and thirty-four minutes, compared with eleven hours for its piston-engine predecessors. Aboard the flight, 111 "jet-age pioneers" (as a Pan Am press release called the passengers) ranged in age from eight to eighty and came from diverse walks of life. Karl Johanson, a seventeen-year-old high school senior from Valhalla, New York, had saved money for five years to purchase a ticket on the inaugural jet flight, but his parents gave him the trip as a graduation present. Noble Hopkins, his wife, and his four children, who lived on a farm in Garrettsville, Ohio, had never been on a train or a plane prior to winning their tickets in a contest cosponsored by Pan Am and Kellogg's Corn Flakes. Elderly Mrs. William Eck of Arlington, Virginia, went in place of her husband, who had passed away before he could use his ticket. Billie Miller, a New York secretary, said that she booked passage on the flight because "I believe that there should be aboard an ordinary workaday person." The flight's more elite passengers included Mrs. Clive Runnels, a member of the

Republican National Finance Committee who planned to lunch in Paris and then return to New York; Talbert Abrams, a former Army Air Corps pilot and chairman of the Abrams Aerial Survey Corporation; Joseph Goodhue, a United Aircraft executive; and the actress Greer Garson, vacationing with her husband, oilman E. E. Fogelson. The forty passengers in first class paid $900 for the round-trip and enjoyed a gourmet dinner with wine and brandy; the seventy-one economy-class passengers paid $489.60 and ate sandwiches.[1]

Pan Am president Juan T. Trippe described the jet inaugural, as he had described previous "firsts" in his company's history, as a patriotic achievement—the latest, greatest expression of America's pioneering spirit and technological ingenuity. At the *Clipper America*'s christening ceremony, in which First Lady Mamie Eisenhower broke a bottle of water from the seven seas over the aircraft's nose, Trippe hailed the jet as "a witness to the American tradition, and a triumph of that tradition." The 707, he said, represented "the clearest and simplest characteristics of all things American"—speed, size, efficiency, and imagination. According to a Pan Am public relations memo, the 707 embodied "Yankee traits of resourcefulness and perseverance," which, the company claimed, had produced an uninterrupted legacy of technological progress from pioneer wagons to nineteenth-century merchant clipper ships to Jet Clippers. Nationalist sentiments also pervaded media coverage of the jet inaugural. The *New York Herald Tribune*, in an editorial typical of many, stated that the jet age had crowned American aviation "the unchallenged queen of the ocean air."[2]

The advent of passenger jets reanimated aviation's cultural significance as a symbol of American national greatness and global power. Jets debuted, moreover, at precisely the moment when developments in the Cold War seemed to threaten the United States' technological preponderance. On October 4, 1957, the Soviet Union launched Sputnik, the world's first artificial satellite, prompting widespread anxieties that the United States had fallen behind in the Cold War air and space race. In response to what President Eisenhower labeled the "Sputnik Crisis," Congress passed the National Defense Education Act, which significantly increased federal funding for all levels of scientific education. In this context, achievements in jet aviation reassured the American public that their country had not lost its technological edge. Just two years after the *Clipper America*'s debut, the United States was producing and flying more passenger jets than any nation in the world.

Yet American supremacy in the jet age was neither preordained nor uncontested. During the early 1950s, U.S. aircraft manufacturers focused on producing military jets, and European firms consequently took the lead in commercial jet aviation. Indeed, Pan Am's *Clipper America* was not the first passenger jet to traverse the Atlantic; that honor went to a British Overseas Airways Corporation (BOAC) Comet IV, which flew from London to New York on October 4, 1958—exactly a year after the launch of Sputnik. BOAC had introduced domestic jet service in the spring of 1952, followed by Air France in 1953 and Soviet national carrier Aeroflot in 1955. Boeing's 707, meanwhile, was not ready for delivery until 1954, prompting newspaper editorials and trade magazines to decry "the sad position of this country with respect to jet transports," as the *Washington Post* lamented as early as 1950.[3]

Between 1958 and 1968, the jet age seemed to indicate the triumphant ascendance of the "American Century," yet ultimately revealed its limits, contradictions, and unintended consequences. Jets instantiated the visions of commercial empire builders such as Trippe and his friend Henry Luce—visions of a world with no distant places, in which U.S. power and influence could expand infinitely. However, the jet age also registered American globalism's internal contradictions and external challenges. Jets were icons and instruments of U.S. global power, but they impelled transnational phenomena that challenged notions of American exceptionalism. Discourse on jets glorified size, speed, abundance, and access but also revealed anxieties about the personal and social costs of jet-setting. And by the end of the 1960s, the troubled state of the American commercial aviation industry suggested that the American Century's onward-and-upward momentum could not be sustained indefinitely.

The Jet Age Takes Off

In 1950, André Priester, Pan Am's chief engineer, sent Christmas cards featuring an illustration of Boston's Old North Church, where Paul Revere's lanterns had signaled the arrival of British troops during the

American Revolution. The card's message read: "The British are coming! One if by land, two if by sea, three if by air." Priester's warning proved accurate. On May 2, 1952, BOAC inaugurated the world's first commercial jet flight from London to Johannesburg. Flying at over 400 miles per hour, BOAC's de Havilland Comet arrived in South Africa in half the time that the trip had previously required. The jet also transformed the experience of air travel. Cruising at 20,000 feet, it soared above the atmospheric turbulence that rocked and rattled piston-engine planes. According to a writer for *Fortune*, jet flight felt "more like hanging still in the air, except for gentle air bumps now and then. It is a new kind of travel." Shortly after BOAC's inaugural flight, Pan Am placed an order for three Comets, scheduled for delivery in 1956, with an option on seven more. Within a year, two other British firms, Vickers and Britannia, were producing their own passenger jets. "Every day, the British lead in commercial jet planes seems to be getting longer and longer," observed *Business Week*.[4]

Britain, along with Germany, had pioneered jet aviation during the interwar years. British engineer Frank Whittle patented a turbojet engine in 1930; six years later, German physicist Hans von Ohain patented a similar, independently invented design. On August 27, 1939—four days before Hitler's armies invaded Poland—a *Luftwaffe* Heinkel He 178, outfitted with Ohain's engine, became the first aircraft to fly purely on turbojet power. The pilot informed Ohain that the flight had been "perfect," lacking turbulence, noise, and vibration.[5]

World War II prompted further advances in jet aviation, with Britain and Germany competing to design and manufacture turbojet-powered fighters and bombers. Drawing on British and German designs, the United States produced two types of jet fighters during the war—the P-59 Airacomet, designed in 1942 by Bell Aircraft Corporation with an engine by General Electric, and Lockheed's P-80 Shooting Star, designed in 1943—but neither was deemed combat-worthy until after the war ended. The first American jet bomber, Boeing's B-47, did not debut until 1947, and its distinctive design feature—drag-reducing swept-back wings—derived from designs created by German engineers during the 1930s.[6]

With the onset of the Cold War, however, the Truman administration authorized unprecedented increases in peacetime military spending that

enabled the United States to quickly become the world's leading innova-
tor in jet aviation. The fastest jet bomber in existence, able to fly at 610
miles per hour, the B-47 confirmed Boeing's stature as the world's leading
designer of jet bombers. In 1952, a year after the B-47 entered service,
Boeing produced an even more impressive successor: the B-52 Stratofor-
tress, an eight-engine behemoth that would serve as the flagship of the
U.S. Air Force through the 1980s.[7]

As U.S. manufacturers concentrated on fulfilling their military contracts, they did not begin developing jets for commercial use until the
mid-1950s. The Air Force did not actively discourage manufacturers from
producing jet transports, yet it nonetheless insisted that the production of
civilian aircraft must not interfere with "critically important" military
manufacturing. Airline leaders, meanwhile, viewed the technology as
simply too expensive: the price tag of a jet airliner would be at least $4
million, compared with $1.5 million for the propeller-driven DC-7, the
mainstay of commercial fleets during the early 1950s. As a result of its
decision to delay jet transport development, the American aviation industry came under sharp criticism in the media. When BOAC began flying
Comets on domestic routes in 1952, editorials in the *New York Times* and
the *Washington Post* lambasted "a significant lag in American aviation
development," while *Newsweek* warned that "American commercial-
transport manufacturers can no longer ignore the threat: they are in danger of losing the lion's share of the lucrative international market." Civil
Aeronautics Board chairman Donald W. Nyrop concurred: in testimony
before Congress in the spring of 1952, he urged the U.S. government to
fund three-quarters of the costs of developing a jet transport.[8]

Exacerbating such concerns, the Soviet Union had also begun to make
rapid strides in the development of jet aviation, both military and commercial. At Moscow's 1954 May Day parade, the star attraction was the
Myasishchev M-4, a four-engine, long-range jet bomber comparable to the
B-52. This "major and fairly chilling Soviet achievement" (as the *Washington Post* described it) was initially believed, though falsely, to enable
Moscow to order atomic air strikes on North American territory. A year
later, the Soviets unveiled the Tupolev Tu-104, a turbojet transport plane
resembling the Comet, which Aeroflot debuted on a flight from Moscow
to London in March 1956. The American media dismissed the Soviet jet

as an "uneconomic fuel-gobbler which only a government could afford to operate, for show purposes"; trade journal *Aviation Week* derided the Tu-104's dowdy cabin décor, which included damask-backed seats, tray tables with crocheted doilies, and baggage racks made of ship netting. Nonetheless, as *Newsweek* stated, the Tu-104 had clearly "delivered more than a slight jolt to any notions of Anglo-American supremacy in commercial jets." American observers worried, moreover, that the Tu-104 was "only the opening gun in Moscow's plans for expanding commercial air service to cities around the world"—plans that included the extension of Aeroflot's routes to nonaligned countries in Asia and the Middle East, where the Soviets "obviously hoped to score a propaganda victory." By 1957, fifty Tu-104s were in service on Aeroflot routes, while U.S. airlines had yet to operate a single jet.[9]

Unbeknownst to the American public, however, Boeing president William M. Allen was developing a plan to launch U.S. commercial aviation into the jet age. Boeing had produced some of the greatest airliners of the 1930s, including Pan Am's signature B-314 "Clippers," but since World War II it had concentrated its resources on developing bombers for the Air Force. Boeing's first postwar transport plane, the B-377 Stratocruiser (an adaptation of the B-29 bomber), proved too large and too expensive for most airlines to operate, resulting in a loss of $15 million. Meanwhile, rival firms Douglas and Lockheed had become the nation's leading manufacturers of commercial aircraft; by 1954, Douglas planes were flying nearly half of all U.S. passenger miles. The successful development of a jet transport, Allen believed, would enable Boeing to surpass its competitors in the commercial market. Between 1952 and 1954, Allen invested $16 million, a quarter of his company's net worth, in "Project X," whose goal was to produce a jet prototype that could serve as a commercial airliner, a military transport, or a refueling tanker.[10]

On May 15, 1954—two weeks after the Soviet Union debuted its Myasishchev M-4 bomber—the result of Project X, the 707 "Dash 80" jet prototype, rolled out of Boeing's football-field-sized factory in Renton, Washington. A brass band played the Air Force anthem, while a crowd of eight thousand Boeing employees and five hundred community leaders cheered and waved American flags. In light of recent headlines about Soviet aviation, the 707 must have reassured the crowd that the United States had not lost its edge in aeronautical innovation. Befitting such patriotic sentiments, the jet prototype's creation had been a truly national endeavor, requiring bolts from El Monte, California, screws from Rockford, Illinois, nylon plastics from Garwood, New Jersey, glass from Pittsburgh, and control cables from Indianapolis. The 707 also embodied the American obsession with size and speed. Its length exceeded that of the Wright brothers' first flight, it could carry 100 to 150 passengers (twice as many as the first Comet), and, with cruising speeds of up to 500 miles per hour, it was the fastest commercial airplane in existence. "Breakfast in London, a mid-morning cup of coffee in New York, lunch in San Francisco and a pre-dinner swim at Waikiki Beach—that's an example of what the Boeing 707 Jet Stratoliner will mean when it goes into commercial operation," promised a company press release. Yet the 707 promised to serve military needs as well as tourist travel. After Strategic Air Command leader General Curtis E. LeMay witnessed its first flight, the Air Force awarded Boeing a contract to build twenty-one refueling tankers modeled on the 707 prototype.[11]

Just months after the 707 rollout, Douglas Aircraft announced that it was building its own passenger jet—the DC-8, which looked nearly identical to the 707 but carried a slightly lower price tag ($5.3 million, compared with $5.6 million for the Boeing jet). In October 1955, Pan Am became the first U.S. airline to invest in jets, purchasing twenty 707s and twenty-five DC-8s at a total cost of $269 million—the most expensive shopping spree in airline history. By October 1958, when Pan Am inaugurated passenger service on the 707, U.S. airlines had invested nearly $3 billion in jets.[12]

Taking to the air a year ahead of the DC-8, the 707 not only crowned Boeing as the leading U.S. manufacturer of commercial aircraft; it also made the United States the world's leading manufacturer of jet transports. During their first year of service, 707s transported over a million passengers, winning rave reviews from the cockpit and the cabin alike. Pilots remarked on the jet's ease of operation; passengers marveled at its relative lack of noise and cabin vibration. By 1960, Boeing jets were flying 54.5 percent of air traffic within the United States and 48.5 percent within the noncommunist world, carrying some 32,000 passengers per day on routes that connected 112 cities in fifty-nine countries. The 707's distinctive swept-back wings and hanging engines became ubiquitous sights at airports around the world, offering striking visual proof that the United States, in spite of its belated start, had decisively captured the lead in the international race to develop jet transportation.[13]

America's leadership in the jet age, however, did not derive from some kind of technological manifest destiny, as Juan Trippe liked to suggest. Rather, it resulted from a variety of historically contingent factors—most important, the Cold War military appropriations that underwrote the costs of aircraft development and production. Unforeseen tragedy also played a role. In March 1953, a BOAC Comet crashed shortly before takeoff, killing all of its passengers and crew. After three additional crashes within a year, the British government grounded the entire Comet fleet. Investigations revealed that the accidents had been caused by metal fatigue, and by the time that de Havilland corrected the problem in later models of the Comet, Boeing and Douglas had already come to dominate the world's commercial transport market.[14]

American manufacturers also proved adept at reproducing and improving on the designs of foreign competitors. In 1959, Air France began flying Sud Aviation's Caravelle, the first jet designed specifically for short-haul flights. The Caravelle was faster and cheaper to operate than the 707, and its innovative design—with the engines located in the tail and on either side of the rear fuselage, instead of suspended below the wings as on the 707—significantly reduced cabin noise. The Caravelle debuted to widespread acclaim in Europe, and in 1960 United Airlines placed an order for twenty of the French jets. Boeing subsequently incorporated elements of the Caravelle's design into its own short-haul jet, the 727, which debuted in 1963 and quickly eclipsed Sud Aviation's model.

Flown by some sixty airlines, the Boeing 727 became the signature passenger plane of the midcentury jet age. Once again, the productive capacities of American manufacturing proved decisive. Nearly two thousand 727s had been produced by the time that Boeing discontinued the line in 1984, compared with only 282 Caravelles. With the introduction of the 727, writes aviation historian Carl Solberg, "everything went the American way . . . it was jets—American jets—all the way on the world's air routes." The 727's only true competitor was its American cousin, Douglas's slightly smaller DC-9, first manufactured in 1965.[15]

Even as the jet age confirmed the global dominance of U.S. commercial aviation, however, it also created new economic and logistical challenges. The airline industry took a significant financial risk by investing in jet technology, as indicated by news articles with titles such as "The Great Jet Gamble" *(Fortune)* and "Gamble in the Sky" *(Time)*. On top of the $2.8 billion that the airlines had expended in acquiring jets, another $220 million would be required to modernize ground facilities and equipment. Airports needed to strengthen and lengthen runways (at the cost of $1 million per 1,000 feet), update air traffic control systems, enlarge passenger terminals, expedite ticketing and baggage procedures, and expand maintenance facilities. The total cost of the jet transition was expected to exceed $3 billion. And because jets had twice as many seats as their piston-engine predecessors, such costs could be recouped only by massive increases in ticket sales. According to trade magazine *Aviation Week*, the jet could be "the biggest money maker the airlines have ever seen," or it could "produce economic disaster."[16]

Increasing ticket sales, moreover, would not be an easy task. As of 1958, two-thirds of Americans had never flown on a commercial airplane. Most alarming to the aviation industry, the jet age seemed to exacerbate fears of flying. In a 1957 survey conducted by the University of Michigan, just 33 percent of all respondents (and only 28 percent of female respondents) expressed positive feelings about jet travel, due overwhelmingly to concerns that the new planes were unsafe.[17]

The airline industry confronted such challenges by doubling its expenditures on advertising between 1959 and 1964. Pan Am's 1958–1959 Jet Clipper campaign exemplified the rhetorical and representational strategies that many airlines used to get people flying during the early jet age. Most notably, Pan Am's advertising aimed to assuage public anxieties about jet travel by incorporating the aircraft into familiar cultural narratives, particularly those involving gendered conceptions of the home and the family. Just as the airline's ads had long domesticated foreign destinations, the new campaign domesticated the jet itself, making flight at 600 miles per hour and 30,000 feet appear as safe and as comfortable as an evening in one's own living room.[18]

The Domestication of Jet Travel

In the summer of 1958, Pan Am commissioned its advertising agency, J. Walter Thompson (JWT), to conceive and implement "the greatest single advertising and sales promotional program the Company has ever mounted," at a cost of $1.5 million. Its objectives were, first, to prove to the public that jets were safe, and, second, to advertise Pan Am as "THE jet airline—the airline responsible for bringing the United States into the Jet Age." In late June, travel agents around the world received a confidential mailing of illustrated brochures, advance ad proofs, and order forms for full-color, 36-by-49-inch window displays (available in English, French, Italian, and German) to announce the inauguration of Jet Clipper service across the Atlantic. At Pan Am ticket offices in the United States and abroad, windows were curtained off to reveal only a sign that told passersby to "Watch This Space." On July 15, the curtains lifted, displaying cutaway models of the 707. That same day, full-page announcements appeared in newspapers in the United States, Europe, Africa, and the Middle East, and cocktail parties were held at Pan Am offices.[19]

During the next several months, Pan Am promoted the 707 with missionary zeal. The campaign targeted travel agencies and ticket sales offices, where first-time leisure travelers would purchase tickets. "It remains with us to educate the unenlightened; to indoctrinate the hesitant; and to prove to all the many advantages inherent in this latest medium of travel," wrote national sales manager John W. Ogilvie in a memo distributed to district sales managers. Consistent with this mission of indoctrination, Pan Am's district offices became evangelical shrines to the jet, their walls and windows filled with 707 dioramas, posters, and catchy, personalized brochures such as "Your Jet Clipper." Visual materials were especially important; colorful, three-dimensional window displays promised to attract the attention of "unenlightened" or "hesitant" passersby. To ensure that Pan Am's own staff would internalize the message, an Internal Communications Working Committee came up with the idea of printing pro-jet slogans on employees' paycheck stubs. The jet gospel also spread far beyond the United States' borders. An initial mailing of 75,000 folders of promotional materials went to forty-one foreign sales offices, including such Asian and African cities as Accra, Dakar, Leopoldville, Baghdad, Karachi, and Calcutta.[20]

Beyond its travel agencies and sales offices, Pan Am appealed to the public directly. During the summer of 1958, the ads created by JWT appeared in some three hundred publications, read by an estimated sixty-three million readers in over forty countries. According to Pan Am advertising manager Donald J. Dougherty, if two members of a family read each copy of a given magazine or newspaper, the ads would be seen by "158,819,025 potential passengers." Full-color, two-page spreads appeared in leading popular publications such as *Life* (circulation 6,000,000), *Time* (2,250,000), *National Geographic* (2,000,000), and the *New York Times Sunday Magazine* (1,227,000). The airline's publicity department also generated "editorial" copy for local newspapers, which public relations director H. B. Miller instructed sales managers to "personally" place in the hands of a features editor for use "under his own byline if he prefers." Subway posters, billboards (including 430 in foreign countries), a direct mail campaign, and commercials on *Meet the Press*, the popular Sunday morning television program, further disseminated the message. And Pan Am partnered with other companies to promote the Jet Clipper via boxes of Kellogg's Corn Flakes (for mailing in a box top, children would receive a plastic model of the plane), coloring books, and even Jantzen swimsuits.[21]

The jet publicity campaign effectively exploited Cold War ideologies of gender, domesticity, and the family. To reach the "unenlightened" and the "hesitant," the advertisements created by JWT identified jets with familiar social spaces: the office and the home. The pilots who occupied the "front office" of a Jet Clipper, stated a Pan Am press release, "have undergone extensive training in the most modern Jet techniques—in addition to their already impressive flying experience in propeller-driven planes." Describing the 707's cockpit as a "front office" likened airline pilots to corporate executives: portrayed as skilled, serious, and buttoned-down, they bore scant resemblance to the reckless, leather-clad flyboys of the early twentieth century. Images of uniformed captains—who tended to be white, male, and middle-aged, just like the CEOs who occupied corporations' real front offices—pervaded airline advertisements in the 1950s, in order to convince nervous passengers that they could confidently entrust their lives to these experienced, fatherly-looking professionals. One of Pan Am's best-known jet-age ads, "This Is the Captain Speaking," featured a pilot who informed readers that the Jet Clipper "is the civilian version of the Boeing KC-135—the *backbone* of the United States Air Force Strategic Air Command." After these ads appeared in magazines such as *Life*, JWT received numerous letters from readers who found the "captain" reassuring and persuasive; one reader claimed that the ad had "converted him from a non-flyer to a flyer." According to a JWT account manager, on a recent flight to Chicago, "the man sitting next to me told me in the course of our conversation that he thought it was a wonderful advertisement and instilled confidence in him in a very unique and re-freshing way."[22]

If the Jet Clipper's cockpit was run like an office, its passenger cabin was made to feel like home. The Boeing 707, according to a Pan Am press release, resembled a "huge living room . . . decorated in a pleasing, eye-resting color scheme of blue and gray pastels and equipped with chairs designed for complete relaxation." Advertisements extolled its "decorator-designed interior," "deep-cushion reclining seats—each with *its own* read-ing light and fresh-air control," and "handsomely appointed lavatories . . . you couldn't feel more at home in your own powder-room." Illustrated brochures such as "Jet Clippers Are Here," distributed free of charge at ticket sales offices and travel agencies, offered full-color proof that the jet resembled "a penthouse in the sky," flying so smoothly and quietly that passengers could play chess, read, or enjoy quality time with their chil-dren, while being waited on by attentive stewardesses (as the brochures' images depicted). When other U.S. airlines introduced jet service in 1959, they too advertised the plane as an airborne home. American Airlines, for example, assured passengers that jet travel would feel "like being carried from coast to coast in [your] own living room." Jets were so "soothing," American claimed, that babies were "usually lulled to sleep in no time."[23]

Boeing's own promotional materials described the 707 in similar terms. The experience of jet travel was "like flying ten miles a minute in my easy chair," claimed a passenger in a *Boeing Magazine* article entitled "Pent-house in the Sky." In fact, Boeing went to great lengths to make the 707 as safe, silent, and comfortable "as your living room at home." Ergonomics experts designed the jet's seats, audio technicians soundproofed its cabin, and dozens of other specialists customized its lavatories, lighting, carpeting, doorways, luggage racks, galleys, and ventilation systems. Accordingly, Boeing's advertisements for the 707 used domestic imagery to emphasize the aircraft's comfort and safety, depicting passengers in familiar, and familial, activities: a mother reads *Alice's Adventures in Wonderland* to her daughter; another cradles her peacefully sleeping baby; a young couple, possibly newlyweds, practice their French en route to Paris; an elderly woman knits (only to realize that the flight is so quick, there is "hardly time to start a sweater").[24]

Likening the 707 to the office and the home made jet travel seem safe, familiar, and all-American. In airline advertising, representations of domesticity assimilated new, potentially fear-inducing technologies into normative American life—in the same fashion as ABC's animated television series *The Jetsons* (originally airing in 1962), whose name reflected the jet's influence on popular culture. Set in the mid-twenty-first century, *The Jetsons* depicted a futuristic society in which people owned robots, inhabited modular houses, and commuted to work in flying cars. Gender roles in this future world, however, mirrored those of early 1960s America. George Jetson, the family breadwinner, worked in an office (even though technological conveniences had reduced the workweek to three three-hour days); his wife, Jane, stayed at home raising their two children. The 707, similarly, flew at death-defying speeds and altitudes, but its cabin resembled a cozy living room, with attractive female flight attendants serving passengers while fatherly male pilots helmed the "front office."

As part of its efforts to domesticate jet travel, Pan Am made special appeals to women, who by 1959 comprised half of all transatlantic travelers. According to *Business Week*, airline advertising should create "a public attitude that someday will induce a housewife to pick up a pair of tickets to Hawaii as casually as she shops for potato chips." Pan Am attempted to appeal to housewives by emphasizing that jet-age technology would simplify their household chores: the infrared ovens designed for the 707, for example, would not only treat passengers to a "cookout above the clouds" but also "might well put an end to many a domestic crisis." Able to cook sixteen steaks in seven minutes, or forty-two portions of scrambled eggs in just two minutes, the high-tech ovens promised to free women from their "'slaving-over-a-hot-stove' routine." Still, while the logic of such advertisements suggested that jet-age technology could make a woman's domestic work easier, they nonetheless implied that her place remained in the kitchen. Other elements of the jet publicity campaign, meanwhile, exploited women's concerns about their looks. In 1959, Pan Am published an international travel guide written expressly for women, "A Woman's Way to See Europe," which offered practical information on hotels, restaurants, and travel regulations—along with makeup tips. The speed of jets "eliminat[ed] the need for bulky cosmetic cases," the booklet stated, since makeup would require only "minimal" refreshing during the six-hour flight. But it also advised female passengers "to use a pinker shade of power," which apparently looked more flattering under the 707's nighttime cabin lighting.[25]

Targeting multiple constituencies and published in diverse forms of media, Pan Am's jet publicity campaign produced its intended effects. During their first six weeks of service in 1958, Jet Clippers transported 12,168 passengers between the United States and Europe, with 94.2 percent of available seats filled. When Pan Am introduced jets on routes between the United States' West Coast and Europe, advance bookings increased by 190 percent; bookings on flights to Hawaii and Tokyo rose by 106 percent. Other U.S. airlines reported similar results. American, for example, filled 90 percent of its seats during its first seven months of flying jets, compared with only 66 percent the previous year. In 1959, a total of fifty-one million U.S. citizens purchased tickets for domestic jet flights and seven million for international flights.[26]

Although U.S. airlines posted net financial losses during the first few years of the jet age, because of their expensive equipment purchases and upgrades, jets finally became profitable for the industry in 1962. That year, Pan Am reported that its gross revenues had reached an all-time high of a half billion dollars, due to "a substantial increase in business volume." The company continued to post record profits and break passenger traffic records in 1963, 1964, and 1965. Such gains were reported across the industry. The total number of airline passengers in the United States nearly quintupled between 1955 and 1972. By 1972, according to Gallup polls, half of all U.S. citizens had flown at least once.[27]

As Boeing and Douglas jets came to dominate world skyways, and as air travel became an increasingly normative aspect of middle-class American life, cultural discourse on jets reaffirmed long-standing nationalist narratives. The United States' victory in the international jet race became the latest evidence of the nation's ostensibly unique pioneer heritage, technological ingenuity, and traditions of free enterprise. According to Pan Am's brochure "Jet Clippers Are Here," the jet age represented "one of America's proudest moments."[28] Materially, jet aviation expanded the United States' military and economic power; culturally, the jet cathected fantasies of a limitless American imperium.

Imagining the American Jet Age

On July 24, 1959, Vice President Richard M. Nixon boarded a Boeing 707-321 Intercontinental, operated by Pan Am but owned by the U.S. government, and flew to Moscow to attend the opening of the American National Exhibition. Sponsored by the State Department, the event showcased American culture, technology, and consumer goods. At the exhibition, while standing in the kitchen of a model American home, Nixon and Soviet leader Nikita Khrushchev held a candid discussion that would subsequently be known as the "Kitchen Debate." Nixon conceded to Khrushchev that "there are some instances where you may be ahead of us, for example in the development of the thrust of your rockets for the investigation of outer space." But, he added, "there may be some instances in which we are ahead of you—in color television, for instance."[29] In Nixon's view, such consumer goods could be as important as rockets in determining the outcome of the Cold War contest.

Nixon's method of travel to Moscow exemplified his arguments in the Kitchen Debate. Although not formally part of the American National Exhibition, the vice president's Boeing 707 Intercontinental—the latest, largest, and fastest model of the jet—was nonetheless a dramatic display of the United States' post-Sputnik technological progress. Along with color televisions, Pepsi-Cola, and Polaroid cameras, the State Department dispatched the 707 to Moscow "to show off all the dazzling creature comforts with which this nation is blessed," as Hearst reporter Bob Considine put it. The jet remained in Moscow for several days, and some four thousand Soviet citizens toured its cabin. According to Pan Am's crew, the Soviet visitors "were particularly interested in the special conveniences of the airplane, such as the softness of the cushions and the completeness of the galley," features that Aeroflot planes conspicuously lacked. Two weeks after Nixon's visit, a second Pan Am–operated 707 Intercontinental, carrying a group of American reporters, flew from New York to Moscow in the record time of eight hours, fifty-four minutes—nearly an hour faster than the Tu-114 that had brought Soviet Deputy Premier Frol R. Kozlov to the United States the week before.[30]

In the context of the Cold War, the United States' victory in the international jet race carried powerful ideological meanings. Achievements in jet transportation provided evidence that the United States had not fallen behind the Soviet Union in the air and space race. "In these days of Sputnik and Lunik, we can all be proud that a great new jet airliner, the largest and fastest in the world, built in the United States by American workmen, owned by an American company operating as a private enterprise, in charge of an American captain and flying the Stars and Stripes, will be first to provide jet service around the world," proclaimed Juan Trippe in a 1959 speech inaugurating his airline's round-the-world jet service.[31] Jets seemed to confirm nationalist narratives about the American aptitude for technological innovation—and they quite literally displayed this image of the United States to the world. In an era when the public flocked to airports to see and tour the latest planes, jets staged the nation, just as it had been staged at the exhibition in Moscow. Compared with the austere Tupolev jets flown by Aeroflot, U.S. airlines' luxuriously appointed Boeing 707s and DC-9s flaunted the material abundance of American society. Their ever-increasing size and speed, meanwhile, signified the nation's military and commercial power.

Jets not only projected the image of U.S. global power but also maintained and expanded its material infrastructure. "The JET AGE—and America's leadership in it—will be important not only to you as a traveler, but to our country," stated an American Airlines passenger brochure. It also emphasized that commercial jets "represent a ready force in case of national emergency" as part of the Civil Reserve Air Fleet (CRAF), created in 1952. Through membership in CRAF, U.S. airlines pledged to make their aircraft and equipment available to the U.S. military in the event of a national or international crisis, such as the recent Soviet blockade of Berlin, which had prompted the Berlin airlift. A similar organization, the Military Air Transport Service (MATS), committed commercial airlines to ferrying military personnel and cargo as part of their regular flight schedules. Pan Am, for example, was by 1965 making forty flights per week to Saigon in support of the United States' escalating war in Vietnam. With MATS functioning as a kind of shuttle service for the Cold War national security state, the airlines flew soldiers and their families between the United States and its hundreds of overseas bases. Commercial jets regularly landed at bases including Dover, McGuide, Torrejon, and Wheelus, although such destinations never appeared on official airline timetables.[32]

Yet even as the commercial airline industry contributed to military activities, jet-age culture reproduced long-standing cultural narratives about the inherently peaceful nature of American democracy. Films, television programs, and magazines such as *Life* tended to focus on commercial rather than military aviation. This emphasis resulted, in part, from Air Force security restrictions. It also, however, reflected the ideological symbolism of jetliners—which, unlike bombers, signified not militarism or imperialism but the more benevolent activities of trade and tourism that contributed both to national prosperity and global integration. In the jet age as in previous eras, the commercial airplane's symbolic currency derived from its ability both to expand American global influence and to sustain Americans' denials of imperial ambition, to assist the fighting of wars while publicly promoting an image of peace, to project power seemingly without force. The vision of globe-girdling airways peacefully facilitating international exchanges of people and products inspired a vision of empire without imperialism—an image of global hegemony minus the unseemly activities of conquest, war, and domination.

Commerce, as always, was central to this vision. The United States had ascended to global power as a "market empire" (to use historian Victoria de Grazia's term), and jets enabled American capital, consumer goods, and corporate executives to circumnavigate the world. "One of the 20th century's greatest romances is between the businessman and the jet," observed *Time* in 1964. Jets expedited global commerce not only by transporting people and products but also by increasing the efficiency of business transactions. As airports became small cities in their own right— exurban enclaves with hotels and chain restaurants—business travelers could fly in, conduct a meeting, and fly home within a single day. By the 1970s, air travel had become as essential as stock tickers to the accumulation and globalization of capital. The airline industry further encouraged American business transactions via such initiatives as Pan Am's World Wide Marketing Service, created in 1962. Formalizing the airline's long-standing practice of advising U.S. overseas corporations, the division offered information on "market conditions and commodity needs" in the 114 cities served by Pan Am, sponsored international trade workshops, published a monthly magazine in five languages (English, Spanish, French, German, and Japanese), and produced a series of booklets such as *Strategies for Penetrating Markets of Eastern Europe and Russia* and *Business Customs in Japan and Hong Kong.* By 1965, Pan Am reported, over 107,000 "buyers and suppliers" had used its World Wide Marketing Service.[33]

As jets expedited the worldwide diffusion of U.S. capital, citizens, and products, the airline industry also invested in the creation of American enclaves abroad. The Intercontinental Hotel Corporation (IHC), a Pan Am subsidiary created in 1946, operated luxury hotels throughout the airline's route network, including establishments in cities in the developing world such as Karachi, New Delhi, Abidjan, and Singapore. In 1967, TWA also entered the hotel business when it acquired Hilton International. Together, jets and hotels ensured that tourists and business travelers would have access to other nations' markets, treasures, and pleasures while remaining insulated from less palatable foreign realities. "Air-Conditioned Luxury Takes Mystery out of Mysterious East" promised a 1965 issue of Pan Am's *Clipper* newspaper. Inside the plush cabin of a Boeing 707 or a posh Hilton dining room, well-off Americans could enjoy the cuisine, services, and high-tech comforts to which they were accustomed. Jets and hotels were therefore as much national prophylactics as agents of globalization—as a 1962 TWA advertisement suggested. Captioned "Coming home from overseas—to get the U.S. kind of care," the ad informed readers that all TWA jets, after returning "from the distant skies of Asia, Africa, and Europe," flew to the airline's overhaul base in Kansas City—in the middle of the American heartland—to be "tested and tuned." Passengers could feel

assured that "TWA maintains your jet this meticulous U.S. way," upholding national standards that might not be available under "distant skies."[34]

As the jet expanded the United States' market empire, it also inspired a
distinct visual style—a look—for U.S. global power. In aircraft design
and airport architecture, the aesthetics of the American jet age displayed
national confidence on a grand scale. Living-room-like aircraft cabins
aimed to reassure nervous passengers but also projected a national fantasy
of consumer abundance. Indeed, this was precisely the self-image that the
United States desired to project internationally during the Cold War. As
Nixon had argued in the Kitchen Debate, the United States could beat
the Soviet Union with refrigerators rather than rockets, and U.S. airliners
likewise displayed affluence and comfort as Cold War assets. "Except for
the seating arrangement, I might have been entering a swanky night
club," wrote a journalist who chronicled a flight to Europe for the *Saturday Evening Post*. "The purser, dressed in white dinner jacket and cum-
merbund, stood by the door like a headwaiter. Soft music from special
tapes flooded into the compartment through the loudspeaker system. The
décor was strikingly modern: pastel grays and blues against white, with
soft, indirect lighting."[35] As airline advertisements emphasized, such luxuries were not limited to passengers in first class; thanks to tourist-class
fares, middle-class American families could also increasingly afford to fly
in style. Jets thus represented the United States as simultaneously affluent
and democratic, powerful yet egalitarian. In the context of the Cold War,
jets visibly projected the benefits of "free world" capitalist democracy.

Nowhere was the jet age more spectacularly displayed than at U.S. airports. Nationwide, the era prompted a boom in airport construction. The
size, speed, noise, and passenger capacity of jets necessitated either radically redesigning airports or building new ones. Furthermore, as jets
diminished the sensory thrill of flight (passengers had little sensation of
speed and often could not see the earth), aviation leaders looked to airports to generate public excitement about air travel. TWA president Ralph
Damon, for example, envisioned an airport "that starts your flight with
your first glimpse of it and increases your anticipation after you arrive"—
and that is exactly what TWA received in its new dedicated terminal at
New York City's Idlewild Airport (now JFK), designed by Eero Saarinen
and opened in May 1962.[36]

The Finnish-born architect had been a passenger on a test flight of the Boeing 707, and he keenly understood the symbolic potency of aviation. Saarinen also believed that in the context of the Cold War, American architecture needed to express power on an epic scale. "Our architecture is too humble. It should be prouder, much richer and larger than we see it today," he stated. Saarinen's design for TWA's Idlewild terminal magnificently synthesized his ideas about aviation and U.S. global power. Saarinen placed ideological value on abstraction; streamlined form, he believed, signified freedom, which he understood as a universal aspiration but a distinctly American achievement. Like the American National Exhibition in Moscow or the international exhibitions of abstract expressionist paintings sponsored by the State Department during the Cold War, Saarinen's TWA terminal staged a specific vision of U.S. democracy for travelers from the world over. Its exterior resembled a giant, swooping eagle; inside, curvilinear spaces encouraged flow and free movement. Passengers boarded their flights after walking down long, windowless tunnels, such that emerging onto the aircraft felt like a kind of rebirth. The overall effect epitomized the onward-and-upward ethos of the American Century. "Everything about Eero's TWA Terminal says . . . it is going to be wonderful up there," opined Yale architectural historian Vincent Scully—and wonderful out there, for those arriving into the United States through the terminal.[37]

Saarinen's design for Dulles International Airport, which he described as the "best thing I have ever done," applied this idiom on an even grander scale. In 1958, President Eisenhower had selected Chantilly, Virginia, an expanse of farmland twenty-five miles west of Washington, DC, as the site for the United States' first airport designed specifically for jets. Unlike Idlewild, where each airline had its own terminal, Dulles would have just one terminal—a single gateway to the nation's capital. And Saarinen's structure, built over four years at a cost of $30 million, was quite a gateway. A three-story cathedral of glass, aluminum, and stainless steel that spanned the equivalent of eleven city blocks, the terminal translated aerodynamic principles into built form. An upwardly slanting roof, suspended on cables, suggested takeoff; curved glass walls and an open, light-filled interior conveyed an exhilarating sense of free movement. To reduce congestion and to maximize the experience of movement, 54-by-16-foot "mobile lounges" transported passengers directly from terminal to jet. And as with his TWA terminal, Saarinen designed Dulles not simply to embody the jet age but to embody the *American* jet age. Constructed in white limestone-aggregate concrete, the building harmonized with the neoclassical federal architecture of Washington, DC, creating a striking visual analogy between the airport and the monuments of the National Mall. Indeed, Dulles itself was a "superb monument to our time," wrote *New York Times* architecture critic Ada Louise Huxtable—a monument that represented the United States to the world, displaying its power, confidence, and soaring ambition. Not by coincidence did the nation's flagship airport bear the name of John Foster Dulles, Eisenhower's globe-trotting secretary of state.[38]

The image of national power expressed in jet-age airport and aircraft design also resonated with fantasies of personal power. Passengers' descriptions of jet travel consistently emphasized sensations of mastery and grandeur: flying aboard a jet is "akin to striding a cloud," opined a writer for the *New York World and Sun*. Jet-age aesthetics, moreover, had a distinctive gendered logic that visually and discursively linked the conquest of aerial frontiers to the conquest of women's bodies. The conflation of these two types of conquest was hardly new; during World War II, Air Force pilots had named their planes after women and painted images of curvy pinups on their hulls. During the jet age, similarly, the U.S. aviation industry marketed air travel by juxtaposing women's bodies with aircraft bodies. In airline ads, corporate publications, and mass-circulation magazines, flight attendants and female passengers posed flirtatiously in or next to aircraft. On a cover of *Boeing Magazine* from 1952, a woman wearing heels and form-fitting shorts posed in front of the B-52 bomber; on the cover of a 1954 issue, a woman in a low-cut blouse leaned forward, preparing to unveil a model jet draped in fabric. The caption read, "You'll See More in '54." The September 1963 issue of Pan Am's *Overseas Clipper* juxtaposed two streamlined bodies: that of the airline's new Dassault ten-passenger jet, shown soaring over a mountain range, and that of Tunisian-born movie star Claudia Cardinale, shown wearing an above-knee skirt with her bare legs crossed and pointed toward the camera. In another issue of the publication, JFK Airport employee Inge Wolf perched, in the same leggy pose as Cardinale, atop a stack of Firestone tires. A related article explained that the tires would be transported by Pan Am jet freighter to Pakistan for use in a U.S.-funded dam project.[39]

These visual juxtapositions of women and aircraft expressed fantasies of unlimited abundance and access: the jet made the world available to the United States just as the women on display made their bodies available to the viewer. Such images aestheticized, sexualized, and naturalized American globalism. Through Inge Wolf's implied offering of her body, U.S. development aid to Pakistan was figured as an act of love, not an expression of power. Claudia Cardinale and the Dassault jet—both, notably, of foreign origin—were rendered as willing and available objects of desire.

During the mid-1960s, airlines notoriously began to sexualize their brands and advertising campaigns. In 1965, Braniff International Airlines launched its "Air Strip" campaign, in which flight attendants performed an airborne striptease. As shown in television commercials set to burlesque-style music, Braniff's flight attendants wore demure skirt suits as they greeted arriving passengers. Then, during the course of the flight, they would "slip into something a little more comfortable" (as a male voiceover explained), removing layers of clothing to reveal increasingly skimpy outfits. Most notoriously, in 1971 National Airlines produced its "Fly Me" series of television ads and billboards, which featured flight attendants proclaiming such lines as "I'm Maggie! Fly me to New York! You'll love my two 747s to Kennedy. Fly me!" By the early 1970s, the airlines extended sexualized marketing strategies to vacation packages, such as "The Thrilling Threesome with TWA." Designed for unmarried travelers under the age of thirty-five, these trips were "perfectly suited to the new lifestyle for singles," as a TWA brochure explained. "Free-swinging and non-escorted," they included air transportation, accommodations, cocktail parties, free drinks at restaurants and nightspots, and even a "Bachelor Party Bonus Booklet," a city guide with coupons for more drink discounts. The "Thrilling Threesome" moniker referred to the trip's triad of destinations—Las Vegas, San Francisco, and a cruise from Los Angeles to Ensenada, Mexico—but the name, an obvious double entendre, also implicitly identified travel with sexual adventure.[40]

The "stewardess mystique" epitomized the jet-age conflation of technological and sexual conquest. Like jet-age airplanes and airports, the sexy stewardess icon conjured fantasies of abundance and access. Her carefully cultivated glamour—the product of a strict regime of age and weight requirements, cosmetics, and comportment—reinforced the jet's equation with optimism, progress, and national greatness. Indeed, the stewardess mystique exemplified the American dream that the aviation industry and the Cold War state alike wished to project. Magazines such as *Life* displayed "glamor girls of the air" (as a *Life* cover story described flight attendants in 1958) as national assets, embodiments of the postwar good life—while the U.S. media simultaneously described Aeroflot flight attendants as "husky, plain, and efficient," as the *Wall Street Journal* stated in 1972. Like the aesthetic differences between the sleek Boeing 707 and the damask-filled Tu-104, the visual contrasts between America's "glamor girls of the air" and the supposedly "husky and plain" Soviet flight attendants were said to reflect the distinctions between capitalism and communism. Jet-age displays of technology, sexuality, youth, and material plenty equated American capitalism with pleasure and desire. Flight attendants' glamour thus took on nationalist connotations as well as sexual ones.[41]

In the jet age, then, the United States appeared to be at the pinnacle of its power. Victory in the international jet race revived the nationalist vision of an American Century, and jets themselves propelled its twin strategies of military and commercial expansion. No less significantly, the jet inspired a look for the American imperium, a distinctive visual style displayed to the world in the sumptuous cabins of 707s, the soaring architecture of Idlewild and Dulles Airports, and the polished glamour of flight attendants. The jet promised to bring U.S. citizens the world, literally and figuratively. And cultural representations of jets connoted personal, national, and global power—unlimited abundance at home, unlimited access to markets and pleasures abroad.

Jet-age realities, however, were more complex than such representations suggested. By the late 1960s, the idea of the American Century faced increasing challenges both within and beyond the United States' borders. And while the jet age expanded the scope of U.S. global power, it also simultaneously underscored the limits of that power.

From Jet Set to Jet Lag

In its very cartography, the jet age yielded a picture of the world that implicitly problematized nationalism and American exceptionalism. Recalling American Airlines' "Air Map" advertisement from 1943, maps of jet

routes from the 1960s depicted a world without nations, a unified planetary space marked only by individual cities. Jet speeds, which brought the
world's major cities within twenty-four hours' reach of one another, made
visions of "one world" real. Nations did not become less significant, of
course; in fact, as a result of increasingly elaborate customs regulations
and border-crossing procedures, international air travelers encountered
more of the nation-state in the jet age than they did during the early 1950s,
when many countries did not require foreign visitors to present passports.
But the nation was symbolically erased from the global imaginary of jet-
age culture, which suggested alternative conceptualizations of space. The
globalization of capital and commerce gave rise to a geoeconomic vision
of the world, whose primary spatial units were cities (sites of commerce)
rather than nations (sites of state formation).

Airline route maps from the 1960s offered striking visual evidence of
the nation's diminishing cartographic legibility. In Pan Am's 1960 annual
report, a map showed the outlines of continents, although only cities were
identified by name. The map depicted air routes from Seattle to Tokyo,
Beirut to Bangkok; nowhere did it indicate, however, that such flights required travelers to pass through entities called the United States, Japan,
Lebanon, or Thailand. The nation had ceased to be a relevant spatial
category. By 1965, even the continents had begun to disappear. That
year's annual report featured a map that projected Pan Am's route network above an image of the earth. Against the background of outer space,
the curvilinear lines of jet routes resembled a constellation, whose "stars"
represented major cities. In the late 1960s, TWA printed maps on its
boarding cards and timetables that likewise depicted a world composed of
cities. In these images, though, the world could scarcely be recognized as
such. Dots representing cities extended horizontally on a white, planar
surface, making TWA's routes look like nothing so much as subway lines.
The earth was flat, according to the United States' two leading international airlines, long before journalist Thomas L. Friedman identified such
spatial "leveling" as the salient feature of globalization.[42]

The global imaginary of the jet age was therefore metropolitan and
geoeconomic, a picture of the world that privileged cities over nations,
commerce over government. International air travel was not yet as fast or
as simple as intracity subway travel, TWA's map notwithstanding, but jets

did transport people and products over vast distances while seeming to remain entirely within metropolitan cultural space. "The Jet Age passenger will eat breakfast in Tokyo and arrive in San Francisco three hours before the meal he just ate," declared a 1959 Pan Am press release, which indicated how jet travel not only disrupted traditional conceptions of time and space but appeared to dissolve the earthly distances between cities of departure and arrival. By its very nature, jet travel diminished passengers' awareness of the nations over which they flew. At 33,000 feet, clouds frequently obstructed views of the earth, and jets' relative lack of vibration and noise made passengers feel as if they were not moving at all. The experience of jet travel in some ways did resemble a very long, transnational subway ride, marked by metropolitan departures and arrivals with minimal awareness of space in between.[43]

The distinctive subjectivity of the jet age, like its spatial imaginary, largely eschewed the category of the nation. The jet-age individual fashioned him- or herself as a cosmopolitan, a world citizen whose habits, style, and persona expressed transnational class and cultural affinities rather than national citizenship. This technologically enhanced brand of cosmopolitanism inspired a new idiom, the "jet set," coined by society columnist Igor Cassini in the mid-1950s. The jet set, Cassini explained, "are people who fly away for weekends. They are the avant-garde, the pace-setters. The jet set is people who live fast, move fast, know the latest thing, and do the unusual and the unorthodox." Cassini's own life embodied his definition of the jet set. Born Igor Loiewski in Sevastopol, Russia, in 1915, Cassini grew up among privilege. His maternal grandfather, Count de Cassini, had been the czar's ambassador to the United States and a prominent figure in Washington high society. When Igor emigrated to the United States in 1937, he adopted his mother's maiden name and, through her social connections, landed a job writing gossip for the Washington *Times-Herald*. His caustic wit, eye for detail, and elite insider's perspective soon attracted the attention of editors at Hearst, who invited him to take over the popular "Cholly Knickerbocker" society column at the *New York Journal-American*. The original Cholly Knickerbocker was Maury Paul, who coined the term "café society" in the 1920s, and Cassini identified the jet set as café society's modern successor. "I used 'jet set' because it seemed appropriate in the age of the jet plane. . . . Café society is outmoded," he stated. Through the 1950s and 1960s, Cassini's column chronicled the exploits of royals, debutantes, millionaire businessmen, models, and fashion designers—including his own brother, Oleg Cassini, who acquired stardom by designing Jackie Kennedy's signature dresses—as they gambled in Monte Carlo, skied in St. Moritz, and danced until dawn at exclusive nightspots such as Manhattan's Le Club (which Cassini partly owned).[44]

The glamorous offspring of advanced technology and high society, the jet set became the object of widespread cultural fascination in the United States. Circulating well beyond the gossip pages, Cassini's idiom occasioned a profusion of media commentary. As a *Washington Post* columnist explained, jet-setters "are men and women in a hurry, who seek to get as much fun out of their nights as they can pay for or get anyone else to pay for." In a major *Sunday Magazine* feature from 1962, the *New York Times* defined the jet set as "a group of worry-proof people, their Louis Vuitton luggage filled to overbrimming with gay apparel like the Pucci dress." (Fittingly, Italian designer Emiliano Pucci—himself a famed member of the jet set—created the uniforms of Braniff flight attendants during the 1960s.) The *Times* further divided the jet set into three classes, distinguished by their wealth and travel habits: the "707 or long-range jet-setter" flew to Europe several times a year, "to say nothing of side trips to Barbados and such"; the "Caravelle or medium-range jet-setter" traveled abroad mostly on business; and the "short-haul jet-setter" rarely used his passport, circulating instead between Palm Beach, Aspen, and Southampton. Living perpetually in search of exotic cross-border adventures, the jet set made cosmopolitanism fashionable. Although its members numbered perhaps no more than a few thousand (according to the *Times*), the coterie's wealth and media visibility ensured that its tastes influenced a much broader public. Trading in a transnational currency of style, mobility, and cultural fluency, jet-setters crafted identities based on leisure instead of work. They did not much care which passport one carried; what mattered, rather, was where one's passport had been stamped.[45]

Nonetheless, the jet set tended to be more European than American, suggesting that the ascendance of the so-called American Century had never fully eclipsed the cultural capital of the "Old World." Regulars in Cassini's columns included Scandinavian royals, scions of Greek shipping fortunes, wealthy Eastern European exiles from communism, and the relatives of Latin American dictators. (Indeed, Cassini's friendship with Rafael Trujillo, the autocratic ruler of the Dominican Republic, prompted Robert Kennedy to authorize an FBI investigation in 1961.) Several of the era's most iconic jet-setters hailed, like the jet itself, from the United Kingdom—including the Beatles, who began the "British Invasion" by arriving at JFK Airport on February 7, 1964, aboard Pan Am Flight 101 from London, as well as James Bond, the globe-trotting spy of Ian Fleming's novels and Hollywood movies, who could be seen taking Pan Am jets in *Dr. No* (1962), *From Russia with Love* (1963), and *Live and Let Die* (1973).[46]

Jets themselves, meanwhile, promoted cosmopolitanism by expediting a rapid, worldwide diffusion of cultures and styles. The major musical trends of the 1950s and 1960s—surf, Latin, Beatlemania, and above all rock-and-roll—were creations of the air and airwaves, as radio, television, and jets elevated sounds across national borders. Fashion, too, benefited from jet transport, which sped the latest styles from designer's atelier to department stores and made possible the yearly circuit of international runway shows that soon became the lifeblood of the fashion industry. By 1964, Parisian couture houses Balmain, Lanvin, Chanel, and Givenchy all held accounts at Pan Am, and their creations were shipped around the world in sky-blue garment bags bearing the airline's logo.[47]

Other cosmopolitan jet-setters, however, did not wear Parisian couture or appear in Igor Cassini's columns. While the Monte Carlo crowd sipped champagne in first class, tourist-class cabins in the 1960s were increasingly filled with students, traveling to study at foreign universities. Junior-year abroad was a jet-age phenomenon: after the commencement of transatlantic jet flights, the number of Americans studying in Europe rose by 150 percent. By 1972, 59 percent of U.S. citizens aged eighteen to twenty-four had flown, compared with just 22 percent a decade earlier. The U.S. aviation industry helped underwrite this educational diaspora. Pan Am, which had sponsored student exchange programs within the Americas since the 1930s, funded numerous study-abroad fellowships and produced a series of international travel guides written specifically for students. Boeing, too, offered "People-to-People Scholarships" to coincide with the 707's debut in 1958. More than six hundred U.S. high-school seniors applied for the scholarships by writing an essay on "how jet-age transportation and communication can speed world peace." Sixteen contestants received money for college, and Boeing treated the top two winners to a ten-day tour of London, Paris, Bonn, Cologne, and the Brussels World's Fair. Thanks to the jet, wrote winner Polly Chase of Newburyport, Massachusetts, "the world is within our reach. For centuries men have fought wars because they misunderstood each other. Now we have an opportunity to meet, know and understand people from far-off nations." Among travelers like Chase, the jet age fostered a new cosmopolitan ethic. Perhaps not by coincidence, the first generation to study abroad in mass numbers was also the generation that led the multinational student rebellions of the late 1960s—which attacked, among other things, what many young people perceived as the narrow-minded provincialism and self-congratulatory patriotism of mainstream national cultures.[48]

Flying in the other direction, jets brought to the United States passengers and products that diversified the nation both culturally and demographically. Pan Am's 1948 introduction of tourist-class fares between New York and San Juan touched off an aerial migration that produced New York's Puerto Rican community. Other aerial migrations occurred in the early 1960s: TWA, for example, transported some 25,000 Basque sheepherders to the United States in a project sponsored by the U.S. Wool Growers Association, which believed that the Basques could improve U.S. wool cultivation. Paul Laxalt, a Republican governor of Nevada (1967–1971) and United States senator (1974–1987), was the son of one of these shepherds. In 1963, Pan Am introduced a special financing program for immigrants to the United States. Available in forty-five countries and based on the "Fly Now, Pay Later" vacation credit plan, it allowed immigrants to finance their travel in their country of origin and make repayment in the United States. They could then book passage on Pan Am flights with only a 10 percent down payment and three months to pay the first installment on the balance. The initial beneficiary of the plan was Antonia Iordanides, a twenty-three-year-old secretary from Athens.[49]

By giving rise to transnational cultures and cosmopolitan communities, the jet age did not simply problematize the concept of nationalism; it challenged the very centrality of the nation-state in the world system. As jets shrunk geographical distances and made possible connections across geopolitical boundaries, they ultimately exposed the artificiality of those boundaries. Nation-states continued to monopolize political legitimacy and military force. But by the late 1960s, they could exert less and less control over worldwide flows of technology, consumer goods, media images, and information.

Cultural and economic globalization even began to produce thaws in the Cold War. Direct flights between New York and Moscow, offered reciprocally by Pan Am and Aeroflot, began on July 15, 1968. The opening of an aerial bridge across the world's major geopolitical divide marked an important shift since 1944, when the Soviet Union had refused even to participate in the International Civil Aviation Conference held in Chicago. In the United States, the flights prompted concerns that commercial airliners "may bring Soviet agents to U.S.A. unsuspectingly," as the Jackson Heights Young Republican Club wrote in a letter to Pan Am. Similar warnings appeared in right-leaning publications such as the *New York Daily News* and the trade magazine *Aviation Week & Space Technology*, and anticommunist protesters held demonstrations in front of the Pan Am Building. More typically, however, the media celebrated the New York–Moscow flights as welcome evidence of aviation's continuing ability to promote international harmony. *Life* chose for the cover of its July 26, 1968, issue a photo of smiling Pan Am and Aeroflot flight attendants in a warm embrace.[50]

As jet-age cosmopolitanism challenged older forms of American exceptionalism and nationalism, jet travel also gave rise to anxieties about the social, economic, and even physiological fitness of the American body politic. Beginning in the mid-1960s, alarming reports on "jet fatigue" or "jet lag" appeared in news and popular science magazines, as well as in jet-set glossies such as *Vogue* and *Holiday*. The human body, it seemed, was not naturally suited for travel at jet speeds. As *Time* explained, "time-zone crossings foul up man's daily physiological cycles, the 'circadian rhythms' that are still one of nature's deepest mysteries." The Federal Aviation Administration commissioned several studies of jet travel's physiological effects, and scientists found that "jet lag" not only produced feelings of tiredness but also measurably impaired travelers' mental acuity. In one study, subjects were asked, before and after jet travel, to punch a telegraph key after seeing a light flash; after flying, their average reaction times doubled. Studies also found that flying increased heart rate, body temperature, and water retention.[51]

Although jet lag was far from life threatening, commentators worried about its effects on productivity, especially since those who frequently traveled by air included key business and government leaders. *Foreign Policy* described the problem thus: "The Great Man lopes across the tarmac, shoulders squared, tummy in, teeth revealed in a jaunty smile. . . . But settled in the limousine, away from the public view and photographers' lenses, the Great Man sags. And well he might. He has been traveling for many hours, and although it is high noon in Cashmania, it is only 3:00AM for him." The article's title, "Fly Now, Pay Later" (a sardonic play on Pan Am's vacation-credit slogan), suggested that jet-setters were paying for their travels with their bodies as well as their cash. And, *Foreign Policy* feared, the United States and the world as a whole might "pay" if leaders made important decisions when exhausted. In fact, Secretary of State John Foster Dulles, a constant international traveler, reportedly blamed travel fatigue for his disappointing performance in diplomatic negotiations over Egypt's Aswan Dam, which led to the Suez Canal Crisis of 1956—and this occurred two years before the introduction of jets made Dulles's overseas trips an even more frequent aspect of his job. Such anxieties about air travel's effects on individual bodies expressed deeper anxieties about the body politic. *Foreign Policy*'s image of the "Great Man" sagging in the back of his limousine personified a loss of potency on both an individual and a national scale.[52]

Thus, while the jet augmented the United States' global power, by the late 1960s it also seemed to jeopardize that power. The American economy, in particular, had begun to suffer from unforeseen consequences of the jet age. On New Year's Day 1968, President Lyndon B. Johnson delivered a televised address on the United States' international balance of payments, which he described as "a subject of vital concern to the economic health and well-being of this nation and the free world." Twenty years ago, the U.S. government had urged citizens to travel abroad, particularly to Europe, in order to stimulate war-ravaged foreign economies. Now Washington urged them to stay home. Due to government and consumer spending, Johnson explained, too many dollars were going to other countries; the United States had posted trade deficits for seventeen of the last eighteen years. Federal foreign aid programs, overseas military bases, and corporations' ballooning foreign investments were all at fault—but so, too, LBJ emphasized, were jet-setting U.S. citizens, whose expenditures abroad far exceeded those of visitors to the United States, resulting in a "travel deficit" of over $2 billion. Among other proposed solutions to the balance-of-payments problem, Johnson asked U.S. citizens "to defer for the next two years all non-essential travel outside of the Western Hemisphere," with the goal of reducing the travel deficit by $500 million. A month later, Treasury Secretary Henry Fowler proposed a tax of 5 percent on airline tickets for travel outside the Western Hemisphere. Overseas travelers' daily expenditures would also be taxed at a graduated rate: 15 percent for those spending between $7 and $15 per day, 30 percent for those spending more than $15 per day. Fowler's proposal, as the *New York Times* explained, would mean that a family that spent $950 during a thirty-day trip to Europe would pay $48 in federal taxes on those expenditures, plus $75 for the airplane-ticket tax.[53]

The balance-of-payments problem was not new. Since 1960, however, Washington had tried to resolve it by bringing more foreign tourists to the United States, not by restricting or taxing the travels of U.S. citizens. After President Eisenhower proclaimed 1960 as "Visit U.S.A. Year," U.S. airlines, hotels, and travel agencies offered special discounts to foreign tourists. The Kennedy administration formalized and expanded on this initiative by creating a government agency, the U.S. Travel Service. And in 1965, Johnson had begun his campaign against overseas travel by urging citizens to voluntarily "See America First"; he even wrote an article for *Parade* magazine extolling the nation's tourist attractions. But LBJ's New Year's Day speech, and especially Fowler's proposed taxes, upped the ante. Washington was now effectively telling citizens how and where they should (or should not) travel and proposing to punish those who did not comply.[54]

Not surprisingly, Johnson's speech met with vociferous opposition from the travel industry, as well as from the media and the public. "If the President can tell me where I cannot go, I can still use my vote to tell him where to go in November," as one man said to a *Los Angeles Times* reporter; the *Washington Post* called the plan "a threat to the freedom of

Americans." A prominent black journalist, meanwhile, pointed out that for African Americans, traveling within the United States involved considerable dangers and hardships. "We love America, and we'd like to see more of it, but we don't have time to go to the Supreme Court just to get into some broken-down motel. . . . *Really* seeing America first could get us arrested or even killed," wrote Ernest Dunbar, senior editor of *Look* and the author of *Black Expatriates: A Study of American Negroes in Exile* (1968). Conversely, Dunbar continued, "for those of us black Americans who can occasionally scrape together enough money to go, a trip abroad is not just a vacation, it's an urgent necessity. . . . Away from the pressures and inanities of the racial struggle in this country, we pull together our shredded psychic garments and gird ourselves for another round or two." To be sure, race-based economic inequalities prevented the vast majority of African Americans from being able to afford international air travel. The University of Michigan survey, for example, found that just 2 percent of black respondents had taken any kind of air trip (foreign or domestic) in 1962. However, for relatively elite African Americans such as Dunbar, international travel could offer a welcome escape from Jim Crow America.[55]

Airline leaders acknowledged that the balance-of-payments problem harmed the U.S. economy, but they urged voluntary solutions rather than punitive measures. TWA chief Charles C. Tillinghast, decrying Fowler's taxes as "discrimination" against lower-income travelers, proposed a more affordable exit fee of $10 to $15 for all international travelers, regardless of their destinations. Pan Am president Juan Trippe, meanwhile, proposed a "Fly American" campaign to reduce U.S. airlines' $580 million "fare deficit." Only half of the U.S. citizens who flew to Europe in 1967 had taken U.S. airlines, Trippe noted, whereas some 80 percent of Europeans patronized their own national flag carriers. Pan Am also sponsored a "See America, Sell American" program to attract foreign visitors to the United States, publishing such guidebooks as *The Pan Am Planning Guide to Travel in the U.S.A.*, available free of charge and in five languages at foreign sales offices. (The campaign included another guidebook for women, *A Woman's Way to See the U.S.A.*, the counterpart to *A Woman's Way to See Europe*.) By 1968, 80 percent of Pan Am's overseas advertising budget funded the promotion of travel to the United States—a major shift for an airline whose mission had always been getting Americans to go abroad. Nonetheless, in 1969 American tourists spent over a billion dollars in Europe, an all-time record.[56]

Even as Americans continued to travel abroad, discourse on the jet age began to reflect a growing critical awareness of the consequences of jet-setting. An October 1968 essay by Richard Johnson, the travel editor at *Esquire*, serves as a case in point. Johnson had been a passenger on Pan Am's inaugural jet flight, and for the anniversary of that event, he penned a dark meditation on the jet's first decade. "Although the jet age has cut travel time by almost one-half and opened up whole new areas of the world for the beautiful people as well as the tourist mob, all is not joy," the article began. Jet travel exhausted passengers, unlike the pre-jet-age flights that allowed them to enjoy "a drink or two" before a "leisurely meal." Contrary to airline ads, jets were far from luxurious, Johnson contended, especially compared with the modes of transportation they had rendered obsolete: the long-distance train and the ocean liner, "one of the last bastions of leisurely and elegant living." And foreshadowing later critics of tourism, Johnson argued that the jet age had spawned cultural homogenization, environmental degradation, and a "resort-industrial complex" that was "international in its ugliness." At tourist destinations everywhere could be seen "the same wanton disregard for natural beauty, the same proliferation of monstrous buildings bearing no relationship to their surroundings, the same emergence of hamburger joints and pizza parlors." Gone was "any remnant" of authentic nature or culture.[57]

Johnson's critique accurately identified some of the challenges that the airline industry faced by the late 1960s. As air travel became a mass activity, it had also become increasingly tedious and unpleasant. Passenger demand outstripped the capabilities of the nation's air traffic system, resulting in overcrowding and delays. "The jet age is not a movement. It's a social illusion," concluded *Los Angeles Times* columnist Art Seidenbaum in a piece entitled "Join the Jet Set—and Keep Running Aground." Not one of Seidenbaum's last eight flights had arrived on time, he complained; half had been delayed for over an hour. When Aeroflot made its first Moscow-to-New-York flight on July 16, 1968, its Ilyushin-62 jet had to circle over JFK Airport for ninety minutes before receiving clearance to land. At a press conference, a Soviet Ministry of Civil Aviation official sarcastically reported that Aeroflot's passengers did not find the fourteen-hour flight "at all fatiguing or exhausting if you exclude the one and a half hours necessary to circle over New York prior to landing." In coverage of the incident, U.S. newspapers noted that delays of this length were typical for flights arriving into New York City during the late afternoon. By 1969, airport congestion in the United States had become so endemic that the FAA commissioned a $35,000 study of proposals to build airports on landfill or on floating pylons in the ocean.[58]

By the late 1960s, the onward-and-upward optimism of the early jet age gave way to a growing awareness of its limits and unintended consequences. As the travel deficit worsened, as mass tourism contributed to cultural homogenization, and as jet-lagged passengers endured delayed flights and overcrowded airports, the costs of jet-setting—both literal and figurative—became ever more apparent. In their advertisements, Pan Am and other airlines would continue to associate flying with glamour and empowerment, but such representations increasingly seemed like defensive strategies designed to conceal less palatable realities. Indeed, by the early 1970s, the American airline industry was on the defensive, as an unprecedented combination of internal challenges and external dangers imperiled the very foundations of its ascendancy. The 1970s would ultimately prove that the United States' aerial empire could not expand infinitely.

Turbulence and Terror

Just as U.S. airlines had begun to pay off their 707s and 727s, in 1966 Boeing announced production of an even bigger, faster jet: the 747, a 550-passenger behemoth that was over twice the size of the 707 and, at $25 million, four times as expensive. In April 1966, Juan Trippe and Boeing president William Allen signed a $525 million contract for Pan Am to purchase twenty-five Boeing 747 "jumbo jets," which could reach speeds of Mach .9. Trippe upped the 747's price tag further by insisting on such amenities as a cocktail lounge and staterooms on the upper deck, which, recalling the Stratocruiser of the late 1940s, would connect to the main cabin by a spiral staircase. Before a single 747 was built—and before the FAA certified the plane as airworthy—Trippe ordered thirty-three of the jumbo jets at a total cost of more than $750 million.[59]

Pan Am leaders understood that filling the jumbo jets to capacity would not be easy, and in 1969 the company instructed JWT to conduct a year-long study of public attitudes toward the 747. After interviewing thousands of potential passengers in the United States, England, France, and Germany, JWT determined that "there just aren't enough people going overseas to fill a fleet of 707s and 727s *and* a fleet of 747s." Many of those interviewed worried that such a large plane would not be safe: "I'm not in favor of these Jumbo Jets; it reminds me a bit of the Titanic," said a respondent from London. To assuage such fears—and to convince the public that the 747 was "Everyman's Airplane," built not only for elite jet-setters but for the masses—Pan Am launched its largest advertising campaign to date, spending $20 million to run ads in ninety-seven countries and thirty-two languages. Created by JWT, the ads featured "warm" and witty cartoons by illustrator Hank Syverson, along with folksy slogans that addressed the work-weary middle class: "Get out of the country. Get into this world"; "What you need is a long vacation on a short island"; "Let us take you out of all this."[60]

On January 21, 1970, Pan Am debuted the 747 on a flight from New York to London. The introduction of the jumbo jet garnered much public excitement, seeming to indicate that JWT's advertising campaign had been successful. Over two thousand people had placed their names on the waiting list for tickets even before a destination for the first flight was specified, and ticket sales continued to be strong throughout the year. Nonetheless, the 747 cost Pan Am far more than it earned. The airline spent $130 million updating airport facilities to accommodate the jumbo jets; in addition, the jumbo fleet incurred $10 million per month in financing charges, causing Pan Am to post a $48 million loss in 1970. When Trippe had purchased the 747s during the flush mid-1960s, he had justified their expense by predicting that air traffic would increase by 17 percent a year. Instead, air traffic increased by only 4.6 percent and international traffic by just 1.5 percent. Meanwhile, Pan Am had nearly 16 percent more seats to fill.[61]

The recession of the early 1970s, which prompted consumers to curtail discretionary travel, took its toll across the aviation industry. But Pan Am, which did not fly domestic routes and thus entirely depended on sales of costly international tickets, suffered disproportionately. The oil crisis of 1973 also delivered a particularly hard blow to Pan Am, which obtained 93 percent of its fuel from foreign sources. Between 1969 and 1976, "the financially troubled Pan Am" (as the press now routinely described the airline) lost some $364 million, while its debt ballooned to over $1 billion. In response, management laid off some fifteen thousand employees, including a third of all pilots and flight engineers, and eliminated routes and overseas bases.[62]

The struggling U.S. airline industry suffered another blow in 1971, with the cancellation of the nation's supersonic transport (SST) program. Able to fly at twice the speed of sound, the SST promised to shrink the flight time between New York and Paris from eight to 3.5 hours. In Washington, discussions concerning SST development began in 1960; three years later, President Kennedy established the National Supersonic Transport Program, which would fund 75 percent of developing a commercial SST. In 1967, contracts were awarded to Boeing, for airframes, and to General Electric, for engines. Meanwhile, both the Soviet Union and an Anglo-French consortium (British Aircraft Corporation and Aérospatiale) began designing their own SST prototypes. As international competition mounted, aviation industry and Air Force leaders urged Washington to increase investment in SST development, arguing that national prestige and national security required nothing less. At stake, declared Juan Trippe in 1966, "is the continued world leadership of the American aviation industry. At issue, is American prestige abroad." U.S. Air Force Lieutenant Colonel Donald I. Hackney concurred in a 1966 paper for the Army War College, contending that "The supersonic transport will have a significant impact on all basic elements of national power—economic, military, political/psychosocial, and technological. Quite conceivably, the nation can derive more benefits in terms of the broad aspects of national power from the supersonic transport than from any other single project undertaken by the Government."[63]

However, just as the United States had lagged behind Europe in the development of commercial jets, European nations debuted their SSTs before Boeing even finished building its prototype, the B-2707. In December 1968, the Soviet Union's Tu-144 made its first flight, followed two months later by BAC-Aérospatiale's Concorde. In the United States, meanwhile, public opposition to the SST mounted, as residents of New

York and other cities with major airports expressed concern about noise pollution from sonic booms. Their protests proved effective. In March 1971, Congress abruptly voted to suspend further funding for SST development, in spite of President Richard Nixon's fervent support of the project. Two months later, the United States discontinued its SST program entirely. In response, the aviation industry raised a sonic boom of an outcry. "The setback of the American SST was a damnable betrayal of history and our national welfare," fumed Wayne Parrish, a leading airline consultant and trade journalist. "The air age got its start in the United States, and we had bloody well keep at it or we, too, will vanish as a super power just as many short-sighted peoples who thought they were on top for good, have declined and all but vanished before us." Recalling arguments previously made by Billy Mitchell, Hap Arnold, Thomas Finletter, and other critics of demobilization after the world wars, Parrish construed aviation as the nation's birthright and the foundation of its superpower status. Any limitations on its development, then, augured broader threats to U.S. national security and global power.[64]

Indeed, the airline industry faced an alarming new external threat in the late 1960s and 1970s: aircraft hijacking. The first recorded instance of an attempted hijacking had occurred in 1931, when revolutionary guerrillas in Arequipa, Peru, surrounded a Panagra Ford Tri-motor and demanded that its pilot surrender the aircraft. Until the jet age, however, such incidents were exceedingly rare; during the 1930s and 1940s, the gravest threats to airline security came from livestock that wandered onto runways. Then, in 1961, the era of "skyjacking" began when a knife- and pistol-wielding Cuban national who called himself "El Pirata Cofrisí" (after a legendary early-nineteenth-century pirate from Puerto Rico) forced a National Airlines pilot to divert to Havana. No crew or passengers were harmed, and the plane returned safely to the United States after dropping its assailant in Cuba. But the incident proved to be the first of many. Between 1968 and 1972, 154 hijackings of U.S. commercial flights occurred. Most of these early hijackers were either Cuban nationals, who commandeered aircraft because they had no means to travel to the island legally, or American citizens—including several members of the Black Panthers and other radical activist groups—who wanted to defect to Cuba for political reasons. None of these incidents resulted in loss of life. They proved to be more of a nuisance than a danger, happening with such regularity that pilots began flying with landing charts to Havana's José Martí Airport.[65]

A new, lethal era of hijacking began in 1968, when the Palestinian Liberation Organization endorsed it as a political tactic. Over four hundred hijackings occurred during the following decade, affecting some 75,000 passengers. Levels of violence also sharply escalated, with hijackers taking hostages, firing on planes, and detonating on-board explosives. Nor was hijacking the only means of aviation terrorism. On December 17, 1973, Pan Am flight 110 was scheduled to fly from Rome to Beirut. As the jet prepared to taxi away from the gate, a group of Palestinian terrorists began spraying machine gun fire through the terminal windows and then proceeded to throw hand grenades at the aircraft, setting it on fire. Twenty-nine passengers and the flight's purser perished in the attack, which Secretary of State Henry Kissinger described as a "moral outrage."[66]

The specter of hijacking could be seen and felt at American airports. Beginning in 1973, the FAA required that passengers pass through metal detectors and have their carry-on baggage searched. Security concerns also, temporarily, put an end to exuberant experiments in airport architecture. The transparent glass walls of Eero Saarinen's terminals "had suddenly become a liability at the airport," writes architectural historian Alistair Gordon. "Posthijack terminals were heavy and grounded, whereas earlier ones had been light and soaring. The sleek and sexy envelopes of the 1960s gave way to blocky concrete and bunkerlike shapes." Inside, Saarinen's flowing and expansive spaces were replaced with "partitions, narrow corridors, single-entry points, and artificial lighting." Even control towers came to resemble penitentiary lookouts, "impregnable . . . isolated and forbidding." If airport architecture had once expressed the optimism and self-confidence of the American Century, it now evoked a nation on the defensive, hunkered down and fearful.[67]

The trio of crises that threatened the U.S. aviation industry beginning in the late 1960s—financial overreach, the recession and oil crisis, and hijacking—received dramatic depiction in Arthur Hailey's novel *Airport*, the fiction best seller of 1968, and its acclaimed 1970 film adaptation. Hailey, a novelist known for conducting extensive firsthand research on his subjects, spent nearly a year interviewing airport personnel, and his novel offered a detailed insider's view that sympathized with airport employees even as it sharply criticized their industry. The film, starring Burt Lancaster, Dean Martin, Jean Seberg, and Jacqueline Bisset, received six Oscars (out of seventeen nominations), inspired three sequels, and defined the 1970s "disaster movie" genre.[68]

Airport is a saga of crisis, both personal and institutional. Set in an unidentified Midwestern city, at fictional Lincoln International Airport (loosely based on Chicago's O'Hare), it chronicles the events that take place on a single, fateful night. During a record winter storm, a 707 gets stuck in the snow, blocking the airport's only jet-sized runway. Aboard another 707 that has just departed for Rome, a passenger detonates a bomb, blowing a hole through the rear fuselage. The explosion kills the bomber and critically injures a flight attendant (who happens to be the captain's pregnant mistress). The half-destroyed jet heads back toward Lincoln but cannot land because of the stuck 707. Mechanics frantically try to dig out the mired jet, and the airport manager considers wrecking it with snowplows. Meanwhile, in front of the terminal, a group of neighboring residents, angry about jet noise, hold a demonstration to demand the airport's closure.

In the midst of these catastrophes, the airport's employees also grapple with personal crises. An air traffic controller, distraught over an accident he believes that he caused, plans to commit suicide. The flight attendant contemplates terminating her pregnancy, while the suave and arrogant captain plans to end their affair. The airport's workaholic manager flirts with a coworker and feuds with his bored socialite wife, who is having an affair of her own; at the end of the novel, the two decide to divorce. In this dark vision of the jet age, aviation is no longer transcendent or liberatory; it is, rather, a source of chaos, anxiety, and failure. The jet itself is no longer a family-friendly living room but a threat to domestic tranquility. In the opening scene of the film, a family of four says grace before dinner, only to be interrupted by the roar of a jet ascending overhead. Airport employees contend with infidelity, unplanned pregnancy, and suicidal inclinations while the airport separates them from their families: "Screw that demanding, stinking, marriage-wrecking airport!" fumes the manager's wife after her husband skips yet another social engagement to deal with problems at work. The airplane bomber represents the most extreme example of familial failure. Financially ruined, unable to provide for his wife and children, the unemployed construction worker rigs a bomb and detonates it—not to make a political statement but so that his wife will receive the money from his flight-insurance policy.[69]

The airport itself functions as a synecdoche for the novel's larger narrative of failure and decline. From the outside, as Hailey describes it, the airport resembles a "brightly lighted, air-conditioned Taj Mahal." But like its employees, the building is inwardly falling apart at the seams. "Judged by its terminal alone, the airport was still spectacular," yet functionally it was "close to becoming a whited sepulcher," with inadequate runways and dangerously overburdened air traffic control systems. The film expresses these contrasts visually, cutting between the terminal, with its warm lighting and calmly stated loudspeaker announcements, the cramped and smoke-filled control tower, where air traffic controllers hunch tensely over radar screens, and the darkened runway, where shivering mechanics agonize over what to do with the snowed-in jet on the ground and the bombed-out jet in the air. In short, the airport is a victim of jet-age hubris. Built during a more modest era of air travel, it proves unable to accommodate the airlines' unbridled expansion. "Lincoln International was obsolescent," Hailey writes. "Much was talked about aviation's growth, its needs, coming developments in the air which would provide the lowest cost transportation of people and goods in human history, the chance these gave the nations of the world to know each other better, in peace, and to trade more freely. Yet little on the ground—in relation to the problem's size—had been done."[70] Gleaming surfaces that conceal inner decay, expansive airborne ambitions versus constrained on-the-ground realities: as of 1968, such tensions afflicted not only the U.S. aviation industry but the global American Century that leaders such as Juan Trippe and Bill Allen had helped to bring about.

On the one hand, the jet age represented the zenith of the American Century. Consistent with *Fortune*'s 1945 prediction of a world in which Americans would "fly everywhere," jets expedited "the free movement of goods and people and ideas, at the lowest possible cost, in the largest possible numbers and amounts, between anywhere and everywhere" (as *Fortune* had stated).[71] By 1968, the United States led the world in aircraft sold, miles flown, and passengers carried. Boeing 707s and 727s lined airport runways from New York to Nairobi; Pan Am's globe-girdling routes stretched even beyond the iron curtain; and as jet-setting became more affordable, a diaspora of U.S.-passport-carrying tourists dispersed throughout the world. The culture of mass air travel both reflected and sustained Cold War ideologies of consumer abundance, just as the Boeing 707 that carried Vice President Richard Nixon to Moscow in 1959 exemplified the arguments he made in the Kitchen Debate. For all of these reasons, the jet became a compelling cultural symbol of midcentury optimism about the United States' national identity and global mission. Airports such as Dulles and JFK translated soaring ambition into built form; representations of aviation in airline advertising and the mass media expressed fantasies of power, both national and individual. Jet-age culture construed the United States as a benign imperium whose borders would stretch ever upward and outward, which would enjoy ever-faster access to new markets and new pleasures, whose ever-increasing prosperity would fulfill all desires. Abundance, access, size, and speed: these were the keywords of the American jet age.

On the other hand, however, the jet age also registered the American Century's limits, contradictions, challenges, and unintended consequences. The jet engine was invented in Britain and Germany; Canada, Britain, and the Soviet Union all flew jet airliners before the first Boeing 707 rolled off production lines; and even the Boeing 727, the signature "American" jet of the era, had a French pedigree, deriving from the Caravelle. Jet-age geography recognized no national entities; airline maps featured only cities, and cities themselves were increasingly cosmopolitan. The so-called jet set, a transnational cohort, defined itself by social class and style rather than citizenship. And in quite literal ways, the jet lowered barriers to movement across national boundaries, enabling American students to study abroad, foreign nationals to emigrate to the United States, and U.S. and Soviet citizens to venture across the iron curtain. But the jet age not only challenged the nation at the level of its ideological coherence; by the late 1960s, it had also begun to challenge more conventionally defined national interests. Jet lag weakened individual bodies, while the balance-of-payments crisis weakened the national body politic. After

1968, even the aviation industry was in trouble. Hijackings, flight delays, outmoded airports, nine-figure debts: all of these phenomena suggested that the onward-and-upward thrust of the jet age had become unsustainable.

Of course, all did not end in failure. In *Airport*, the stuck 707 is finally liberated, the bombed 707 lands safely, and the injured flight attendant lives, as does the suicidal air traffic controller. In real life, too, jets continued carrying American power, in all of its myriad forms, to places no longer distant. During the 1970s, the United States would continue to lead the world in numbers of planes produced and passengers carried. Henceforth, however, the U.S. aviation industry had no choice but to reckon with diminished expectations and shrunken horizons. Europe, not the United States, would lead the new era of supersonic jet transport. Britain and France jointly produced the Concorde, which began flying in 1976; plans to develop a U.S. supersonic passenger jet were canceled in 1971. And when President Jimmy Carter signed the Airline Deregulation Act of 1978, he struck a fatal blow to the midcentury corporatism that had allowed otherwise unprofitable airlines to thrive as subsidized "chosen instruments." Reorganized around neoliberal principles, the U.S. airline industry split into a multitude of smaller companies, none of which could claim to be *the* American flag carrier, as Pan Am had been for half a century. Deregulation not only changed the economic rules of the game; it transformed the very meanings of air travel. In this new era, aviation would cease to capture Americans' imagination as the circuitry of a new kind of empire or the harbinger of a world with no distant places. No longer of transcendent cultural significance, flying would become a business like any other.

Conclusion:
"Empires Rise and
Empires Fall"

At 7:00 a.m. on the morning of December 4, 1991, Pan American World Airways *Clipper Goodwill* departed New York's JFK Airport on its regularly scheduled flight to Barbados. The five-hour journey was smooth and uneventful, and the Boeing 727-200 touched down under clear, sunny skies and a pleasant temperature of 80°F. To the passengers, all must have seemed well. Little did they know that while they had been enjoying a hot in-flight breakfast, Pan Am had quietly gone out of business. "We ceased operations at 9 o'clock this morning!" shouted the Barbados station manager as he ran across the tarmac to meet purser Exie Soper, the first crew member to exit the aircraft. Disbelieving her ears, Soper grabbed the telegram from the station manager's hands and returned to the cockpit, where the crew gathered to read the verdict.

SUSPENSION OF SERVICES—STATUS NBR 1
    PAN AM PRESIDENT & CEO REGRETS TO ANNOUNCE
THAT PAN AM CORP HAS WITHDRAWN ITS MOTION
FOR CONFIRMATION OF THE CHAPTER 11 PLAN OF
REORGANIZATION FOR PAN AM AND ITS AFFILIATED

COMPANIES AND THAT AS A RESULT IT IS CEASING
FLIGHT OPERATIONS EFFECTIVE IMMEDIATELY.[1]

For Pan Am's employees, the news came as devastating but not entirely
unexpected. Following two decades marked by escalating debt, unfortunate management decisions, and the sale of valuable assets, their company
had been reduced to a shadow of its former self. In the summer of 1991,
Delta Airlines had provided temporary life support by purchasing Pan
Am's European and eastbound routes and agreeing to reorganize the
bankrupt airline with $140 million in financing. After Delta's bailout,
however, losses continued to mount. On December 3, Delta executives
informed a federal bankruptcy court that they were pulling the plug, leaving Pan Am with only enough cash to operate for another day. Consequently, after a tearful farewell ceremony at the Barbados airport, Exie
Soper and her fellow crew members, along with 111 passengers, boarded
the *Clipper Goodwill* for one last time. During the flight back to Miami,
flight attendants gave away pillows, blankets, and "as much liquor as people
wanted," recalled stewardess Jeanne Katrek. "The ship was going down
and Fate had decreed that we were the crew to go down with the ship. We
would do it with class." Passengers, meanwhile, took up a collection for
the crew so that they could have Christmas dinner if they did not receive
their paychecks. Even the logistics of the flight were poignant. As the
aircraft descended into Miami International Airport, the tower instructed
it to follow a flight path that passed directly over Dinner Key, where Pan
Am's original terminal still stood.[2]

Just as Pan Am's first four decades epitomized the ascendance of the
American Century, the great airline's decline and fall between 1968 and
1991 embodied broader challenges to the United States' global hegemony.
Some of these challenges originated from within: by the late 1970s, both
Pan Am and the U.S. government had overstretched their financial resources, suffered from crises of leadership, and lost much of their public
legitimacy. Other challenges were external: the oil crises of 1973 and
1979, the deregulation of the airline industry, escalating foreign economic competition, and the rise of aerial terrorism. The airline's final
years, then, offer a cautionary tale about the limits of power and the
challenges of the American Century.

As Pan Am struggled to recoup the losses it had incurred in purchasing its fleet of Boeing 747s—which were delivered right as the oil crisis steeply increased fuel costs and reduced voluntary air travel—a second, political, crisis was brewing in Washington. The Civil Aeronautics Board (CAB), the federal agency charged with regulating commercial aviation, had come under increasing fire from economists and politicians from both major parties. Under existing CAB guidelines, airlines could not fly new routes, make changes to existing routes, or alter fares without first applying for CAB permission to do so, and the process of receiving such permissions could be long and cumbersome. Critics of this system (including some prominent liberals who typically favored regulating commerce, notably Democratic senator Edward Kennedy) argued that it kept fares artificially high, encouraged airlines to maintain unprofitable routes, and violated free-market principles. As criticism and demands for reform mounted, in 1976 the CAB voluntarily began to liberalize its own policies, for the first time allowing individual airlines to offer promotional discounts, such as American Airlines' "SuperSaver" fare. The following year, President Jimmy Carter signaled his support for further reforms by appointing economist Alfred E. Kahn, a vocal critic of regulation, as CAB chairman. These efforts ultimately culminated in the landmark Airline Deregulation Act, signed by Carter on October 24, 1978. The legislation dramatically reduced the CAB's authority and mandated its complete abolition by the end of 1984. Henceforth, U.S. airlines would determine their own routes and fares. The immediate effect of the act was a competitive feeding frenzy, as airlines raced to attract passengers by slashing fares and adding new flights to popular travel destinations in the United States and abroad.[3]

For Pan Am, deregulation was nothing short of disastrous. Since its inception, Juan Trippe's airline had served as a corporate partner of the U.S. government—its "chosen instrument" in commercial aviation, whose vast international operations had long depended on postal subsidies, regulated fares, restrictions on competition, and diplomatic support from the State Department, as well as War Department contracts during World War II. With deregulation, these privileges disappeared, subjecting Pan Am to unlimited domestic competition on international routes. Moreover, the fact that Pan Am was not authorized to fly domestically gave its competitors a distinct advantage, as their extensive routes within the United States could funnel passengers to such international hubs as New York and Miami. Domestic airlines also did not fly unprofitable "prestige routes" to places such as Lagos, Pago Pago, or Rabat, which attracted relatively few customers but had long been deemed as essential to Pan Am's self-image as a truly worldwide carrier.[4]

A crisis of leadership compounded the problems caused by deregulation and the 747 purchases. By the late 1960s, Pan Am's management structure was bloated to the point of redundancy: an inside joke held that the company "had almost as many vice presidents as it had secretaries," and foreign outposts that saw only one flight per day were fully staffed with station managers, travel agents, and public relations representatives. Meanwhile, from his office on the forty-sixth floor of Manhattan's Pan Am Building, Juan Trippe presided over the company as his "personal fiefdom" (as one critic put it). He selected its board of directors, ensuring little opposition to his decisions and priorities. Most of these directors, furthermore, were men of his generation, whose business philosophies had been molded during the "chosen instrument" era.[5]

As an unintended consequence of his leadership style, Trippe's retirement on May 7, 1968, set off shock waves from which Pan Am never fully recovered. For more than a decade, a revolving door of CEOs created ongoing turbulence within the company. Harold Gray, Trippe's personally chosen successor as CEO and chairman of the board, was a quintessential company man who had been hired as a pilot at age twenty-three. Employees widely respected Gray, in spite of his aloof demeanor, as a technocrat who understood the aviation business from the cockpit to the boardroom. Unbeknownst to his colleagues, however, Gray had been diagnosed with terminal cancer, and he resigned in November 1969 after little more than a year in office. Gray's replacement was Najeeb Halaby, a former test pilot of mixed cultural heritage (his father was a Lebanese businessman, his mother a Texan socialite) who had served as FAA administrator under President Kennedy. Yet Halaby had no experience running an airline, and in March 1971, Pan Am's board asked for his resignation. Next up was William T. Seawell, an Air Force general, graduate of West Point and Harvard Law School, former commandant of the Air Force Academy, and senior vice president for operations at American Airlines. With such sterling credentials, Seawell seemed like a sure bet—he had been nominated by Trippe, after all, who still presided over Pan Am's board—and he managed to stay in office for ten years. But employee morale declined during Seawell's tenure as a result of his imperious attitude, volatile temper, and controversial decisions (notably the ill-fated 1980 acquisition of National Airlines). In August 1981, after unsuccessfully trying to recoup his losses by selling off major corporate assets, Seawell chose to take early retirement. Pan Am's final president was C. Edward Acker, an imposing, colorful Texan with "a handshake as hard as granite." A respected airline industry veteran, Acker had previously headed Braniff International Airways and regional carrier Air Florida, whose revenues increased twenty-three-fold during his four-year tenure. Brought in to do for Pan Am what he had done for Air Florida, Acker instead ended up presiding over its dissolution.[6]

Pan Am's post-Trippe era was plagued not only by leadership turmoil but also by a series of managerial decisions that, in hindsight, seem remarkably misguided. The first of these was Bill Seawell's acquisition of domestic carrier National Airlines. Adding National's domestic routes to Pan Am's international ones, Seawell reasoned, would bring Pan Am feeder traffic to fill seats on its 747s. Yet the takeover, which cost Pan Am $437 million, did little to improve its position in the domestic market—and turned out to be completely unnecessary. By the time the merger became official on January 7, 1980, deregulation had allowed Pan Am to fly within the United States. Merger logistics, furthermore, proved cumbersome and costly. Separate manufacturers had built the two airlines' aircraft and engines, creating a "maintenance and supply nightmare"; differences in corporate culture, not to mention competition for jobs, caused friction among employees. And in a final, bitter irony, National's routes, which ran mostly north-south, failed to provide adequate feeder traffic to Pan Am's hubs on the East and West Coasts. Among employees, the merger became known as "Seawell's Folly."[7]

In a desperate bid for financial solvency, Pan Am began selling off its major corporate assets. After the fateful merger with National, in 1980 Seawell negotiated the sale of the Pan Am Building to Metropolitan Life Insurance for $400 million, then the highest price ever paid for a Manhattan office building. The following year, he relinquished the Intercontinental

Hotel Corporation for $500 million. But even these controversial sell-offs paled in comparison to Ed Acker's abrupt and unexpected decision, in April 1985, to sell Pan Am's entire Pacific network—routes, airport bases, offices, and equipment—to United Airlines for the sum of $750 million. The decision shocked Pan Am employees and industry analysts, who widely viewed it as a coup for United. Acker had considered no rival bids or merger proposals; he and United CEO Richard Ferris reportedly sealed the deal on the golf course. Many employees viewed the sale of the Pacific routes as especially galling, since these routes, once plied by the *China Clipper*, had first made Pan Am into a world-class airline. Adding insult to injury, within several years the Pacific Rim's booming economy made United's investment extremely profitable. By 1984, the United States did a greater volume of business with Asia than with Europe, and some aviation experts have claimed that Pan Am might have survived if it had held on to its transpacific routes. Instead, the company's financial troubles continued to worsen. By 1987, Pan Am had a negative net worth and just $200 million in liquid assets. Two years later, it sold off the last of its "crown jewels," Pan Am World Services, the division that had once managed Cape Kennedy. By 1991, when Pan Am sold its transatlantic routes to United and Delta, its route map looked much the way it had in 1931: the flagship carrier that had once encircled the globe now flew only within the Americas. Later that year, United acquired several of Pan Am's key Latin American routes.[8]

Although Pan Am's tumultuous last decades contained an element of farce, tragedy ultimately sealed its fate. On December 21, 1988, as Flight 103 proceeded from Heathrow to JFK, a bomb exploded in the 747's forward cargo hold. The explosion killed 270 people—all 243 passengers and sixteen crew members, along with eleven residents of Lockerbie, Scotland, where the burning wreckage of *Clipper Maid of the Seas* plummeted to earth, creating a 140-foot-long crater and vaporizing several houses. The bombing was later determined to have been masterminded by Abdelbaset Ali Mohmed al-Megrahi, a former Libyan intelligence officer, whom a Scottish court sentenced to life in prison in 2001. One of the most lethal acts of aerial terrorism in history, the bombing of Flight 103 caused travelers who already knew of Pan Am's financial problems to further lose confidence in the "World's Most Experienced Airline." Bookings declined

sharply after the Lockerbie crash, with many travelers citing safety concerns as their reason for not flying overseas. After all of Pan Am's other
recent troubles—its 747 debts, its revolving door of CEOs, its ill-timed
acquisition of National Airlines, the selling of its Pacific routes and other
assets—Lockerbie proved to be the final nail in its coffin. It was, said one
pilot, "the day the heart of Pan American died."[9]

Yet Pan Am was not the only major U.S. airline to self-destruct in the
post-deregulation era. By the early 1990s, Eastern and TWA, both of
which dated to the 1930s, had also folded their wings. Eastern, predominantly a domestic carrier, had purchased an expensive fleet of Boeing 757
and Airbus A300 jets in 1977 and 1978 and could not repay its debts ($2.5
billion by the end of 1985) amid rising oil prices, the recessionary down-
turn in air travel, and competition from new, low-cost carriers and non-
unionized airlines such as Delta. Drastic pay cuts and layoffs then resulted
in years of hostile labor relations and a crippling strike in March 1989. By
1991, Eastern was out of business. TWA declared Chapter 11 bankruptcy
in 1992 and managed to continue flying only after eliminating many of its
routes. And as in the case of Pan Am, tragedy compounded TWA's financial troubles: on July 17, 1996, TWA Flight 800 exploded off the coast of
Long Island, killing all 230 people on board. Mechanical failure caused
the explosion; at the time, TWA's fleet was among the oldest in the
industry.[10]

In April 2001, when TWA ceased to exist after being absorbed by
American Airlines, only two of United States' great mid-twentieth-
century airlines, American and United, remained in business—and within
ten years, even they would be facing grave troubles. In May 2010, United
acquired Continental in a controversial $3 billion merger, which was
completed in the spring of 2012. Although the long-term results of the
merger remain to be seen, its initial effects included a 37 percent quarterly
decline in profits and widespread consumer and employee complaints
about unsatisfactory customer service, a glitch-ridden new computer system, and poorly maintained equipment. In October 2012, a Google search
for "United Continental merger problems" yielded 2,580,000 results,
nearly twice as many as "United Continental merger" (just 1,320,000 results). Meanwhile, American Airlines filed for bankruptcy in November
2011 and has since been plagued by conflict with its pilots' union and the

highest rates of delayed flights in the U.S. airline industry, as well as by more bizarre problems, such as seats becoming detached from its aircraft.[11]

Yet in spite of the airline industry's ongoing crisis, more people were, and are, flying than ever before. By 1977, 63 percent of the American population had flown commercially, compared with just 10 percent in the late 1940s, and U.S. airlines were operating some seven hundred flights per hour, the same volume they had flown per day in 1932. Deregulation, by creating competition that lowered fares—by as much as 40 percent in real dollars, according to the Air Transport Association—further democratized the skyways. Passenger traffic on domestic and international routes has steadily increased every year since 1978. That year, U.S. airlines carried some 254 million passengers; in 1998, they carried nearly 657 million, and in 2011, 726,007,934. In March 2012, the FAA predicted that passenger traffic would double within the next twenty years. And even as the older flagship carriers filed bankruptcy or disappeared altogether, low-cost carriers such as Jet Blue, Frontier Airways, and Southwest Airlines have emerged to take their place. Dallas-based Southwest, in particular, has been the U.S. airline industry's greatest post-deregulation success story, with 2012 its thirty-ninth consecutive year of profitability. In the second quarter of that year, it posted a record 42 percent increase in profit, which more than offset the rising cost of jet fuel. Southwest also receives the industry's lowest ratio of complaints per passenger, according to Department of Transportation statistics, and has earned industry acclaim for high rates of employee and customer loyalty.[12]

However, as air travel became more accessible, it lost much of its romantic allure. Increasingly routine and mundane, aviation largely ceased to captivate the imagination, let alone inspire expansive visions of national greatness and global unity as it had at midcentury. Flying increasingly felt no more glamorous than traveling by Greyhound or Amtrak. In fact, as airlines lowered fares by cramming more seats into cabins and eliminating in-flight meals and other amenities, Amtrak trains, with their spacious seating and dining cars, could be more comfortable than planes. As early as 1976, a travel industry trade journal stated, "From the nature of airline advertising one may reasonably deduce that airline executives and ad copy writers live in some kind of fantasy world, one in which flying is a glamorous, exciting, fulfilling experience. . . . In fact flying is a bus-ride." Air travel became more of a "bus ride" after deregulation, which made the skyways cheaper but more crowded, resulting in routine delays, missed connections, lost baggage, and surly passengers who occasionally exploded in outbursts of "air rage." As airlines further reduced costs by funneling traffic to regional airport hubs (often located in the Midwest or South, inconvenient to major destinations on the East and West Coasts), even domestic flights often required multiple legs and layovers.[13]

As the U.S. airline industry grapples with financial and logistical challenges, the mantle of luxury air travel has been taken up by such airlines as Singapore Airlines, Cathay Pacific, Qatar Airways, and Emirates, reflecting the broader economic ascendancy of the Pacific Rim and the oil-producing Persian Gulf states. Tellingly, every company on the 2012 Top Ten List of World Airlines, annually compiled on the basis of passenger surveys by the U.K.-based research firm Skytrax, hails from Asia or the Persian Gulf region; not one U.S. or European carrier made the cut. These "five-star" carriers have invested heavily in the newest jumbo jets, Boeing's 787 Dreamliner and the Airbus A380, and they pride themselves on offering amenities and standards of service that long ago vanished (or never existed) on U.S. airlines. Aboard Emirates' A380s, first-class cabins feature private suites replete with minibars, desks, and showers. On South Korea's Asiana Airlines, first- and business-class passengers can recline on lie-flat seats with duvet covers and tuck-in service. Qatar Airways treats its first- and business-class passengers to eight-course dinners that include sushi plates, foie gras, and caviar; in Doha, they enter and leave the airport through a private "Premium Terminal" that offers expedited check-in, along with sleeping cabins and showers. Such luxuries come at extremely high cost, of course, but even these airlines' economy-class cabins feature wide-body seats, individual seatback screens with a dizzying array of entertainment options, and high-quality meals. Fittingly, their advertisements market air travel not simply as an efficient mode of transportation but as a glamorous and romantic experience, just as Pan Am and TWA ads once did. (Many of these ads, such as Singapore Airlines' "Singapore Girl," also reprise the gendered and sexualized imagery that pervaded U.S. airline advertising during the 1960s and 1970s.) Superiority in commercial aviation, then, has become integral to the national self-image of ascendant powers in Asia and the Middle East, just as it once was to that of the United States.[14]

In this context, recent American discourse on aviation has linked the struggles of the U.S. airline industry to a perceived crisis of U.S. international power and legitimacy more generally. Indeed, some writers have explicitly linked the decline and fall of the nation's flagship airlines to the decline and fall of American empire. In September 2012, novelist Gary Shteyngart penned a humorous yet blistering op-ed for the *New York Times*, entitled "A Trans-Atlantic Trip Turns Kafkaesque." Chronicling a calamitous American Airlines flight from Paris to New York, which ended thirty hours after it began and involved baggage compartments held together with masking tape, a broken altimeter, a lengthy detour to Heathrow, and a sweltering airport shuttle bus whose doors would not open, the editorial begins: "You, American Airlines, should no longer be flying across the Atlantic. You do not have the know-how. You do not have the equipment. And your employees have clearly lost interest in the endeavor. Like the country whose name graces the hulls of your flying ships, you are exhausted and shorn of purpose." Although American Airlines receives the brunt of Shteyngart's criticism, he clearly intends to indict the United States as a whole, using the airline's mishaps as a metaphor for a broader sense of national malaise, exhaustion, and loss of purpose. In the editorial's last paragraph, Shteyngart quotes a fellow passenger, a wistful "old-timer" who says, "This used to be a great airline." Addressing American once again, Shteyngart writes, "I know you were. And I know you are not alone in failure. . . . Empires rise and empires fall. A metaphor you may need to consider closely." Just as the United States' commercial "empire of the air" once augured an ascendant American Century, the troubles of American and other airlines now become harbingers of imperial decline.[15]

It is a grim yet fitting irony, then, that the deadliest act of violence ever to occur on U.S. soil involved commercial airplanes. In spite of the many ways that aviation's meanings have changed since the late 1960s, the terrorist attacks of September 11, 2001, testified to its enduring significance as a symbol of U.S. global power. Al Qaeda used commercial airplanes, the chosen instruments of America's market empire, to attack the market empire at its heart, the World Trade Center. "Ground zero," the name given to the Twin Towers' smoldering ruins, bears a double meaning, for those buildings had also been ground zero of neoliberal globalization, from whence flowed the capital and commodity trades that had made the United States the wealthiest and most powerful nation on earth—and which had infuriated those, including the 9/11 hijackers, who perceived themselves as the American Century's enemies or casualties. The attacks also emphatically demonstrated that the communication and transportation technologies that propelled the American ascendancy could be used against the United States. According to the 9/11 Commission Report, "It should now be apparent how significant travel was in the planning undertaken by a terrorist organization as far-flung as al Qaeda. The story of the plot includes references to dozens of international trips. Operations required travel, as did basic communications and the movement of money." The very infrastructure of the United States' aerial empire could be weaponized and deployed at the expense of American lives.[16]

Yet the "empire of the air" was never simply a market empire. It both derived from and depended on the maintenance of a formidable military arsenal, the scope of which has far exceeded any other in history. In its decision to attack the Pentagon as well as the World Trade Center, al Qaeda implicitly recognized the military foundations of the American ascendancy. If the Twin Towers were ground zero of U.S. economic power, the Pentagon has been ground zero of the worldwide network of air, naval, and army bases, aircraft carriers, submarines, nuclear missiles, and satellites that has ensured American preponderance since World War II. The "open skies" envisioned by Adolf Berle have also been fortified skies. And since 1927, when Hap Arnold, Carl Spaatz, and Jack Jouett founded Pan Am to help protect the security of the Panama Canal, commercial airlines have played a critical role in building, managing, and expanding this global military network—as demonstrated by Pan Am's subsequent history as a military contractor, from its Latin American Airport Development Program and trans-African airlift to its Guided Missile Range Division.

If air travel had lost much of its allure long before 2001, the attacks of September 11 cast an even darker shadow over the experience of flying and the cultural meanings of the airplane. With the war on terrorism a seemingly permanent condition of American life, passengers have become accustomed to the additional inconveniences of waiting in lengthy security lines, removing their shoes, and undergoing full-body scans. In the post-9/11 era, air travel is as likely (or perhaps more likely) to evoke anxiety or frustration than exhilaration or awe. Although the average frequent flier may not think about the burning Twin Towers every time she boards a plane, airport security announcements and questions from Transportation Security Administration agents make it nearly impossible to ignore the fact that commercial airplanes are potential terrorist targets and weapons. If airliners once connoted the power and greatness of the United States, they now also ineluctably conjure its vulnerability.

Meanwhile, America's empire of the air has been militarized to an unprecedented extent. Even as U.S. airlines struggle to remain profitable, leading aircraft manufacturers have thrived, as they did during the Cold War, on Defense Department contracts. And in the United States' wars in Iraq and Afghanistan, the administrations of both George W. Bush and Barack Obama have relied heavily on unmanned aerial vehicles, or "drones." Drone warfare takes Anne Morrow Lindbergh's concerns about the lofty hubris of the "air observer"—"the curious illusion of superiority bred by height . . . the illusion of irresponsible power"—to a whole new level. Sitting at computer stations in the Pentagon or CIA headquarters, thousands of miles from their faceless targets, drone operators are radically insulated from the human consequences and moral implications of their actions. As critics have argued, the drone program creates the illusion of war without cost, in which calculations of efficiency take precedence over consideration of human lives and in which civilian casualties are conveniently explained away as "collateral damage."[17]

Aviation, in the aftermath of 9/11, thus carries a complex array of cultural meanings. On the one hand, aviation has largely lost its transcendent connotations. As air travel has become increasingly routine and inconvenient, it rarely evokes glamour, excitement, or expansive visions of "one world"; the commercial airplane, furthermore, now signifies the vulnerability of the United States as much as its national and global power. On the other hand, in the era of drone warfare, aviation continues to serve as the foundation of national security and the war on terrorism. Thus, even as the United States' commercial air empire has declined since its early jet-age apex, U.S. military infrastructure—which Pan Am and other commercial airlines played a crucial role in building and maintaining—remains, more extensive and more expansive than ever and increasingly normalized as a permanent, inevitable aspect of American life.

From Pan Am to *Pan Am:*
The "Golden Age of Air Travel" Reimagined

In the era of Predator drones, bankrupt airlines, escalating foreign competition, and the ever-present threat of terrorism—not to mention security lines, baggage fees, cramped seating, and few in-flight amenities aside from "meals available for purchase"—it is scarcely surprising that many Americans would feel nostalgic for the mid-twentieth-century "golden age of air travel," as the 1930s through 1960s have sometimes been called. And, in fact, deeply nostalgic currents run through many contemporary cultural representations of aviation. At the Smithsonian Institution's National Air and Space Museum (NASM) in Washington, DC, the exhibition "America by Air," chronicling the history of U.S. commercial aviation from the first airmail flights to the Boeing 787 Dreamliner, comments on but also taps into this nostalgia. Reopened in 2007 after a major redesign, the exhibition questions whether the mid-twentieth century truly *was* a golden age of air travel, pointing out that flying was then more hazardous, costly, and time-consuming than it is today. But with its gorgeously restored vintage aircraft, stylish crew uniforms, and colorful Art Deco airline posters, "America by Air" still ends up romanticizing a bygone era of commercial aviation (as visitors' remarks suggest). After seeing the exhibition, visitors can purchase their own piece of aviation nostalgia in NASM's gift shop, which sells a whole line of Pan Am memorabilia—posters, postcards, T-shirts, and replicas of the airline's famous blue-and-white flight bags.[18]

Beyond museums, nostalgic representations of air travel seem to be everywhere these days, especially in fashion. Faux vintage flight bags are widely available on the Internet (for example, at www.airlinesoriginals. com), and one can obtain the real items on eBay, a thriving market for every type of airline memorabilia. In 2006, haute-hipster designer Marc Jacobs created a line of "Pan Am" travel bags, which were nearly identical to the originals except for the presence of the designer's logo opposite Pan

Am's. The limited-edition bags, retailing for over $500, sold out so quickly that in 2007, Jacobs introduced a second line, priced at around $60, under his less-expensive Marc by Marc Jacobs label. The boutique store Flight 001, meanwhile, sells high-end luggage and travel accessories for the modern jet-setter, including retro luggage tags, passport covers, and eye masks inspired by 1960s designs. Presumably named after Pan Am's famous round-the-world route, Flight 001 was founded aboard Air France Flight 023, "somewhere between New York and Paris," by two business travelers who "had spent far too much time preparing for their trip" and "envisioned a travel store as streamlined as flight itself." The store has been remarkably successful, branching out from its original location in Manhattan's West Village to five additional outposts in Brooklyn, Chicago, Los Angeles, Berkeley, and San Francisco.[19]

Since 2000, Hollywood has produced several lavish homages to the golden age of air travel, most notably *The Aviator* (2004), Martin Scorsese's Academy Award–winning biopic of eccentric TWA magnate Howard Hughes, and Steven Spielberg's *Catch Me If You Can* (2002). Based on a true story, *Catch Me If You Can* portrays the life of con artist Frank Abagnale Jr. (played by Leonardo DiCaprio), who courted wealth and women in the early 1960s by impersonating, among other high-powered professionals, a Pan American World Airways captain. The film brilliantly captures the aura of power, glamour, and sexuality that surrounded air travel, and Pan Am specifically, in the early jet age. In a scene that takes place in an opulent Manhattan restaurant, for example, Abagnale dines with his father (Christopher Walken), a modest salesman, who marvels at the attention his son attracts with his navy-and-gold-striped uniform: "See these people staring at you? These are the most powerful people in New York City. And they keep peeking over their shoulders wondering where you're going tonight," Abagnale Sr. says. As the scene suggests, Abagnale chose to impersonate a Pan Am pilot for strategic reasons. The airline's iconic visual signifiers—its blue globe logo on a check, a gold-braided uniform—allowed the young con artist to turn heads, seduce flight attendants and bank clerks, and forge checks.

Retro aviation aesthetics have even appeared in movies that are not historical. In the light-hearted comedy *View from the Top* (dir. Bruno Barreto, 2004), Gwyneth Paltrow starred as Donna Jensen, a showgirl's daughter who escapes small-town Nevada by training to be a flight attendant. She takes a job at a budget regional airline, endures its sexually exploitative culture, and ultimately works her way "to the top" on international carrier Royalty Express. Though the film ostensibly takes place in the present, costume designer Mary Zophres clearly looked to the past for inspiration, as her flight attendant uniforms are straight out of the mod era. In the film's poster, Donna's uniform—an orange minidress with a vertical white strip running down the center—replicates almost exactly the uniforms worn by United Airlines flight attendants between 1968 and 1970, created by designer Jean Louis, which were displayed at Seattle's Museum of Flight in its 2008 exhibition "Style in the Aisle."[20]

On the small screen, AMC's acclaimed television series *Mad Men*, about the professional and private lives of New York advertising executives in the 1960s, has prominently featured commercial aviation in key plot lines of several episodes. During the show's second season, boutique ad agency Sterling Cooper persistently tries to acquire an airline account, considered a de rigueur asset of any blue-chip agency. Sterling Cooper signs Mohawk Airlines, a small regional carrier, only to drop the account in an ill-fated attempt to sign American Airlines after one of its jets crashed into Jamaica Bay in March 1962, killing all of the ninety-five people aboard; the agency again picks up Mohawk in Season Five not because the account is especially prestigious but simply because it is an airline. "Smile! We've got an airline!" exclaims partner Roger Sterling after the agency re-signs Mohawk. Air travel also prominently figures into the personal life of the show's charismatic, enigmatic lead character, Sterling Cooper's creative director Don Draper (played by Jon Hamm). "You want to get on a plane to feel alive," says Don during a meeting about the Mohawk account, and his own life certainly corroborates the statement. In "The Jet Set" (Season Two), Don flies to Los Angeles to attend an aerospace industry convention. After listening to a dreary presentation on the virtues of MIRV nuclear warheads, he skips the conference to accompany a seductive young woman, the appropriately named Joy, to Palm Springs. There, in a glass-walled California contemporary estate—whose architectural aesthetics recall Eero Saarinen's airport designs—he meets Joy's cohort of wealthy jet-setters: "We're nomads together," she tells him. A gray-flannel-suit man who worked his way up from rural poverty, Don is simultaneously perplexed, repulsed, and fascinated by these high-flying aristocrats who disdain employment ("I should take something up," muses one jet-setter—not for income but because "I'm smarter than I've ever been in my life") and spend their days poolside in Palm Springs and Capri, sipping champagne. Although Don, who has spent his whole life escaping from himself in one way or another, considers running off with Joy and her fellow "nomads," he ultimately returns to his wife and children in a suburb of New York. But in the inaugural episode of Season Three, entitled "Out of Town," Don again boards a plane, flying to Baltimore for meetings with London Fog. Mixing business with pleasure, he has a tryst with a flirtatious TWA flight attendant, whose forwardness ("I've never seen a stewardess *that* game," says Don's coworker) embodies the jet-age conflation of access to women and access to the world. Several episodes later, Don and his wife, Betty, become jet-setters themselves, taking a spontaneous trip to Rome at the invitation of hotelier Conrad Hilton.[21]

A third-season plot line involving Hilton, whose worldwide network of hotels was as integral as Pan Am's route network to the United States' aerial empire, best embodies *Mad Men*'s take on jet-age hubris. In a late-night meeting over bootleg liquor, the hotel magnate (played by Chelcie Ross) advises Don on the ad campaign that he wants Sterling Cooper to create. "It's my purpose in life to bring America to the world, whether they like it or not," Hilton says, speaking slowly and emphatically. "You know, we are a force of good, Don. . . . Generosity. The Marshall Plan. Everyone who saw our ways wanted to be us." Hilton insists that he "doesn't want politics in my campaign," but "there should be goodness, and confidence." To highlight the extent of his ambitions, he declares, "I want Hilton on the moon. That's where we're headed." Don interprets the statement as metaphorical, but Hilton meant it quite literally, expressing displeasure when the ads that the agency creates do not include explicit references to hotels on the moon. Like the Smithsonian's "America by Air" exhibition, *Mad Men* adopts a critical take on Hilton's unabashed confidence in the United States' goodness and global influence—and the series consistently underscores the darker side of American society in the 1960s, including its racism, sexism, and the extramarital affairs and alcoholism that belied the era's ideology of placid suburban domesticity. Yet

Mad Men also clearly participates in nostalgia for a time when the United States seemed to be at the pinnacle of its global power and influence and when American aspirations reached all the way to the moon.[22]

The apotheosis of early millennial jet-age nostalgia was ABC's highly anticipated yet brief-lived television series *Pan Am*, which premiered in September 2011 and ran for fourteen episodes before a sharp decline in ratings and viewership prompted its cancellation (over eleven million viewers watched its premiere, while just 3.8 million tuned into the finale). A Technicolor confection of beautiful people, beautiful planes, and beautiful places, *Pan Am* exuberantly reenacted the onward-and-upward buoyancy of the early jet age. Taking place in 1963, the ensemble drama centered around six main characters: four flight attendants, a captain, and a first officer. Brash, bohemian purser Maggie Ryan (Christina Ricci), who faked a degree from Berkeley to obtain her job, flouts Pan Am's dress code rules, campaigns for John F. Kennedy, and eventually takes up smuggling in partnership with a loutish older captain. Serious, trilingual Kate Cameron (Kelli Garner) leads a double life as an undercover spy for the CIA, which includes a doomed love affair with a Yugoslavian asset (Goran Visjnic). Kate's naive yet plucky younger sister Laura (Margot Robbie) becomes a stewardess—and lands on the cover of *Life* magazine, under the caption "Welcome to the JET AGE"—after leaving her fiancé at the altar. French-born Collette Valois (Karine Vannasse) has a penchant for rescuing orphans and falling in love with the wrong men. Dashing, daredevil pilot Dean Lowrey (Mike Vogel), promoted to captain at an improbably young age after getting stuck in an elevator with Juan Trippe, skillfully helms his 707s through numerous near disasters and spends his off hours piloting flashy sports cars and pursuing women. First officer Ted Vanderway (Michael Mosely), the son of a wealthy aircraft manufacturing firm executive, is a former naval test pilot who was discharged after a controversial crash. The series follows the crew's high-flying personal and professional adventures, both on board and off, from Pan Am's Idlewild terminal to such cosmopolitan world capitals as Paris, Berlin, Monte Carlo, and Rio.

With its lush cinematography, jaunty soundtrack (Buddy Greco's "Around the World"; Bobby Darin's "Mack the Knife"), and painstakingly recreated period details, *Pan Am* paid loving homage to its namesake—and to the "golden age of air travel" that the airline continues to signify. Creator Jack Orman, who had previously written and produced episodes of *ER* and *JAG*, explicitly intended the series as an exercise in nostalgia. "Part of the draw of the show is the glamour of the jet age, and the really interesting element for me is that they got it right the first time," Orman said in an interview. The show, he continued, aimed to resurrect an era when "the journey was as important as the destination . . . in a more innocent time before all the security and all the rigamarole that you [now] have to go through to get through the airport." Like *Mad Men* creator Matthew Weiner, Orman went to great lengths to accurately capture the look and feel of the early jet age. The show's pilot episode cost a staggering $10 million, in part because it entailed the recreation, from scratch, of a Boeing 707—"which meant a lot of pictures, a lot of digging around the California desert for pieces of galley and cockpit and whatnot," Orman recalled. Executive producer Nancy Hult Ganis, who had worked as a Pan Am flight attendant from 1968 to 1976, "dug through boxes and boxes" of Pan Am's records at the University of Miami, distributed a twenty-page questionnaire to members of World Wings International (an organization of former Pan Am flight attendants), conducted interviews with dozens of retired flight attendants and pilots, and advised the show's actors, writers, and directors on everything from uniform design to how coffee was served. Some of the plot lines, including a memorable scene in which Christina Ricci stabs a passenger with a serving fork after he drunkenly gropes her, were based on Ganis's own experiences.[23]

Pan Am portrayed serious issues and historical events—sexual harassment, racism, espionage, John F. Kennedy's June 1963 "Ich bin ein Berliner" speech, and his assassination five months later—and its characters grappled with various forms of emotional drama, from Ted's bitterness over his Navy discharge to Kate's stressful secret life to Colette's recurring tendency to have her heart broken. Yet the series' dominant tone conveyed confidence, exhilaration, and optimism. As Ella Fitzgerald sings in "Blue Skies" (featured in Episode Four, "Eastern Exposure"): "Blue skies / Smiling at me / Nothing but blue skies / Do I see." Recalling Pan Am's sky-blue jet-age logo, which depicted a world divided only by latitudes and longitudes, *Pan Am* reproduced the ideology of American nationalist globalism that identified the entire world as the United States'

rightful sphere of influence, open for the taking. The show's most memo-
rable image, which appeared on its promotional posters, depicted the four
stewardesses, clad in trim sky-blue uniforms and clutching Pan Am bags
in their white-gloved hands, proudly sauntering through an airport terminal with winsome smiles and heads held high. In the pilot episode, as
the *Clipper Majestic* lifts off on its maiden voyage from New York to London, Maggie says to Laura, who is working her first international flight:
"Better buckle up. Adventure calls!" Later, as the women toast pint glasses
at a London pub, Colette muses, "Tomorrow another plane takes off to
someplace new, vanishing today's mistakes." Such sentiments pervaded
the series, identifying Pan Am with opportunity and optimism, romance
and renewal, glamour and adventure. In Episode Four, Maggie admon-
ishes Kate for being overprotective of Laura, telling her: "Pan Am is the
ultimate new beginning! Let her begin!" And when Maggie wants to meet
her idol, JFK, in Berlin, Kate suggests that she simply wear her uniform
when approaching the president's security detail: "We're Pan Am! And
who can turn down Pan Am?"[24]

What does *Pan Am*—along with the many other recent reimaginings of
the golden age of air travel, from the Smithsonian's "America by Air" exhibition to Marc Jacobs's $500 Pan Am bags—indicate about our current
cultural moment? Are such expressions of mile-high nostalgia mere
flights of fantasy, wistful yearnings for "a more innocent time" (as Jack
Orman put it) before shoe bombers and checked-baggage charges? Or do
they signify something more profound? Numerous critics have charged
that retro television shows such as *Mad Men*, *Pan Am*, and NBC's *The
Playboy Club* (which, like *Pan Am*, premiered in the fall of 2011 and ran for
a single season) are little more than indulgences in sexism, which "seem to
long for the days when women could be freely and openly objectified," to
quote one critic. As if to corroborate this interpretation, *Esquire*'s Tom
Chiarella wrote, in reference to *Pan Am*, "I still call them 'stewardesses'
and I don't give a damn what anybody thinks. . . . I like the implication of
a sisterhood that puts up with the horseshit harassment of flight officers in
hotels in St. Louis for the chance to later land a spot on an intercontinental route to London." Yet while such patently offensive statements do
make it tempting to read *Pan Am* and its ilk as manifestations of antifemi-
nist backlash, a historically contextualized analysis of aviation nostalgia

yields a more nuanced explanation. Although recent portrayals of the mid-twentieth-century jet age have certainly reanimated the "stewardess mystique," these have also emphasized the inequities and daily forms of harassment that flight attendants experienced—and by no means simply "put up with." (Chiarella obviously did not watch the *Pan Am* episode with the serving fork.) Furthermore, such depictions have glamorized and sexualized male pilots as well as stewardesses and have lavishly fetishized vintage aircraft, airports, and airline iconography. On *Pan Am*, Idlewild Airport and the Boeing 707 practically play roles in their own right, and each episode begins with a lingering close-up shot of the Pan Am logo.[25]

In the context of aviation's century-long salience as both symbol and instrument of the American ascendancy, nostalgia for the "golden age" of air travel ultimately expresses nostalgia for a perceived golden age of U.S. global power. It is no accident or surprise that, in the wake of 9/11 and the 2008 financial crisis, when the United States' international clout and economic strength appear to be waning, and when book titles ask *Are We Rome?*, popular culture romanticizes the heyday of the American Century—and the empire of the air that derived from and expanded the United States' global dominance. In a review of *Pan Am*, Slate.com critic Troy Patterson astutely comments, "The cool of Pan Am is a cover for a fanciful trip to a happy hegemony." A happy hegemony indeed, when even the disastrous Bay of Pigs invasion of Cuba—portrayed, in a flashback, in the show's premiere episode—becomes an example of American triumphalism, with Captain Dean Lowrey proposing to his stewardess girlfriend on the tarmac in Havana right before he flies a cabin full of celebrating Cuban exiles to safety back in the United States. Nostalgia for the golden age of air travel, then, is about much more than fantasies of sexy stewardesses. On a broader scale, it conjures a past in which the United States, with its booming economy and ideological credibility as undisputed leader of the "free world," seemed to face few limits on its ambitions, desires, and access to the world.[26]

The golden age of air travel is more imagined than real—or, at least, the reality has always been less glamorous than the fantasy. Recall *Fortune*'s 1946 complaints: "To travel by plane, a passenger must now sacrifice his comfort, his sleep, and often his baggage. He must endure inconveniences that rise to the level of punishment." Or Arthur Hailey's dystopian novel *Airport*, published in 1968 during the height of the jet age, which depicted inadequate runways, crowded terminals, surly passengers, burnt-out air traffic controllers, and a bomb explosion on board. And, of course, air travel, and especially international air travel, remained prohibitively expensive for a vast majority of the American population during its supposed golden age. However, the popular appeal of air travel, in its ostensible golden age as well as in twenty-first-century representations, reflected more than its realities. As *Atlantic* writer Virginia Postrel noted in 2007,

> Airline glamour was never on the planes themselves. It was in the imagination, especially in the imagination of the people who could only dream of flying. Airline glamour never promised anything as mundane as elbow room, much less a flatbed, a massage, or an arugula salad. *It promised a better world.* . . . Airline glamour was not about the actual experience of flying but about the idea of air travel—and the ideals and identity it represented. (emphasis added)

A better world: a world of blue skies, divided only by latitudes and longitudes, fully open to U.S. expansion, eagerly awaiting the arrival of American tourists and technocrats, commodities and popular culture.[27]

Nostalgia for the golden age of air travel is ultimately nostalgia for Henry Luce's vision of the American Century—a vision of unlimited national promise and global power, in which Americans "accept[ed] wholeheartedly our duty and our opportunity as the most powerful and vital nation in the world and in consequence . . . exert[ed] upon the world the full impact of our influence, for such purposes as we see fit and by such means as we see fit." Yet as the history of commercial aviation vividly demonstrates, there always were limits to U.S. power, both internal and external: from the Latin American governments that resisted Pan Am's initial expansion, to aerial terrorism, to the airlines' own financial inability to sustain their investments in ever-larger, ever-more-expensive fleets of jets. Just as the realities of air travel have always been more complex than its idealized representations, the notion of the American Century that men such as Henry Luce, Juan Trippe, and Conrad Hilton believed in and propagated obscures the many challenges to and contradictions within the American ascendancy, even when the United States stood at the pinnacle of its global influence.[28]

Empires rise and empires fall. The spectacular success of Pan American World Airways during its first four decades, followed by its precipitous twenty-year demise, can be read as a metaphor for the fate of empires in general, as Gary Shteyngart suggests in reference to another of the nation's troubled flagship carriers. In an era of limits and diminished expectations, when flying (at least on U.S. airlines) often seems no more exciting or convenient than traveling by bus, it is tempting to conclude that America's empire of the air—and, by extension, the United States itself—is indeed in decline. To be sure, no empires in history have been permanent. And rather than arising from an inevitable process of technological manifest destiny, the United States' triumphs in aviation during the twentieth century—and the broader American imperium that the airplane both symbolized and sustained—depended on a historically specific set of factors: the achievements of such individuals as the Wright brothers and Charles Lindbergh, extensive government support for the aviation industry (including postal subsidies, investments in aircraft manufacturing, and State Department diplomatic assistance), the phenomenal expansion of U.S. air routes and bases during World War II, postwar consumer affluence and the international influence of the "market empire," and, not least, the compelling culture of aviation that prompted Americans to envision and desire a heightened role for the United States in a globalized world.

However, even as the U.S. airline industry faces serious challenges, and even as aviation's meanings have changed since the days of Pan Am's Clippers, it seems premature to conclude that the empire of the air is on the brink of demise. Although Pan Am no longer exists, its legacy endures in the worldwide infrastructure of air routes, bases, and communications facilities that continue to enable the global circulation of people and products, capital and commodities—not to mention its iconic status in popular culture. No-frills budget airlines such as Southwest are thriving, and more U.S. citizens than ever are flying. The war on terrorism, like the Cold War, has funneled defense contracts to the aerospace industry, and unmanned aerial vehicles have significantly expanded and expedited the United States' ability to launch military strikes on other countries (not to mention the increasing use of surveillance drones by the CIA as well as by domestic law enforcement agencies). Finally, perhaps in response to current anxieties about the future of U.S. global leadership, nostalgic renderings of the golden age of air travel have revived the visions of national greatness and exceptionalism that propelled the original American ascendancy. In short, the United States' empire of the air may be in *relative* decline, but its infrastructure—especially its military infrastructure—continues to serve as the backbone of the nation's power.

Yet the history of commercial aviation also suggests other ways of imagining the United States' role in the world. In contrast to Luce's conception of the American Century, which made globalism coterminous with nationalism and conflated the world's interests with those of the United States, it is possible to discern, in the "logic of the air," a more expansive and genuinely democratic kind of globalism, consistent with Wendell Willkie's vision of "one world." American culture's current fascination with the original jet set may express a longing not simply for greater national power but for more meaningful kinds of international engagement—a desire to "get into this world," as Pan Am's ads used to say, instead of attempting to stand above it in the imperious position of Anne Morrow Lindbergh's "air-observer." For those who value contact and connection across national borders, the twentieth-century culture of aviation can thus function as what critic Van Wyck Brooks famously defined as "a usable past": a historical moment rich with possibilities for transforming the present and future. In this interpretation, the logic of the air was not necessarily misguided or misleading; its implications simply remain unrealized.

Sources and Abbreviations

FDRPL	Franklin D. Roosevelt Presidential Library
AAB	Adolf A. Berle Papers
FDR	Franklin D. Roosevelt Papers
MP	Map Room File
OF	Official File
PP	Papers as President
PSF	President's Secretary's File
SF	Safe File
SWB	Sumner B. Welles Papers
HHPL	Herbert Hoover Presidential Library
HH	Herbert Hoover Papers
PP	Papers as President
SC	Papers as Secretary of Commerce
WPM	William P. MacCracken Papers
WP	Wayne Parrish Papers
IU	Indiana University, Lilly Library
WW	Wendell Willkie Papers
JWHC	John W. Hartman Center for Sales, Advertising, and Marketing History, Duke University
JWT	J. Walter Thompson Archives
AF	Account Files
AVF	Advertising Vertical File
DA	Domestic Advertising Collection
DS	Dan Seymour Papers

RB	Review Board Records
WE	Wallace Elton Papers
WT	Winfield Taylor Papers
LOC	Library of Congress
CBL	Clare Boothe Luce Papers
HRL	Henry R. Luce Papers
HHA	Henry H. Arnold Papers
NARA	National Archives and Records Administration
RG	Record Group
CDF	Central Decimal File (for RG 59)
RG 18	U.S. Army Air Forces
RG 59	Department of State
RG 84	Department of State, Foreign Service Post Files
RG 107	Department of War
RG 197	Civil Aeronautics Board
RG 224	Office of Inter-American Affairs
RG 234	Reconstruction Finance Corporation
RG 237	Federal Aviation Administration
RG 255	National Aeronautics and Space Administration
RG 469	U.S. Foreign Assistance Agencies
NASA	National Aeronautics and Space Administration Headquarters, Historical Reference Collection
NASM	National Air and Space Museum, Smithsonian Institution
ATC	Air Transport Collection
AWF	Army World Flight Collection
JTT	Juan Terry Trippe Papers
UM	University of Miami, Richter Library, Special Collections Division
PAWA	Pan American World Airways, Inc., Records
WSU	Wright State University, Special Collections and Archives
AEC	Aeronautical Ephemera Collection
HRH	Harold R. Harris Papers
WB	Wright Brothers Collection
Yale	Yale University, Sterling Memorial Library, Division of Manuscripts and Archives
HB	Hiram Bingham III Papers
CAL	Charles A. Lindbergh Papers
RAL	Robert A. Lovett Papers

Notes

Introduction

1. "Report of the City of Miami Planning, Building and Zoning Department to the Historic and Environmental Preservation Board on the Potential Amended Designation of the Pan American Seaplane Base and Terminal Building, 3500 Pan American Drive, as a Historic Site," November 16, 1993, egov.ci.miami.fl.us (accessed 10/22/12).

2. Ibid.; Gene Banning, *Airlines of Pan American since 1927* (McLean, VA: Paladwr Press, 2001), 54–57; William E. Brown Jr., "Pan Am: Miami's Wings to the World," *Journal of Decorative and Propaganda Arts* 23 (1998): 148–153.

3. "Maps and Globes," *Classroom Clipper*, June 1945; Owen Edwards, "Sky King," *Smithsonian Magazine*, November 2007; Noel F. Busch, "Juan Trippe: Pan American Airways' Young Chief Helps Run a Branch of U.S. Defense," *Life*, October 20, 1941, 111.

4. On the importance of cartography in shaping the American global imagination and U.S. claims to global power, see Susan Schulten, *The Geographical Imagination in America, 1880–1950* (Chicago: University of Chicago Press, 2001); Neil Smith, *American Empire: Roosevelt's Geographer and the Prelude to Globalization* (Berkeley: University of California Press, 2003); and Denis Cosgrove, *Apollo's Eye: A Cartographic Genealogy of the Earth in the Western Imagination* (Baltimore: Johns Hopkins University Press, 2001), 205–255.

5. "The Logic of the Air," *Fortune*, April 1943, 72–74, 188, 190, 192, 194.

6. Adolf A. Berle Jr. to John Philip Sousa III, April 19, 1943, CDF 800.796/296, RG 59, NARA; Juan T. Trippe, speech opening Boy Scout Week, February 7, 1947, Box 1, JTT, NASM.

7. Henry Luce, "The American Century," *Life*, February 7, 1941, reprinted in *Diplomatic History* 23 (1999): 159–171. I define "discourse" as a set of communicative practices—or "communicative action," to use Jürgen Habermas's term—that constitutes knowledge and meaning. As Michel Foucault has argued, discourse both reflects the power relations of the social institutions in which it is embedded and produces power in the form of truth claims. See Jürgen Habermas, *The Theory of Communicative Action* (Boston: Beacon Press, 1984); and Michel Foucault, *The Archaeology of Knowledge* (New York: Pantheon, 1972). The literature on American empire—and on the question of whether the United States even "has" or "is" an empire—is vast. Three texts, in particular, have influenced my use of the term and my understanding of the United States' imperial history: Paul A. Kramer, "Power and Connection: Imperial Histories of the United States in the World," *American Historical Review* 116 (2011): 1348–1391; Charles S. Maier, *Among Empires: American Ascendancy and Its Predecessors* (Cambridge, MA: Harvard University Press, 2006); and Smith, *American Empire*.

8. "New Look in Our Lapels," *Pan Am Clipper*, July 1963, 10, NASM Library.

9. Robert Wohl, *A Passion for Wings: Aviation and the Western Imagination, 1908–1918* (New Haven, CT: Yale University Press, 1994), 1.

10. Benedict Anderson, *Imagined Communities: Reflections on the Origin and Spread of Nationalism* (New York: Verso, 1991), 22, 44–45; Wendell Willkie, text for the Museum of Modern Art exhibition "Airways to Peace: An Exhibition of Geography for the Future," 1943, Box 7, WW, IU; Franklin D. Roosevelt, text for "Airways to Peace" exhibit in the Department of Commerce Auditorium, November 11, 1943, PSF 1025, PP, FDR, FDRPL; Franklin D. Roosevelt, "Annual Message to Congress," January 6, 1941, in Samuel I. Rosenman, ed., *The Public Papers and Addresses of Franklin D. Roosevelt, 1940* (New York: Random House, 1938–1950), 672.

11. Stuart Banner, *Who Owns the Sky? The Struggle to Control Airspace from the Wright Brothers On* (Cambridge, MA: Harvard University Press, 2008); Victoria de Grazia, *Irresistible Empire: America's Advance through Twentieth-Century Europe* (Cambridge, MA: Belknap Press of Harvard University Press, 2005), 3; Luce, "The American Century."

12. Luce, "The American Century."

13. John Fousek, *To Lead the Free World: American Nationalism and the Cultural Roots of the Cold War* (Chapel Hill: University of North Carolina Press, 2000), 2. See also David Edgerton, "The Contradictions of Techno-Nationalism and Techno-Globalism: A Historical Perspective," *New Global Studies* 1 (2007), http://www.degruyter.com/view/j/ngs. On American exceptionalism, see Dorothy Ross, *The Origins of American Social Science* (New York: Cambridge University Press, 1991).

14. Anne Morrow Lindbergh, "Airliner to Europe," *Harper's*, September 1948, 45–46; Mary Louise Pratt, *Imperial Eyes: Travel Writing and Transculturation* (New York: Routledge, 1992), 9–10; Bruce Robbins, *Feeling Global: Internationalism in Distress* (New York: New York University Press, 1999), 3; Michael S. Sherry, *The*

Rise of American Air Power: The Creation of Armageddon (New Haven, CT: Yale University Press, 1987).

15. Walter Kirn, *Up in the Air* (New York: Anchor Books, 2002); Michael Shapiro, "Moral Geography," *Public Culture* 6 (1994): 479–502.

16. Lindbergh, "Airliner to Europe." On the "intimacies" of empire and the cultural logic of integration in U.S. foreign relations, see, especially, Ann Laura Stoler, ed., *Haunted by Empire: Geographies of Intimacy in North American History* (Durham, NC: Duke University Press, 2006); and Christina Klein, *American Orientalism: Asia in the Middlebrow Imagination, 1945–1961* (Berkeley: University of California Press, 2003).

17. A. D. P, "A Chapter on Travelers," *The Knickerbocker*, September 1835, 253; Henry P. Tappan, *A Step from the New World to the Old, and Back Again: With Thoughts on the Good and Evil in Both* (New York: D. Appleton, 1852), 32; Daniel C. Eddy, *Europa: Or, Scenes and Society in England, France, Italy, and Switzerland* (Boston: Higgins, Bradley and Dayton, 1836), 32–33. On previous globalizing technologies, see, for example, Jurgen Osterhammel and Niels P. Petersson, *Globalization: A Short History* (Princeton, NJ: Princeton University Press, 2009); Paul Kennedy, *The Rise and Fall of British Naval Mastery* (Atlantic Highlands, NJ: Ashfield Press, 1986); and Mary Greenfield, "A Generation of Steam: Ships, States, and Statelessness in an Industrializing Pacific, 1836–1922," Ph.D. dissertation, Yale University, 2013.

18. On other turn-of-the-century communications technologies, see Stephen Kern, *The Culture of Time and Space, 1880–1918* (Cambridge, MA: Harvard University Press, 2003); Douglas B. Craig, *Fireside Politics: Radio and Political Culture in the United States, 1920–1940* (Baltimore: Johns Hopkins University Press, 2005); Ruth Vasey, *The World According to Hollywood, 1918–1938* (Madison: University of Wisconsin Press, 1997); and Daniel R. Headrick, *Invisible Weapon: Telecommunications and International Politics, 1851–1945* (New York: Oxford University Press, 1991). Important scholarship on material and consumer culture as mechanisms of internationalism includes Kristin Hoganson, *A Consumers' Imperium: The Global Production of American Domesticity, 1865–1920* (Chapel Hill: University of North Carolina Press, 2007); Helen Delpar, *The Enormous Vogue of Things Mexican: Cultural Relations between the United States and Mexico, 1920–1935* (Tuscaloosa: University of Alabama Press, 1992); and Mari Yoshihara, *Embracing the East: White Women and American Orientalism* (New York: Oxford University Press, 2003). On American internationalism during the Progressive and Popular Front eras, see Daniel T. Rodgers, *Atlantic Crossings: Social Politics in a Progressive Age* (Cambridge, MA: Belknap Press of Harvard University Press, 1998); Frank Ninkovich, *The Diplomacy of Ideas: U.S. Foreign Policy and Cultural Relations, 1938–1950* (New York: Cambridge University Press, 1981); Erez Manela, *The Wilsonian Moment: Self-Determination and the Origins of Anti-Colonial Nationalism, 1917–1920* (New York: Oxford University Press, 2007); and Michael Denning, *The Cultural Front: The Laboring of American Culture in the Twentieth Century* (New York: Verso, 1997). On the global color line—and resistance to it—see, for example, Glenda Gilmore, *Defying Dixie: The Radical Roots of Civil Rights, 1919–1950*

(New York: W. W. Norton, 2008); Gerald K. Horne, *End of Empires: African Americans and India* (Philadelphia: Temple University Press, 2008); James H. Meriwether, *Proudly We Can Be Africans: Black Americans and Africa, 1935–1961* (Chapel Hill: University of North Carolina Press, 2002); Penny Von Eschen, *Race against Empire: Black Americans and Anti-Colonialism, 1937–1957* (Ithaca, NY: Cornell University Press, 1997); and Brenda Gayle Plummer, *A Rising Wind: Black Americans and Foreign Affairs, 1935–1960* (Chapel Hill: University of North Carolina Press, 1996). Key works on the history of international institutions include Elizabeth Borgwardt, *A New Deal for the World: America's Vision for Human Rights* (Cambridge, MA: Belknap Press of Harvard University Press, 2005); Akira Iriye, *Global Community: The Role of International Organizations in the Making of the Contemporary World* (Berkeley: University of California Press, 2002); and Patrick Hearden, *Architects of Globalism: Building a New World Order during World War II* (Fayetteville: University of Arkansas Press, 2002).

19. Political scientist Joseph S. Nye Jr. was among the first to theorize the distinction between "soft" and "hard" power; see, for example, *Soft Power: The Means to Success in World Politics* (New York: Public Affairs, 2005). On how Americans reconciled increasingly destructive methods of warfare with cultural narratives of national innocence and benevolence, see Sahr Conway-Lanz, *Collateral Damage: Americans, Noncombatant Immunity, and Atrocity after World War II* (New York: Routledge, 2006).

20. Peter Fritzsche, *A Nation of Fliers: German Aviation and the Popular Imagination* (Cambridge, MA: Harvard University Press, 1994); Scott W. Palmer, *Dictatorship of the Air: Aviation Culture and the Fate of Modern Russia* (New York: Cambridge University Press, 2006); Willie Hiatt, "The Rarefied Air of the Modern: Aviation and Peruvian Participation in World History, 1910–1950," Ph.D. dissertation, University of California, Davis, 2009; Daniel R. Headrick, *Power over Peoples: Technology, Environments, and Western Imperialism, 1400 to the Present* (Princeton, NJ: Princeton University Press, 2010), 302–333; Joanne London, *Fly Now! The Poster Collection of the Smithsonian National Air and Space Museum* (Washington, DC: National Geographic, 2007).

21. Department of Defense, Base Structure Report, Fiscal Year 2012 Baseline, www.acq.osd.mil/ie/download/bsr/BSR2012Baseline.pdf (accessed 10/22/12); Joseph Corn, *The Winged Gospel: America's Romance with Aviation, 1900–1950* (New York: Oxford University Press, 1983).

22. Michael Adas, *Dominance by Design: Technological Imperatives and America's Civilizing Mission* (Cambridge: Belknap Press of Harvard University Press, 2006), 129–118; John Krige, *American Hegemony and the Postwar Reconstruction of Science in Europe* (Cambridge, MA: The MIT Press, 2006), 4–6. See also Sheila Jasanoff, ed., *States of Knowledge: The Co-Production of Science and Social Order* (London: Routledge, 2004).

23. Thomas Bender, *A Nation among Nations: America's Place in World History* (New York: Hill and Wang, 2006), 4. On American and world historiography after the transnational turn, see, especially, "Diplomatic History Today: A Roundtable," *Journal of American History* 95 (2009): 1053–1091; "*AHR* Conversation: On

Transnational History," *American Historical Review* 111 (2006): 1441–1464; Thomas Bender, ed., *Rethinking American History in a Global Age* (Berkeley: University of California Press, 2002); Bruce Mazlish, ed., *The New Global History* (New York: Routledge, 1996); and Michael Geyer and Charles Bright, "World History in a Global Age," *American Historical Review* 100 (1995): 1034–1060. Scholarship that effectively integrates aviation into international history includes Jeffrey A. Engel, *Cold War at 30,000 Feet: The Anglo-American Fight for Aviation Supremacy* (Cambridge, MA: Harvard University Press, 2007); Alan P. Dobson, *Peaceful Air Warfare: The United States, Britain, and the Politics of Aviation* (New York: Oxford University Press, 1991); and Eric Paul Roorda, "The Cult of the Airplane among U.S. Military Men and Dominicans during the U.S. Occupation and the Trujillo Regime," in Gilbert Joseph, Catherine LeGrand, and Ricardo Salvatore, eds., *Close Encounters of Empire* (Durham, NC: Duke University Press, 1998), 269–310.

1. The Americanization of the Airplane

1. Roger E. Bilstein, *Flight in America: From the Wrights to the Astronauts* (Baltimore: Johns Hopkins University Press, revised edition, 1994), 12; telegram, Orville Wright to Bishop Milton Wright, December 17, 1903, in Marvin W. McFarland, ed., *The Papers of Wilbur and Orville Wright* (New York: McGraw-Hill, 1953), 397.

2. R. E. G. Davies, *A History of the World's Airlines* (New York: Oxford University Press, 1964), 11–38, 57.

3. Quoted in "Current Comment," *Popular Aviation*, May 1928, 28.

4. Editorial, *Aeronautical Digest*, August 1922, 11; Frederick Jackson Turner, "The Significance of the Frontier in American History," address to the American Historical Association, 1893, http://xroads.virginia.edu/~Hyper/TURNER/ (accessed 8/8/12). For further analysis of aviation as a form of frontier conquest, see David Courtwright, *Sky as Frontier: Adventure, Aviation, and Empire* (College Station: Texas A&M University Press, 2005).

5. On aviation and nationalism in countries other than the United States, see Robert Wohl, *A Passion for Wings: Aviation and the Western Imagination, 1908–1918* (New Haven, CT: Yale University Press, 1994); Peter Fritzsche, *A Nation of Fliers: German Aviation and the Popular Imagination* (Cambridge, MA: Harvard University Press, 1992); Scott Palmer, *Dictatorship of the Air: Aviation Culture and the Fate of Modern Russia* (New York: Cambridge University Press, 2006); and Willie Hiatt, "The Rarefied Air of the Modern: Aviation and Peruvian Participation in World History, 1910–1950," Ph.D. dissertation, University of California, Davis, 2009.

6. Bilstein, *Flight in America*, 13–14; Tom D. Crouch, *The Bishop's Boys: A Life of the Wright Brothers* (New York: W. W. Norton, 1989), 258–263.

7. Statement by Wilbur and Orville Wright to the Associated Press, January 5, 1904, in Peter L. Jakab and Rick Young, eds., *The Published Writings of Wilbur and Orville Wright* (Washington, DC: Smithsonian Institution Press, 2000), 14–15; Crouch, *Bishop's Boys*, 280–281.

8. Crouch, *Bishop's Boys*, 284–285; A. I. Root, "Our Homes," *Gleanings in Bee Culture*, January 1, 1905, memory.loc.gov/cgi-bin/query/r?ammem/wright:@field(DOCID+@lit(wright002536) (accessed 3/9/13); "The Wright Aeroplane and Its Fabled Performances," *Scientific American*, January 13, 1906, 40; "Fliers or Liars," *New York Herald*, February 10, 1906.

9. Wilbur Wright to Octave Chanute, June 1, 1905, *Papers of Wilbur and Orville Wright*, 494; Crouch, *Bishop's Boys*, 293.

10. Wilbur and Orville Wright to the Secretary of War, October 9, 1905, *Papers of Wilbur and Orville Wright*, 514–515; Wilbur Wright to Chanute, June 1, 1905; Bishop Milton Wright's Diary, December 28, 1905, *Papers of Wilbur and Orville Wright*, 540; agreement between Wilbur and Orville Wright and Arnold Fordyce and Henri Bonel, April 1906, Box 2, WB, WSU; Crouch, *Bishop's Boys*, 289–310.

11. Wohl, *A Passion for Wings*, 38; contrat entre Mr. Lazare Weiller et Mr. Wilbur Wright et Orville Wright, March 3, 1908, Box 2, WB, WSU.

12. Wilbur Wright to Orville Wright, August 15, 1908, *Papers of Wilbur and Orville Wright*, 912; Wilbur Wright to Orville Wright, October 4, 1908, in Fred C. Kelly, ed., *Miracle at Kitty Hawk: The Letters of Wilbur and Orville Wright* (New York: Farrar, Straus, 1951), 321; letters from Wilbur Wright, August–October 1908, *Miracle at Kitty Hawk*, 294–295, 310, 323.

13. U.S. Army Signal Corps Specification No. 486, "Advertisement and Specification for a Heavier-than-Air Flying Machine," December 23, 1907; Articles of Agreement between Charles S. Wallace, Captain, U.S. Army Signal Corps, and Wilbur and Orville Wright, February 10, 1908, Box 2, WB, WSU; "A Creature of the Air," *Washington Post*, September 11, 1908; Crouch, *Bishop's Boys*, 371–381.

14. Arthur Ruhl, "Up in the Air with Orville Wright," *Collier's*, July 2, 1910.

15. Arthur W. Page, "How the Wrights Discovered Flight," *The World's Work*, August 1910; "The Wright Flyer" advance announcement, 1911, Box 2, WB, WSU.

16. Joseph Corn, *The Winged Gospel: America's Romance with Aviation, 1900–1953* (New York: Oxford University Press, 1983), 24–30; Bilstein, *Flight in America*, 29.

17. Courtwright, *Sky as Frontier*, 30–34.

18. Dominick A. Pisano, "The Greatest Show Not on Earth: The Confrontation between Utility and Entertainment in Aviation," in Pisano, ed., *The Airplane in American Culture* (Ann Arbor: University of Michigan, 2003), 44–51; Alex Roland, *Model Research: The National Advisory Committee for Aeronautics, 1915–1958* (Washington, DC: National Aeronautics and Space Administration, 1985), 25, 29, 394–395.

19. John B. Rae, *Climb to Greatness: The American Aircraft Industry, 1920–1960* (Cambridge, MA: The MIT Press, 1968), 1, 24; Herbert A. Johnson, *Wingless Eagle: U.S. Army Aviation through World War I* (Chapel Hill: University of North Carolina Press, 2001), 55; Joseph Ames to W. F. Duran, August 10, 1918, Folder 18781, NASA.

20. Committee of the Board of Regents of the Smithsonian Institution, "Memorial on the Need of a National Advisory Committee for Aeronautics," February 1, 1915, Folder 19197, NASA; Henry H. Arnold, *Global Mission* (New York: Arno Press, 1972 [1949]), 52; Rae, *Climb to Greatness*, 1; Johnson, *Wingless Eagle*, 5, 90–115, 187–188.

21. Roland, *Model Research*, 41–51; Rae, *Climb to Greatness*, 2–3; John Morrow, *The Great War in the Air: Military Aviation from 1909 to 1921* (Washington, DC: Smithsonian Institution Press, 1993), 268, 336–343; Courtwright, *Sky as Frontier*, 46–52; Corn, *Winged Gospel*, 12–13.

22. Joseph Ames to John J. Pershing, September 9, 1921, Folder 18780, NASA; George de Bothezat to NACA, "Short Note on My Trip to Langley Field," June 4, 1919, Folder 18833, NASA; Stephen Budiansky, *Air Power: The Men, Machines, and Ideas That Revolutionized War, from Kitty Hawk to Gulf War II* (New York: Viking, 2004), 148–149; William Mitchell, *Our Air Force, the Keystone of National Defense* (New York: E. P. Dutton, 1921) and *Winged Defense: The Development and Possibilities of Modern Air Power—Economic and Military* (New York: G. P. Putnam's Sons, 1925); Count Jean de Strelecki, "Russia's Air Force Is Growing," *Aero Digest*, June 1925, 295–296.

23. Editorial, *Aero Digest*, December 1924, 360; National Advisory Committee on Aeronautics, "A National Aviation Policy," November 15, 1920, Box 4, Entry 1, RG 255, NARA; Ralph W. Cram, "Rallying Public Opinion," *Aero Digest*, April 1924, 224; Edward Marshall, "Daniel Guggenheim Says We Must Not Let Air Supremacy Slip," *Aero Digest*, August 1925, 407–409; editorial, *Aeronautical Digest*, November 1922, 190; editorial, *Aeronautical Digest*, December 1922, 260.

24. Henry Woodhouse, "The Wonderful Prospects of American Aeronautics," *Aerial Age Weekly*, March 22, 1915, 5; Harry F. Guggenheim, *The Seven Skies* (New York: G. Putnam's Sons, 1930), 208–209; Arnold, *Global Mission*, 97.

25. Ernest A. McKay, *A World to Conquer: The Epic Story of the First Around-the-World Flight by the U.S. Army Air Service* (New York: Arno, 1981), 5–10. On the Great White Fleet, see Henry J. Hendrix, *Theodore Roosevelt's Naval Diplomacy: The U.S. Navy and the Birth of the American Century* (Annapolis, MD: Naval Institute Press, 2009); and James R. Reckner, *Teddy Roosevelt's Great White Fleet* (Annapolis, MD: Naval Institute Press, 1988).

26. Memorandum, J. E. Fechet to Major F. L. Martin, December 22, 1923, Box 1, Entry 140, RG 18, NARA; Carroll V. Glines and Stan Cohen, *The First Flight around the World* (Missoula, MT: Pictoral Histories, 2000), 9.

27. McKay, *World to Conquer*, 18–49; www.boeing.com/history/mdc/dwc.htm (accessed 8/8/12); Glines and Cohen, *First Flight around the World*, 16.

28. Memorandum, State Department Division of Far Eastern Affairs, August 30, 1923, and letters between the U.S. Ambassador to Japan and the Secretary of State, Box 1, AWF, NASM.

29. McKay, *World to Conquer*, 25; Lowell Thomas, *First World Flight* (Boston: Houghton Mifflin, 1925), 4.

30. McKay, *World to Conquer*, 60–74, 147–150.

31. Fechet to Martin; McKay, *World to Conquer*, 176–188; Glines and Cohen, *First Flight around the World*, 137.

32. "World Flight Now History," *Aero Digest*, November 1924, 263; Thomas, *The First World Flight*, 194; Commander, Destroyer Division 38, to Commander in Chief, U.S. Asiatic Fleet, June 13, 1924, Box 1, AWF, NASM; U.S. Consulate Amoy to Secretary of State, July 21, 1924, Box 2, AWF, NASM.

33. Commander, Destroyer Division 45, to Commander in Chief, U.S. Asiatic Fleet, July 22, 1924, Box 1, AWF, NASM; U.S. Consulate Bushire to Secretary of State, July 12, 1924; U.S. Legation Vienna to Secretary of State, July 14, 1924; and U.S. Legation Bogotá to Secretary of State, March 4, 1925, Box 2, AWF, NASM.

34. Woods to Secretary of State, May 24 and May 26, 1924, Box 1, AWF, NASM; U.S. Consulate Nagasaki to Secretary of State, June 7, 1924, Box 2, AWF, NASM; Thomas, *The First World Flight*, 126, 137, 147.

35. Thomas, *First World Flight*, 221; Mason M. Patrick, "The World from Above," *Aero Digest*, January 1925, 47; Clifford Albion Tinker, "Modern Magellans," *Aero Digest*, May 1924, 276–277; "Magellans of the Air," *New York Times*, September 7, 1924; "Everybody Soon to Fly around the World," *Washington Post*, October 12, 1924; covers, *Aeronautical Digest*, January 1924, and *Aero Digest*, October 1924. Note that this is the same magazine; it changed its name in mid-1924.

36. "Around the World," *Washington Post*, February 11, 1924; "Magellans of the Air"; "Development of Aviation," *Washington Post*, April 1, 1924.

37. Thomas, *First World Flight*, 187, 194–195, 220; "Thrilling Adventures of 'Round the World Flyers," *Hartford Courant*, January 9, 1925. On American orientalism and the vogue for Asian culture in the United States during this era, see Mari Yoshihara, *Embracing the East: White Women and American Orientalism* (New York: Oxford University Press, 2002); and Kristin Hoganson, *A Consumers' Imperium: The Global Production of American Domesticity, 1865–1920* (Chapel Hill: University of North Carolina Press, 2007), 13–56.

38. See Nayan Shah, *Contagious Divides: Epidemics and Race in San Francisco's Chinatown* (Berkeley: University of California Press, 2001), 1–76.

39. Richard P. Hallion, *Legacy of Flight: The Guggenheim Contribution to American Aviation* (Seattle: University of Washington Press, 1977), 45, 69, 85–86; Rae, *Climb to Greatness*, 23; *Report of President's Aircraft Board* (Washington, DC: U.S. Government Printing Office, 1925).

40. Roger E. Bilstein, *Flight Patterns: Trends of Aeronautical Development in the United States, 1918–1929* (Athens: University of Georgia Press, 1983); F. Robert van der Linden, *Airlines and Airmail: The Post Office and the Birth of the Commercial Aviation Industry* (Lexington: University Press of Kentucky, 2002); "Uses of the Air Mail Service," *New York Times*, August 15, 1924; Bilstein, *Flight Patterns*, 41.

41. National Advisory Committee on Aeronautics, "A National Aviation Policy"; Lieutenant Commander Clifford A. Tinker, "Is America First in the Air?" *Aeronautical Digest*, January 1924, 28–29, 62; Stuart Banner, *Who Owns the Sky? The Struggle to Control Airspace from the Wright Brothers On* (Cambridge, MA:

Harvard University Press, 2008), 144–168; Bilstein, *Flight in America*, 51; "Opening the National Airways," *New York Times*, August 16, 1926.

42. Scott Berg, *Lindbergh* (New York: Berkley Books, 1998), 81–111; R. E. G. Davies, *Charles Lindbergh: An Airman, His Aircraft, and His Great Flights* (McLean, VA: Paladwr Press, 1997), 10.

43. Charles A. Lindbergh, *We* (New York: Grosset & Dunlap, 1927), 246.

44. Charles A. Lindbergh, *Spirit of St. Louis* (New York: Scribner, 1953), 444–451; Edwin L. James, "LINDBERGH DOES IT! To Paris in 33 1/2 Hours; Flies 1,000 Miles through Snow and Sleet; Cheering French Carry Him off Field," *New York Times*, May 22, 1927; Berg, *Lindbergh*, 136, 144, 150–151, 158, 160; Courtwright, *Sky as Frontier*, 80–81; Corn, *Winged Gospel*, 22. Sheet music: Box 6, AEC, WSU, and Box 500, CAL, Yale.

45. Richard J. Beamish, *The Boy's Story of Lindbergh, the Lone Eagle* (Chicago: John C. Winston, 1928), 251; editorial, *Aero Digest*, June 1927, 558; Ross H. Rohrer to Lindbergh, June 7, 1927, and Billy Darr to Lindbergh, February 29, 1928, Box 44, CAL, Yale.

46. Berg, *Lindbergh*, 112; John W. Ward, "The Meaning of Lindbergh's Flight," *American Quarterly*, Spring 1958, 6; Walter Hixson, *Charles A. Lindbergh: Lone Eagle* (New York: Harper Collins, 1996), 40–43; Michael Adas, *Machines as the Measure of Men: Science, Technology, and Ideologies of Western Dominance* (Ithaca, NY: Cornell University Press, 1989), 381.

47. George Buchanan Fife, *Lindbergh: The Lone Eagle* (New York: A. L. Burt Company, 1927), 46; James E. West, *The Lone Scout of the Sky* (New York: Boy Scouts of America, 1927), 142.

48. Earl Reeves, *Lindbergh Flies On!* (New York: Robert M. McBride and Company, 1929), 5; Fife, *Lindbergh*, 41; West, *Lone Scout of the Sky*, frontispiece, 143–146, 150–152; Berg, *Lindbergh*, 33, 42; Lindbergh, *Spirit of St. Louis*, 200–201.

49. Beamish, *Boys' Story of Lindbergh*, 130, 132, 224; Fife, *Lindbergh: The Lone Eagle*, 35; Reeves, *Lindbergh Flies On!*, 22.

50. Beamish, *Boys' Story of Lindbergh*, 193; employees of Wauseon, Ohio, Post Office to Charles Lindbergh, June 9, 1927, Box 44, CAL, Yale.

51. Thomas Bentley Mott, *Myron T. Herrick, Friend of France: An Autobiographical Biography* (Garden City, NY: Doubleday, 1929), 352; Frank C. Costigliola, *Awkward Dominion: American Political, Economic, and Cultural Relations with Europe, 1919–1933* (Ithaca, NY: Cornell University Press, 1984), 180–181.

52. Myron Herrick, quoted in Lindbergh, *We*, 246; U.S. Embassy France to Secretary of State, CDF 800.51 W8 France/4951/2, RG 59, NARA; Marguerite Dumay to Lindbergh, May 23, 1927, Box 44, CAL, Yale; Leonard Mosley, *Lindbergh: A Biography* (New York: Doubleday, 1976), 113–115; Hixson, *Charles A. Lindbergh*, 37; Brooke Blower, *Becoming Americans in Paris: Transatlantic Culture and Politics between the World Wars* (New York: Oxford University Press, 2011), 49–51.

53. Hixson, *Charles A. Lindbergh*, 44; Vicente Botero, Eduardo Arango, and P. A. Martinez to Lindbergh, Mme. Legros, Garabed Minassian, and Sophi Kvisnizky to Lindbergh, and other letters, Box 44, CAL, Yale.

54. Quoted in Berg, *Lindbergh*, 137.

55. Sybil Bryant Poston, "Lindbergh and the Negro Problem," *New York Amsterdam News*, June 15, 1927; letter from George Wallace Hunter, *New York Amsterdam News*, June 22, 1927; "What Will the Negro Contribute to Aviation?" *New York Amsterdam News*, June 22, 1927; speech by Josephine Baker, Keil Opera House, February 3, 1952, www.sheldonconcerthall.org/pdf/BakerSpeechKiel.pdf (accessed 8/10/12). Ironically, as Brooke Blower notes, Baker was celebrating Lindbergh's flight at a chic Parisian restaurant when a white couple asked the headwaiter to throw her out, stating, "in America this is not done . . . a nigger woman belongs in the kitchen." The restaurant manager, however, allowed Baker to stay. See Baker's Keil Opera House speech and Blower, *Becoming Americans in Paris*, 51. On Garvey and the UNIA, see Colin Grant, *Negro with a Hat: The Rise and Fall of Marcus Garvey* (New York: Oxford University Press, 2008).

56. NACA 13th Annual Report, December 8, 1927, Box 4, Entry 1, RG 255, NARA; Courtwright, *Sky as Frontier*, 82; Rae, *Climb to Greatness*, 49; Bilstein, *Flight Patterns*, 42, 50–53, 130; Davies, *Charles Lindbergh*, 39–55; John Dos Passos, *Airways, Inc.* (New York: MacAulay Company, 1928), 77.

57. Guggenheim, *Seven Skies*, 111; Berg, *Lindbergh*, 161–162; Corn, *Winged Gospel*, 23. On Lindbergh's complex attitudes toward his fame, see Charles L. Ponce de Leon, "The Man Nobody Knows: Charles A. Lindbergh and the Culture of Celebrity," in Pisano, ed., *The Airplane in American Culture*, 75–101.

58. NACA 13th Annual Report.

2. Good Neighbors Are Close Neighbors

1. Cordell Hull, *Memoirs: Volume I* (New York: Macmillan, 1948), 319–332. For further analysis of the film and its relation to the history of both aviation and Hollywood, see Rosalie Schwartz, *Flying Down to Rio: Hollywood, Tourists, and Yankee Clippers* (College Station: Texas A&M Press, 2004).

2. See Greg Grandin, *Empire's Workshop: Latin America, the United States, and the Rise of the New Imperialism* (New York: Metropolitan Books, 2006).

3. R. Leslie Cizek, report of trip, June 30, 1930, Box 20, Series I, PAWA, UM; Roger Bilstein, *Flight in America: From the Wrights to the Astronauts* (Baltimore: Johns Hopkins University Press, 1994), 33; Wesley Phillips Newton, "The Role of Aviation in Mexican-United States Relations, 1912–1919," in Eugene R. Huck and Edward H. Moseley, eds., *Militarists, Merchants, and Missionaries: United States Expansion in Middle America* (Tuscaloosa: University of Alabama Press, 1970), 107–130; Panagra advertisement, "Panagra Makes Good Neighbors Close Neighbors," 1943, Ad*Access Online Project #T2294, library.duke.edu/digitalcollections/adaccess.

4. See Eric Paul Roorda, *The Dictator Next Door: The Good Neighbor Policy and the Trujillo Regime in the Dominican Republic, 1930–1945* (Durham, NC: Duke University Press, 1998); Frederick B. Pike, *FDR's Good Neighbor Policy: Sixty Years of Generally Gentle Chaos* (Austin: University of Texas Press, 1995); and Bryce Wood, *The Making of the Good Neighbor Policy* (New York: Columbia University Press, 1961).

5. On the crisis in U.S.-Mexican relations during the 1920s, see, for example, Lorenzo Meyer, *Mexico and the United States in the Oil Controversy, 1917–1942* (Austin: University of Texas Press, 1977); Alan Knight, *U.S.-Mexican Relations, 1910–1940: An Interpretation* (San Diego: University of California Press, 1987); John M. Hart, *Empire and Revolution: The Americans in Mexico since the Civil War* (Berkeley: University of California Press, 2002); Daniela Spenser, *The Impossible Triangle: Mexico, Soviet Russia, and the United States in the 1920s* (Durham, NC: Duke University Press, 1999).

6. On the Nicaraguan civil war and its implications for U.S.–Latin American relations, see Walter LaFeber, *Inevitable Revolutions: The United States in Central America* (New York: W. W. Norton, 1984), 64–69; and Wood, *The Making of the Good Neighbor Policy*, 13–47.

7. "Calles and Morrow Eat Ham and Eggs at Former's Ranch," *New York Times*, November 3, 1927.

8. Charles A. Lindbergh, *Autobiography of Values* (New York: Harcourt, Brace, Jovanovich, 1977), 83.

9. "Lindbergh Plans Trip," *Los Angeles Times*, December 9, 1927; "A Good-Will Flight," *New York Times*, December 10, 1927; "Mexicans Await Lindbergh," *Washington Post*, December 11, 1927.

10. A. Scott Berg, *Lindbergh* (New York: Berkley, 1998), 172–173; Lindbergh, *Autobiography of Values*, 87–88; "Calles Stirred by Emotion of the Occasion," *New York Times*, December 15, 1927; "Calles Says Flight of Lindbergh Brings Priceless Good Will," *Chicago Daily Tribune*, December 15, 1927; "President Calles Takes Aero Flight; Lindbergh Is Pilot," *Washington Post*, December 21, 1927; "Parade of 150,000 before Lindbergh Today Is Expected," *Washington Post*, December 18, 1927; Calvin Coolidge to Charles Lindbergh, December 14, 1927, CDF 811.79612 L64/30F, RG 59, NARA; Manuel Tellez to Secretary of State, December 14, 1927, CDF 811.79612 L64/14, RG 59, NARA; Arthur Frost to Secretary of State, December 23, 1927, CDF 811.79612L64/61, RG 59, NARA.

11. "Mexican Congress Acclaims Coolidge and Col. Lindbergh," *Washington Post*, December 16, 1927; "Mexico's Welcome," *New York Times*, December 15, 1927; Russell Owen, "Amity with Mexico Is Brought Closer by Lindbergh Visit," *New York Times*, December 19, 1927; "Air Diplomacy Next," *Popular Aviation*, February 1928, 30.

12. Lindbergh, *Autobiography of Values*, 88–93; Berg, *Lindbergh*, 174–175; articles by Lindbergh, *New York Times*, December 14, 1927, through February 16, 1928; George T. Summerlin to Secretary of State, January 6, 1928; C. Van H. Engert to Secretary of State, January 31, 1928, and February 7, 1928; Russell to Secretary of State, February 8, 1928; Arthur Bliss Lane to Secretary of State, December 19, 1927; Samuel H. Piles to Secretary of State, January 28, 1928; Piles to Secretary of State, February 3, 1928; Roy Davis to Secretary of State, January 7, 1928, all in CDF 811.79612L64/, RG 59, NARA.

13. Roy Davis to Secretary of State, January 7, 1928, CDF 811.79612L64/115, RG 59, NARA; G. R. Taggart to Secretary of State, February 3, 1928, CDF 811.79612L64/181, RG 59, NARA.

14. Charles A. Lindbergh, "Lindbergh Writes for the *Times* Why He Is Flying to Mexico City," *New York Times*, December 14, 1927; "Lindbergh Tells of Warm Greeting and Honors Paid to Him in Salvador," *New York Times*, January 3, 1928; Lindbergh, *Autobiography of Values*, 91.

15. Speech by Enrique Martínez Sobral, enclosure in Stanley Hawks to Secretary of State, December 31, 1927, CDF 811.7962L64/131, RG 59, NARA.

16. "Salutation to Lindbergh," enclosure in Evan E. Young to Secretary of State, January 17, 1928, CDF 811.79612L64/155, RG 59, NARA; speech by Horacio Vásquez, enclosure in Evan E. Young to Secretary of State, February 3, 1928, CDF 811.79612L64/173, RG 59, NARA; Mario Ribas, "Honduras Y Lindbergh," *Renacimiento*, January 30, 1928.

17. "Lindbergh and Haiti, the Country of a Great Liberator," *New York Amsterdam News*, January 18, 1928.

18. Nicaraguan Autonomous Association to Lindbergh, January 1, 1928, enclosure in Samuel S. Dickson to Secretary of State, January 5, 1928, CDF 811.79612 L64/145, RG 59, NARA; letter from forty-eight Salvadoran students to Lindbergh, enclosure in Samuel S. Dickson to Secretary of State, January 9, 1928, CDF 811.79612L64/146, RG 59, NARA.

19. "Plane Brings out Wounded Marines under Enemy Fire," *New York Times*, January 8, 1928; "Serious Strike Occurs in Port of Nicaragua for Which American Marines Leave Tomorrow," *Atlanta Constitution*, January 8, 1928; "Wounded Marines in Planes Fired on by Sandino Rebels," *Washington Post*, January 8, 1928; "Corinto Strike Draws Marines," *Los Angeles Times*, January 8, 1928; Harold N. Denny, "Lindbergh Flies Safely to Nicaragua," *New York Times*, January 6, 1928; Harold N. Denny, "Lindbergh Waves Adios to Managua," *New York Times*, January 8, 1928; Munro to Secretary of State, January 5, 6, 7, and 12, 1928, CDF 811.79612L64/, RG 59, NARA.

20. Memorandum, J. McDermott to Hengstler, April 21, 1928, CDF 811.79612L64/210, RG 59, NARA.

21. John Hambleton to Lindbergh, February 20, 1928, Box 103, Series III, CAL, Yale; Marilyn Bender and Selig Altschul, *The Chosen Instrument: Pan Am, Juan Trippe, the Rise and Fall of an American Entrepreneur* (New York: Simon and Schuster, 1982), 97–101; Gene Banning, *Airlines of Pan American since 1927* (McLean, VA: Paladwr Press, 2001), 12, 33; Juan T. Trippe to Lindbergh, May 15, 1928, Box 103, Series III, CAL, Yale.

22. "Hundreds of Aerial Lines Could Be Established Immediately in South and Central America," *Aerial Age*, February 14, 1916, 518; R. E. G. Davies, *A History of the World's Airlines* (New York: Oxford University Press, 1964), 71–76; Bilstein, *Flight in America*, 79; H. Case Wilcox, "Air Transportation in Latin America," *Geographical Review* 20 (October 1930): 587–604. See also Dan Hagedorn, *Conquistadors of the Sky: A History of Aviation in Latin America* (Washington, DC: National Air and Space Museum, in association with the University Press of Florida, 2008).

23. "General Patrick Reaches Panama," *New York Times*, January 6, 1924; memorandum, H. H. Arnold to Col. Margetts, April 1, 1925, Box 4, HHA, LOC. On

the Army missions, see Wesley Philips Newton, *The Perilous Sky: U.S. Aviation Diplomacy and Latin America, 1919–1931* (Coral Gables, FL: University of Miami Press, 1978).

24. Memorandum, Mason Patrick to Adjutant General, January 9, 1925, and memorandum, William Lassiter to Adjutant General, June 5, 1925, Box 4, HHA, LOC.

25. Henry H. Arnold, *Global Mission* (New York: Arno, 1972 [1949]), 115, 122; Bender and Altschul, *Chosen Instrument*, 84.

26. Bender and Altschul, *Chosen Instrument*, 22–32, 42–43, 49–51, 59–63.

27. Ibid., 71–88; Banning, *Airlines of Pan American*, 3–7.

28. Banning, *Airlines of Pan American*, 12, 18–20; "Antilles Air Mail Starts from Miami," *New York Times*, January 10, 1929; "West Indian Airways," *Washington Post*, January 11, 1929; "The Eagle Extends His Wings," *Christian Science Monitor*, February 6, 1929.

29. "Invitation to the Christening of the American Clipper by Mrs. Herbert Hoover," October 12, 1931, and Root, Clark, Buckner, Howland, and Ballantine to Henry Breckenridge, January 25, 1929, Box 103, CAL, Yale; Harold Bixby to Lindbergh, March 30, 1939, Box 5, CAL, Yale; "PAA Financial and Traffic Data, Route Mileage, Subsidy, Aircraft, Airports, Radio Stations, etc., 1927–1934 (Comptroller's Black Book)," Box 71, Series I, PAWA, UM; "Argus-Eyed Argonaut," *Time*, February 5, 1940; "Merchant Aerial," *Time*, July 31, 1933; William E. Brown Jr., "Pan Am: Miami's Wings to the World," *Journal of Decorative and Propaganda Arts* 23 (1998): 151–153.

30. Pan American Airways Annual Report, 1929, Box 23, Series I, PAWA, UM; "International Air Transport," report to Federal Aviation Commission, 1934, Box 1, JTT, NASM. On the significance of postal subsidies to the development of the U.S. aviation industry, see F. Robert van der Linden, *Airlines and Airmail: The Post Office and the Birth of the Commercial Aviation Industry* (Lexington: University Press of Kentucky, 2002).

31. See Emily Rosenberg, *Spreading the American Dream: American Economic and Cultural Expansion, 1890–1945* (New York: Hill and Wang, 1982), 138–160.

32. Harold R. Harris to Jill LaPierre, January 25, 1984, Box 4, HRH, WSU; Francis White to Charles C. Eberhardt et al., March 17, 1928, CDF 810.79611PAA/50–53, RG 59, NARA; Morgan to Johnson, January 14, 1928, CDF 810.79611PAA/202, RG 59, NARA; Clarence M. Young, "Linking the Americas by Air," *Aero Digest*, April 1931, 37; Bender and Altschul, *Chosen Instrument*, 125; Rosenberg, *Spreading the American Dream*, 59, 105–107.

33. Statement by Pilot Leo Terletzky, August 14, 1933, Box 20, Series I, PAWA, UM; correspondence between George T. Summerlin and Secretary of State, July 1928 through March 1929, CDF 810.79611PAA, RG 59, NARA; Henry S. Villard to Secretary of State, January 12, 1937, CDF 810.79611PAA/1465, RG 59, NARA; "La Concesión a la Pan American Grace Airways," *El Mercurio*, July 23, 1929. For more on Pan Am's negotiations with Latin American governments, see Newton, *Perilous Sky*, 157–330.

34. Cizek, report of trip; Arthur Geissler to Secretary of State, August 20, 1929, CDF 810.79611PAA/624, RG 59, NARA; Newton, *Perilous Sky*, 169–171.

35. Geissler to Secretary of State, April 14 and April 24, 1928, CDF 810.79611PAA/59 and 810.79611PAA/63, RG 59, NARA.

36. "Lindberghs to Take Four on Flight South," *New York Times*, September 18, 1929; Bender and Altschul, *Chosen Instrument*, 135; Jefferson Caffery to Secretary of State, February 7, 1929, CDF 810.79611PAA/242, RG 59, NARA; Ferdinand Lathrop Mayer to Secretary of State, September 14, 1929, CDF 810.79611PAA/650, RG 59, NARA; Warren Robbins to Secretary of State, October 3, 1929, CDF 810.79611PAA/672, RG 59, NARA. On Lindbergh's courtship of and marriage to Anne Morrow, see Berg, *Lindbergh*, 174–177, 186–188, 192–202.

37. "Air Mail Creating Pan American Ties," *New York Times*, August 24, 1930; Davies, *A History of the World's Airlines*, 150. Compañía Mexicana was a classic example of this strategy, as Pan Am's majority ownership of stock in this "Mexican company" enabled it to circumvent national laws requiring airlines to be operated by Mexican citizens. See Bender and Altschul, *Chosen Instrument*, 107–115.

38. Newton, *Perilous Sky*, 181, 292; R. Henry Norweb to Secretary of State, October 31, 1933, CDF 810.79611PAA/1274, RG 59, NARA; John D. MacGregor to Harold R. Harris, December 19, 1930, Box 1, HRH, WSU.

39. W. I. Van Dusen, "The First Million Miles," *Aeronautic Review*, September 1929, 19–21; William Van Dusen, "Wings over Three Americas," *Scientific American*, October 1931, 234–236; Evan E. Young, "Wings for Our Foreign Trade," *Aero Digest*, November 1932.

40. "Our Real Job," *Pan American Air Ways*, April 19, 1930; "Rush Aid to Managua by P.A.A. Planes during Disaster—Trippe Offers Facilities to U.S. and Latin Governments—P.A.A. Office Wrecked," *Pan American Air Ways*, May 1, 1931; "P.A.A. Radio Stands by in Belize Hurricane—Plane Brings Aid," *Pan American Air Ways*, November 1931; "Fellowships," *New Horizons*, October 1, 1940; "Southbound Students," *New Horizons*, March 1942; "Spanish Papers Reprint Pan American Air Ways," *Pan American Air Ways*, November 15, 1930, all in Box 14, Series I, PAWA, UM; "Air Clinic Great Force for South American Good," *Pan American Air Ways*, February 24, 1930, Box 20, Series I, PAWA, UM.

41. "Do You Know?," and "The Airport Is a Community Center," *Pan American Air Ways*, May–June 1936, Box 14, Series I, PAWA, UM; "Ox-Cart to Air Now Commonplace," *Pan American Air Ways*, May–June 1934, NASM Library.

42. Cizek, report of trip; "Good Neighbor Class," *New Horizons*, December 1942; menu from first sales meeting of Pan American Airways Eastern Division, Miami, October 17–20, 1938, Box 201, Series I, PAWA, UM.

43. "Trade and Tolerance," *New Horizons*, November 1942; "Inter-American Relations," *New Horizons*, May 1941; "Good Neighbor Skyways," *Aviation*, March 1939, 19–20.

44. Noel F. Busch, "Juan Trippe: Pan American Airways' Young Chief Helps Run a Branch of U.S. Defense," *Life*, October 20, 1941, 117; Bender and Altschul, *Chosen Instrument*, 101–102; Trippe, "One America" speech, *Herald Tribune* High

School Forum, March 8, 1947, Box 5, JTT, NASM; Trippe acceptance speech at the Americas Foundation Award Dinner, October 12, 1944, Box 1, JTT, NASM; "Americas Award," *New Horizons*, January–March 1945; Texas Quality Network, "Guardians of Freedom" Broadcast #1: "Challenge to the Guardians of Freedom," February 2, 1942, Box 8, JTT, NASM.

45. Francis and Katharine Drake, "Atlantic Laboratory," *Atlantic Monthly*, November 1933, 592–598; Forrest Wilson, "Eagle Wings," Hearst's *International Cosmopolitan*, August 1933, 24–27.

46. "Displays," *New Horizons*, June 1944, 29. On U.S. fascination with Latin American culture during the interwar years, see Helen Delpar, *The Enormous Vogue of Things Mexican: Cultural Relations between the United States and Mexico, 1920–1935* (Tuscaloosa: University of Alabama Press, 1992).

47. Pan American Airways advertisement, "The Good Neighbor Who Calls Every Day," 1941, Ad*Access Online Project #T2301, JWHC, library.duke.edu/digitalcollections/adaccess.

48. See Lars Schoultz, *Beneath the United States: A History of U.S. Policy toward Latin America* (Cambridge, MA: Harvard University Press, 1998); and Frederick B. Pike, *The United States and Latin America: Myths and Stereotypes of Civilization and Nature* (Austin: University of Texas Press, 1992).

49. "Travel Year," *New Horizons*, June 1941; 1937 Pan American Airways timetable and William Van Dusen's "Señor" Letters, December 1938 and January 1939, Box 119, CAL, Yale.

50. Panagra advertisements, "In the Words of a Passenger on a Panagra Skyway Cruise to South America," 1941, Ad*Access Online Project #T1575; "It's as Easy as This," 1940, #T1570; "It's Springtime in Chile, South America," 1940, #T1571; and "To Rio or New Zealand," 1940, #T1569; all at JWHC, library.duke.edu/digitalcollections/adaccess.

51. John D. MacGregor to Harold R. Harris, December 19, 1930, Box 1, HRH, WSU.

52. It is telling that Woods mistakenly described South America as a "country" instead of a continent, for articles such as his—and, indeed, Pan Am's own advertising strategies—tended to elide differences between various nations and regions, constructing "Latin America" as a homogeneous cultural and geographical entity.

53. Rufus Woods, *Riding the Wings over South America* (Wenatchee, WA: 1944), Box 201, Series I, PAWA, UM.

54. Basil Brewer, "Pan-American Airlines Is Putting Good Will in Good Neighbor Movement of Hemisphere," *Standard-Times Mercury*, February 28, 1941, and "Latin-American Answers—The New North and South Problem," *Standard-Times Mercury*, March 19, 1941; letter from Sumner Welles, March 20, 1941, Box 201, Series I, PAWA, UM.

55. Alice Rogers Hager, *Wings over the Americas* (New York: Macmillan, 1940), 23.

56. Eugene Staley, "The Myth of the Continents," *Foreign Affairs* 19 (1941): 484–494.

57. Special Interdepartmental Committee on the Development of Aviation in the Western Hemisphere, Memorandum for the President, "Plan for Aeronautical

Development in the Western Hemisphere," August 4, 1939; G. Grant Mason Jr. to Hull, July 7, 1939; William D. Leahy to Mason, June 17, 1939; Harry K. Woodring to Mason, May 25, 1939; Cordell Hull to Mason, May 19, 1939; all in OF 3760, PP, FDR, FDRPL.

58. "The Airport Development Program of Pan American Airways," undated report, Box 132, Series II, PAWA, UM; memorandum of conversation re: proposed contract between Pan American Airways and War Department, September 14, 1940, CDF 810.79611-PAA/2000, RG 59, NARA.

59. War Department Contract No. W-1097-eng-2321, November 2, 1940, and Executive System Memoranda No. 52, November 16, 1940, Box 15, Series I, PAWA, UM; L. P. O'Connor to Samuel F. Pryor, June 1, 1944, Box 132, Series I, PAWA, UM; undated Pan American Airways memorandum on ADP contracts and charges, Box 6, Series II, PAWA, UM; "The Airport Development Program of Pan American Airways"; remarks by John C. Leslie at Clipper Pioneers' Reunion, October 2, 1975, Box 29, Series I, PAWA, UM.

60. Arnold, *Global Mission*, 204; Welles to Hackworth, June 27, 1942, CDF 810.79611-PAA/6-2542, RG 59, NARA; Adolf Berle to Robert Lovett, July 10, 1943, CDF 810.79611-PAA/3395, RG 59, NARA; "ADP Development," *New Horizons*, June 1942; Basil Brewer, "Lindbergh Omitted Something—When You Say 'That'—Smile," *Standard-Times Mercury*, March 1, 1941; "The Airport Development Program of Pan American Airways"; Banning, *Airlines of Pan American*, 504–505.

61. Van Dusen to Graham Grosvenor [undated] and press release, "Inter-American Cooperation Builds Strategic Air Field," Box 259, Series I, PAWA, UM; O'Connor to ADP Engineers, August 14, 1942, Box 6, Series II, PAWA, UM; "Jungle School," *New Horizons*, April 1942.

62. "French Guiana Airport Construction Report," Box 43, Series II, PAWA, UM; "Scale of Daily Wages Paid by PAA-ADP to Local Labor in Foreign Countries," Box 136, Series I, PAWA, UM.

63. "Chronological Record of Building Amapa Airport," Box 259, Series I, PAWA, UM; ADP Public Relations, "Report, Cochabamba—Bolivia," Box 43, Series II, PAWA, UM.

64. O'Connor to Van Dusen, November 13, 1943, Folder 12, Series II, PAWA, UM; "Tropical Triumph," *New Horizons*, November 1942.

65. Philip O. Chalmers to Philip Bonsal, September 28, 1943, CDF 810.79611-PAA/9-2843, RG 59, NARA; Bender and Altschul, *Chosen Instrument*, 333–335.

66. William Van Dusen, "The Jungle Airport That Stopped the Nazis," *True*, August 1962.

67. Van Dusen, "Señor Letter: Review of 1937/Preview of 1938," Box 119, CAL, Yale; Hull, *Memoirs*, 309, 321, 349.

3. Global Visions, National Interests

1. Henry Luce, "The American Century," *Life*, February 7, 1941, reprinted in *Diplomatic History* 23 (1999): 159–171.

2. Ibid., 165, 169–170. Clare Boothe Luce, "America in the Post-War Air World," *Congressional Record*, House of Representatives, 78th Session, February 9, 1943, 759–764.

3. Matthew Josephson, *Empire of the Air: Juan Trippe and the Struggle for World Airways* (New York: Harcourt, Brace, 1944), 13; Marilyn Bender and Selig Altschul, *The Chosen Instrument: Pan Am, Juan Trippe, the Rise and Fall of an American Entrepreneur* (New York: Simon and Schuster, 1982), 335–337, 344, 374; W. A. Swanberg, *Luce and His Empire* (New York: Charles Scribner's Sons, 1972), 85, 214; John Kobler, *Luce: His Time, Life, and Fortune* (Garden City, NY: Doubleday, 1968), 85, 108; James L. Baughman, *Henry R. Luce and the Rise of the American News Media* (Boston: Twayne, 1987), 2, 5–7, 53, 99; Alan Brinkley, *The Publisher: Henry Luce and His American Century* (New York: Vintage, 2011), 28, 172, 214.

4. Alan P. Dobson, *FDR and Civil Aviation: Flying Strong, Flying Free* (New York: Palgrave Macmillan, 2011), 4, 29–43, 74–75; Roger E. Bilstein, *Flight in America: From the Wrights to the Astronauts* (Baltimore: Johns Hopkins University Press, 1994), 96–97, 100, 104, 173; T. A. Heppenheimer, *Turbulent Skies: The History of Commercial Aviation* (New York: J. Wiley & Sons, 1995), 124; Frederick Graham, "Travel Transformed by Wings," *New York Times*, August 2, 1938; Frederick Graham, "America Spreads Her Wings Wider over Land and Sea," *New York Times*, October 15, 1939.

5. Bender and Altschul, *Chosen Instrument*, 226.

6. Pan American Airways, "Chronology of Transatlantic Air Service," Box 120, CAL, Yale.

7. Thomas J. McCormick, *China Market: America's Quest for Informal Empire, 1893–1901* (Chicago: Quadrangle Books, 1967), 129; William Appelman Williams, *The Tragedy of American Diplomacy* (New York: W. W. Norton, 1972 [1959]), 50. On the origins of the Open Door policy, see Michael Hunt, *The Making of a Special Relationship: The United States and China to 1914* (New York: Columbia University Press, 1983), 152–154. On U.S. ambitions in the Pacific, see, for example, McCormick, *China Market*; Bruce Cumings, *Dominion from Sea to Sea: Pacific Ascendancy and American Power* (New Haven, CT: Yale University Press, 2009); Arthur Power Dudden, *The American Pacific: From the Old China Trade to the Present* (New York: Oxford University Press, 1992); and John R. Eperjesi, *The Imperialist Imaginary: Visions of Asia and the Pacific in American Culture* (Hanover, NH: Dartmouth University Press, 2004).

8. Bender and Altschul, *Chosen Instrument*, 202–226; Carl Solberg, *Conquest of the Skies: A History of Commercial Aviation in America* (Boston: Little, Brown, 1979), 229–232; Charles J. Kelly Jr., *The Sky's the Limit* (New York: Coward-McCann, 1963), 133.

9. Bender and Altschul, *Chosen Instrument*, 233.

10. Bender and Altschul, *Chosen Instrument*, 234–235; Robert Gandt, *Skygods: The Fall of Pan Am* (New York: Morrow, 1995), 74–75, 108; James Trautman, *Pan American Clippers: The Golden Age of Flying Boats* (Erin, ON: Boston Mills Press, 2007), 103, 115, 133; "Pacific Past," *New Horizons*, March 1943, 22.

11. Gandt, *Skygods*, 97; Roy Allen, *The Pan Am Clipper* (London: David & Charles, 2000), 109; "Clipper off Today on Pacific Mail Hop," *New York Times*, November 22, 1935; "Regular Air Route to the Orient Opens," Associated Press, November 17, 1935; Solberg, *Conquest of the Skies*, 232–234; Bender and Altschul, *Chosen Instrument*, 248–250; Juan T. Trippe, "Remarks upon Take-Off of *China Clipper*," November 22, 1935, Box 10, JTT, NASM.

12. Waldo Drake, "How Cloud-Riding Boats Will Span Broad Pacific," *Los Angeles Times*, November 17, 1935; "Regular Air Route to the Orient Opens," Associated Press, November 17, 1935; "The Conquest of the Pacific," *Los Angeles Times*, November 22, 1935.

13. William Stephen Grooch, *Skyway to Asia* (New York: Longmans, Green, 1936), 205; "Plane Trip around World Now Available to Public," Associated Press, November 23, 1936; Gene Banning, *Airlines of Pan American since 1927* (McLean, VA: Paladwr Press, 2001), 426–432; Bender and Altschul, *Chosen Instrument*, 251.

14. Bender and Altschul, *Chosen Instrument*, 284–288; Solberg, *Conquest of the Skies*, 236; "China Clipper Greeted by Throng at Harbor," *Los Angeles Times*, November 11, 1935.

15. Bender and Altschul, *Chosen Instrument*, 300–301.

16. William Van Dusen, "Señor Letter," November 13, 1939, Box 119, CAL, Yale; Pan American Airways, "Chronology of Transatlantic Air Service"; Gandt, *Skygods*, 139–140; Kelly, *The Sky's the Limit*, 138–139. Pan Am's flights to France terminated in the port city of Marseille rather than Paris because the airline then used seaplanes, not landplanes.

17. Trippe to Henry H. Arnold, February 24, 1939, Box 20, HHA, LOC; Harrison Forman, "Trippe around the World," *Today*, October 19, 1935, 5. On Clipper ships, see Carl C. Cutler, *Greyhounds of the Sea: The Story of the American Clipper Ship* (Annapolis, MD: United States Naval Institute, 1961).

18. Solberg, *Conquest of the Skies*, 241; Trautman, *Pan American Clippers*, 110; Pan American Airways press release, June 12, 1939, Box 120, CAL, Yale.

19. Noel F. Busch, "Juan Trippe: Pan American Airway's Young Chief Helps Run a Branch of U.S. Defense," *Life*, October 1941, 112; Bender and Altschul, *Chosen Instrument*, 344–345.

20. Busch, "Juan Trippe," 111. These figures include Pan Am's foreign subsidiaries AVIANCA, Compagne Mexicana de Aviation, China National Airways Corporation, Panagra, Panair, Pan American Airways-Africa, and Pan American Air Ferries. Banning, *Airlines of Pan American*, 506.

21. John Zukowsky, ed., *Building for Air Travel: Architecture and Design of Commercial Aviation* (Chicago and Munich: Art Institute of Chicago and Prestel-Verlag, 1996), 110; Solberg, *Conquest of the Skies*, 237–245; Trautman, *Pan American Clippers*, 225.

22. Bender and Altschul, *Chosen Instrument*, 252–253; Gandt, *Skygods*, 106; Larry Weirather, *The China Clipper, Pan American Airways, and Popular Culture* (Jefferson, NC: McFarland, 2007), 66–74.

23. Trautman, *Pan American Clippers*, 203–212; Weirather, *China Clipper*, 16, 91–95, 164.

24. Weirather, *China Clipper,* 149–160.

25. Gandt, *Skygods,* 98.

26. Bilstein, *Flight in America,* 127–128; John B. Rae, *Climb to Greatness: The American Aircraft Industry, 1920–1960* (Cambridge, MA: MIT Press, 1968), 104, 108.

27. Orville H. Bullitt, ed., *For the President, Personal and Secret: Correspondence between Franklin D. Roosevelt and William C. Bullitt* (Boston: Houghton Mifflin, 1972), 288; Steve Casey, *Cautious Crusade: Franklin D. Roosevelt, American Public Opinion, and the War against Nazi Germany* (New York: Oxford University Press, 2001), 9; Robert Dallek, *Roosevelt's Diplomacy and World War II* (New York: Holt, Rinehart, and Winston, 1970), 171–175; John Buckley, *Air Power in the Age of Total War* (Bloomington: Indiana University Press, 1999), 114–124; Bilstein, *Flight in America,* 128–129; Michael S. Sherry, *The Rise of American Air Power: The Creation of Armageddon* (New Haven, CT: Yale University Press, 1982), 82. See also Jeffrey S. Underwood, *The Wings of Democracy: The Influence of Air Power on the Roosevelt Administration, 1933–1941* (College Station: Texas A&M University Press, 1991).

28. Rae, *Climb to Greatness,* 113–117, 139.

29. Bilstein, *Flight in America,* 159–161; Rae, *Climb to Greatness,* 172; Dominick A. Pisano, *To Fill the Sky with Pilots: The Civilian Pilot Training Program, 1939–46* (Urbana: University of Illinois Press, 1993).

30. American Airlines, "Here, There, Anywhere: How an Airline Became a Lifeline around the World," 1944, Folder F1A-500000-13, ATC, NASM Library; "The World's Greatest Airline," *Fortune,* August 1945, 159, 210; Bilstein, *Flight in America,* 162; Anthony Sampson, *Empires of the Sky: The Politics, Contests, and Cartels of World Airlines* (New York: Random House, 1984), 62; Kelly, *The Sky's the Limit,* 150–151. See also Oliver LaFarge, *The Eagle in the Egg* (Boston: Houghton Mifflin, 1949).

31. Joseph J. Corn, *The Winged Gospel: America's Romance with Aviation, 1900–1950* (New York: Oxford University Press, 1983), 113–134; "Education: High Schools, Air-Conditioned," *Time,* October 12, 1942, 74; "New School Texts Based on Aviation," *New York Times,* December 21, 1942; "Air Education," *Life,* August 20, 1945, 69–72; Lucien Aigner, "Education for the 'Air Age,'" *New York Times Sunday Magazine,* May 31, 1942.

32. J. Parker Van Zandt, *America Faces the Air Age* (Washington, DC: Brookings Institution, 1944), 74. Other social scientific analyses of aviation included T. P. Wright, "Aviation's Place in Civilization" (Washington, DC: U.S. Government Printing Office, 1945); William F. Ogburn, *The Social Effects of Aviation* (New York: Houghton Mifflin, 1946); and the journal *Air Affairs,* which began publication in 1946.

33. American Airlines advertisement, "Teach Them the World," Ad*Access Online Project #T2391, JWHC, library.duke.edu/digitalcollections/adaccess; Elizabeth Borgwardt, *A New Deal for the World: America's Vision for Human Rights* (Cambridge, MA: Harvard University Press, 2005), 85–86.

34. Wendell Willkie, *One World* (New York: Limited Editions Club, 1944), 1; Howard Rushmore, "An Interview with Wendell Willkie: Man from Indiana

'Gulliver-Travels' around the Globe," *Skyways*, February 1943; Steve Neal, *Dark Horse* (New York: Doubleday, 1984); Charles Peters, *Five Days in Philadelphia: The Amazing 'We Want Willkie!' Convention of 1940 and How It Freed FDR to Save the Western World* (New York: Public Affairs, 2005).

35. Willkie, "Report to the People," October 26, 1942, Box 6, WW, IU; William Allen White, "Willkie Again," *Emporia Gazette*, October 30, 1942; "Wendell Willkie's Challenging Report," *Louisville Courier-Journal*, October 28, 1942; "Around the World with Willkie," *Christian Science Monitor*, October 27, 1942; Neal, *Dark Horse*, 264; David Reynolds, *One World Divisible: A Global History since 1945* (New York: W. W. Norton, 2001), 1.

36. "Topics of the Times," *New York Times*, January 31, 1947; Henry A. Wallace, "Freedom of the Air—A Momentous Issue," *New York Times Sunday Magazine*, June 27, 1943; Wallace, *The Price of Vision: The Diary of Henry A. Wallace, 1942–1946*, ed. John Morton Blum (Boston: Houghton Mifflin, 1973), 157, 217–219; Wallace to FDR, June 5, 1942, OF 1571, PP, FDR, FDRPL.

37. "Europe Overnight," *Fortune*, November 1945, 116, 119, 286; letter from Margaret Saunders, *New Horizons*, July 1941, 6.

38. Susan Schulten, *The Geographical Imagination in America, 1880–1950* (Chicago: University of Chicago Press, 2001), 209; J. Parker Van Zandt, "Looking down on Europe Again," *National Geographic*, June 1939, 791–822.

39. Brinkley, *The Publisher*, 253–260; Swanberg, *Luce and His Empire*, 214; "Life Flies the Atlantic: America to Europe in 23 Hours by Clipper," *Life*, June 3, 1940, 17; "Pan American Is Diplomats' Choice," *New Horizons*, July–September 1945, 12–13.

40. David Courtwright, *Sky as Frontier: Adventure, Aviation, and Empire* (College Station: Texas A&M University Press, 2005), 163–166; Kathleen M. Barry, *Femininity in Flight: A History of Flight Attendants* (Durham, NC: Duke University Press, 2007), 16; A. Bowdoin Van Riper, *Imagining Flight: Aviation and Popular Culture* (College Station: Texas A&M University Press, 2004), 96.

41. Eastern Airlines advertisement, "It's a Smaller World Now!," *Newsweek*, March 1, 1943, 29; Wright Engines advertisement, "Modern Atlas," *Skyways*, September 1943, 49; Consolidated Vultee Aircraft advertisement, "No Spot on Earth Is More than 60 Hours from Your Local Airport," *Skyways*, July 1943, 8–9. For additional examples of global imagery in aviation advertisements, see John Fousek, *To Lead the Free World: American Nationalism and the Cultural Roots of the Cold War* (Chapel Hill: University of North Carolina Press, 2000), 91–102; and Douglas K. Fleming, "Cartographic Strategies for Airline Advertising," *Geographical Review* 74 (1984): 76–93.

42. American Airlines advertisement, "Air Map," *Skyways*, January 1943, 50; O. M. Mosier to James M. Barnes, June 6, 1944, Box 8, OF 249, PP, FDR, FDRPL; "New Look in Our Lapels," *Pan Am Clipper*, July 1963, NASM Library.

43. W. Weigard and Vilhjalmur Stefansson, eds., *Compass of the World: A Symposium on Political Geography* (New York: Macmillan, 1944), 6–7; "Great Circle Airways," *Fortune*, May 1943; "One World, One War" map supplement, *Fortune*,

March 1942; "Global Maps," *Life*, August 3, 1942; Millicent Taylor, "Through the Editor's Window: The New Geography," *Christian Science Monitor*, February 6, 1943; "The Logic of the Air," *Fortune*, April 1943, 73; "World Airways: Control of Bases Nub of Rivalry Shaping up as Powers Scan Postwar Prospects," *Newsweek*, March 1, 1943, 26; Schulten, *Geographical Imagination*, 204–206.

44. Van Zandt, *America Faces the Air Age*, 4; George T. Renner, "Air Age Geography," *Harper's*, June 1943, 38–41.

45. Louis Powell, "New Uses for Globes and Spherical Maps," *Geographical Review* 35 (1945), 49–58; "Globes on Parade," *Time*, December 22, 1941.

46. Chicago and Southern Air Lines advertisement, 1942, Ad*Access Online Project #1101, JWHC, library.duke.edu/digitalcollections/adaccess; Jennings Randolph, "Scanning the Skyways," *Skyways*, November 1943, 6.

47. Sherry, *Rise of American Air Power*, 117, 358; "Air: The Limitless Sky," *Time*, May 17, 1943, 61. On British anxieties about aviation, see Uri Bialer, *The Shadow of the Bomber: The Fear of Air Attack and British Politics, 1932–1939* (London: Royal Historical Society, 1980); Tami Davis Biddle, *Rhetoric and Reality in Air Warfare: The Evolution of British and American Ideas about Strategic Bombing, 1914–1945* (Princeton, NJ: Princeton University Press, 2002); and A. M. Gollin, *No Longer an Island: Britain and the Wright Brothers, 1902–1909* (Stanford, CA: Stanford University Press, 1984).

48. "Airways to Peace: An Exhibition of Geography for the Future," *Bulletin of the Museum of Modern Art*, August 1943, Box 7, WW, IU; Museum of Modern Art press release, "Airways to Peace Exhibition with Text by Wendell L. Willkie Opens at Museum of Modern Art," June 30, 1943, www.moma.org/docs/press_archives/886/releases/MOMA_1943_0038_1943-06-30_43630-36.pdf (accessed 10/5/12).

49. Sherry, *Rise of American Air Power*, 185; Alexander de Seversky, *Victory through Air Power* (New York: Simon and Schuster, 1942), 101, 120; "Air: The Limitless Sky," *Time*, May 17, 1943, 61. On conceptions of American benevolence in U.S. airpower doctrine, see Sahr Conway-Lanz, *Collateral Damage: Americans, Noncombatant Immunity, and Atrocity after World War II* (New York: Routledge, 2005), 8–12, 199–206, and 227–230.

50. Trippe interview with NBC, September 29, 1939, Box 1, JTT, NASM.

51. Trippe, "America in the Air Age," address delivered at Charter Week ceremonies, UCLA, March 23, 1944, Box 1, JTT, NASM; Trippe, "America Unlimited" speech, Box 1, JTT, NASM. On Lindbergh's opposition to World War II and involvement with the "America First" movement, see A. Scott Berg, *Lindbergh* (New York: Berkley Books, 1999), 384–432.

52. Pan American Airways advertisements, "America Meets the Challenge of a Changing World," 1940, Ad*Access Online Project #T2299; "There *Are* No Distant Lands by Flying Clipper," 1941, #T1580; "The Most Watched-for Ship in the World!" 1941, #T1578; "America's Outposts of Security and Defense," 1941, #T2300; all JWHC, library.duke.edu/digitalcollections/adaccess.

53. Trippe, speech for Boy Scout Week, October 7, 1947, Box 1, JTT, NASM; J. D. Ratcliff, "Modern Magellan," *Who*, October 1943, 38–40, 59; Matthew

Josephson, "Columbus of the Airways," *Saturday Evening Post*, August 14, 21, and 28 and September 4 and 11, 1943; Josephson, *Empire of the Air*, 16.

54. "Busman's Holiday," *New Horizons*, June 1941, 30; "Return," *New Horizons*, September 1941, 21.

55. Clare Boothe, "Destiny Crosses the Dateline: Report on a Flight across the Pacific," *Life*, November 3, 1941, 99–100. Although Boothe used her maiden name as a *Life* writer, I refer to her as "Luce" in order to maintain consistency over the course of her career, since she used "Luce" as a congresswoman.

56. Ibid., 104–109.

57. Louis P. Mouillard, "The Empire of the Air" (1881), in Eugene Emme, ed., *The Impact of Air Power: National Security and World Politics* (Princeton, NJ: D. Van Nostrand, 1959), 19; Wilbur Wright, "What Mouillard Did," *Aero Club of America Bulletin* 1 (1912): 3–4.

58. Daniel R. Headrick, *Power over Peoples: Technology, Environments, and Western Imperialism, 1400 to the Present* (Princeton, NJ: Princeton University Press, 2010), 302–333; Oliver Lissitzyn, "The Diplomacy of Air Transport," *Foreign Affairs*, October 1940, 169.

59. Hanson W. Baldwin, "America Builds an Empire of the Air," *New York Times Sunday Magazine*, August 28, 1938; August Loeb, "A New Empire of Air Travel," *New York Times*, June 30, 1940.

60. James G. Stahlman, *Wings to the Orient* (Nashville: *Nashville Banner*, 1936), 10.

61. "Air Transport," *Life*, November 30, 1942, 78–79; Richard Edes Harrison, "World-Wide View," *New Horizons*, January 1943, 18–19; "Map Index of the Month," *New Horizons*, November 1943–February 1944; letters from M. J. Thompson and Mary M. Waring, *New Horizons*, June 1943. For further analysis of Harrison's cultural influence during World War II, see Schulten, *Geographical Imagination*, 204–238.

62. Robert E. Peary, "The Future of the Airplane," *National Geographic*, January 1918, 107.

63. Memorandum, Livingston Satterthwaite to Laurence Duggan, May 26, 1942, CDF 810.796/918–1/2, RG 59, NARA; Joseph M. Jones, "A Modern Foreign Policy," *Fortune*, October 1943, 183–185.

64. Samuel Flagg Bemis, *A Diplomatic History of the United States* (New York: Holt, 1936); Henry Luce, "The American Century," 165.

65. Clare Boothe Luce, "America in the Post-War Air World."

66. Wallace, "Freedom of the Air—A Momentous Issue"; "Freedom of the Skies," *The Nation*, February 20, 1943, 255–256.

67. "The World's Greatest Airline," *Fortune*, August 1945, 208–210.

68. Adolf Berle Diary, January 10, 1944, Frame 307, Reel 5, AAB (microfilm); Juan Trippe, "Post-War Air Transport" speech, May 1943, Box 10, JTT, NASM; S. Roger Wolin, "The International Airline of the Americas," Folder F1P-167000-69, NASM Library.

69. Mosier to Barnes, June 6, 1944.

4. "America's Lifeline to Africa"

1. "President's Trip to War Parleys," *New York Times*, February 3, 1943; Roland Nicholson, "4 Pilots Tell How President Followed Flight by Landmarks," *Washington Post*, February 4, 1943.

2. Deborah W. Ray, "Pan American Airways and the Trans-African Air Base Program of World War II," Ph.D. dissertation, New York University, 1973; Deborah W. Ray, "The Takoradi Route," *Journal of American History* 62 (1975): 340–358; Tom Culbert and Andy Dawson, *Pan Africa: Across the Sahara in 1941 with Pan Am* (McLean, VA: Paladwr Press, 1998).

3. Willis Lowe to PAA-Africa Construction Representative, February 16, 1943, Box 25, Series II, PAWA, UM.

4. On liberalism and empire, see, especially, Jennifer Pitts, *A Turn to Empire: The Rise of Imperial Liberalism in Britain and France* (Princeton, NJ: Princeton University Press, 2005); Alice Conklin, *A Mission to Civilize: The Republican Idea of Empire in France and West Africa, 1895–1930* (Palo Alto, CA: Stanford University Press, 2000); and Uday Singh Mehta, *Liberalism and Empire: A Study in Nineteenth-Century British Liberal Thought* (Chicago: University of Chicago Press, 1999).

5. Amy Kaplan, "'Left Alone with America': The Absence of Empire in the Study of American Culture," in Amy Kaplan and Donald Pease, eds., *Cultures of United States Imperialism* (Durham, NC: Duke University Press, 1993), 3–21; Perry Miller, *Errand into the Wilderness* (Cambridge, MA: Harvard University Press, 1978 [1956]), vii–viii.

6. "Flight 262," *New Horizons*, March 1941; Gene Banning, *Airlines of Pan American since 1927* (McLean, VA: Paladwr Press, 2001), 461–498; Robert E. Sherwood, *Roosevelt and Hopkins: An Intimate History* (New York: Harper, 1948), 262; memorandum, Franklin D. Roosevelt to Sumner Welles, June 19, 1941, Box 6, SF, PSF, FDR, FDRPL; memorandum, James Rowe Jr. to FDR, July 30, 1940, and Admiral H. R. Stark to Civil Aeronautics Board Chairman, December 3, 1940, Box 5, OF 2955, PP, FDR, FDRPL.

7. Winston Churchill to Roosevelt, September 24, 1940, in Francis L. Loewenheim et al. eds., *Roosevelt and Churchill: Their Secret Wartime Correspondence* (New York: Da Capo Press, 1975), 115; Joint Strategical Planning Committee, "Study of the Occupation of a Base in West Africa," May 7, 1941, Box 6, SF, PSF, FDR, FDRPL.

8. Major General J. H. Burns to Assistant Secretary of State, July 19, 1941, CDF 880.796/10, RG 59, NARA.

9. Culbert and Dawson, *Pan Africa*, 1–2; Ray, "Pan American Airways and the Trans-African Air Base Program," 9; Derek Gratze, "The Takoradi Run," *Aeroplane*, November 2001, 55–59.

10. Ray, "The Takoradi Route," 350; Culbert and Dawson, *Pan Africa*, 2–3; Marilyn Bender and Selig Altschul, *The Chosen Instrument: Pan Am, Juan Trippe, the Rise and Fall of an American Entrepreneur* (New York: Simon and Schuster, 1982), 348–350; Pan American Airways, Executive (System) Memoranda Nos. 58 and 59, July 1941, Box 15, Series I, PAWA, UM; memorandum, Bureau of the

Budget Director to Roosevelt, September 10, 1941, Box 5, OF 2955, PP, FDR, FDRPL.

11. PAA Inc.—Truman Report, "U.S. Government Contracts with PAA, Inc., Sept. 3, 1939–1942," Box 71, Series I, PAWA, UM; "Outline of General Purposes of Contracts with War and Navy Departments for Operation of Services," January 14, 1942, Box 15, Series I, PAWA, UM; Associated Press, "America Will Ferry Planes to Middle East," *Los Angeles Times*, August 19, 1941; Arthur Sears Henning, "New Moves toward an AEF!" *Chicago Tribune*, August 19, 1941.

12. "Africa and the War," *Chicago Defender*, July 12, 1941; "The Liberian Republic," *Chicago Defender*, September 6, 1941.

13. "United States Will Expedite Delivery of Fighting Planes," *West African Pilot*, August 20, 1942; "West Africa Is Vital Air Route to Middle East War Theatre," *West African Pilot*, October 26, 1942; A. M. Wendell Malliet, "News [sic] Day Dawning for Liberian Folk," *African Nationalist*, November 8, 1941. On alliances between anticolonial intellectuals and activists in the United States and Africa, see Penny Von Eschen, *Race against Empire: Black Americans and Anticolonialism, 1937–1957* (Ithaca, NY: Cornell University Press, 1997).

14. Samuel Rosenman, ed., *Public Papers and Addresses of Franklin D. Roosevelt*. Vol. 10, *1938–1950* (New York: Harper, 1950), 314.

15. "Swarthy Bird-Men of U.S.A.," *West African Pilot*, November 15, 1941; "Traveling by Air," *West African Pilot*, October 17, 1942.

16. George Padmore, "British Erase Racial Bar in Royal Air Force," *Chicago Defender*, February 22, 1941; Azikiwe, "Nigerian Youth and the RAF," *West African Pilot*, September 18, 1941; Azikiwe, "Nigerian 'Bird-Men' Abroad," *West African Pilot*, October 29, 1942; George Padmore, "RAF Men Tell of Abuse and Jim Crow in Africa," *Chicago Defender*, June 17, 1944.

17. Daniel R. Headrick, *Power over Peoples: Technology, Environments, and Western Imperialism, 1400 to the Present* (Princeton, NJ: Princeton University Press, 2009), 306, 309–311, 321–328; David Killingray, "'A Swift Agent of Government': Air Power in British Colonial Africa, 1916–1939," *Journal of African History* 25 (1984): 429–444; Robert McCormack, "Airlines and Empires: Great Britain and the 'Scramble for Africa,' 1919–1939," *Canadian Journal of African Studies* 10 (1976): 99; Beryl Markham, *West with the Night* (San Francisco: North Point Press, 1983 [1942]), 175.

18. Culbert and Dawson, *Pan Africa*, 35–83; Ray, "Pan American Airways and the Trans-African Air Base Program," 123–127; Horace Brock, *Flying the Oceans: A Pilot's Story of Pan Am, 1935–1955* (Lunenberg, VT: Stinehour, 1978), 180–184; Bender and Altschul, *Chosen Instrument*, 352–353.

19. Voit Gilmore, "African Report," Box 54, Series I, PAWA, UM; Major Al Williams, "Modern Clipper," *Pittsburgh Press*, December 9, 1941; "Foreign Trade," *Classroom Clipper*, May 1945, 6.

20. Gilmore, "African Report."

21. *Labor News*, October 30, 1942; Bill Newport, "African Affairs," *New Horizons*, February 1941, 5; Gilmore, "African Report"; "Revised List of Buildings Constructed in Africa," November 30, 1943, Box 259, Series I, PAWA, UM.

22. H. E. Baldwin, "Pan Africa and the Natives," *Africa News Letter*, August 1, 1942, Box 258, Series I, PAWA, UM.

23. Civil Aeronautics Board Docket No. 5818, January 1954, Exhibit No. PA-301, "Early History of Pan American's Service to Africa," Box 54, Series I, PAWA, UM; *New York Herald Tribune*, October 11, 1942; "Miracle in Africa," *New Horizons*, March 1942; W. Clifford Harvey, "Fresh Foods Served to Air Pilots," *Christian Science Monitor*, September 16, 1942; "Lessons in Africa," *New Horizons*, December 1942; Gilmore, "African Report."

24. Alfred Sporrer, "Bomber to the East," *New Horizons*, August 1942.

25. "African Pleasure," *New Horizons*, July 1942; Harold B. Whiteman to William Van Dusen, March 5, 1942, April 16, 1942, and May 18, 1942, Box 54, Series I, PAWA, UM; Gilmore, "African Report"; John Yeomans to Public Relations Director, "Movies for Africa," April 18, 1942, Box 25, Series II, PAWA, UM.

26. Philip Curtin, "'The White Man's Grave': Image and Reality, 1780–1850," *Journal of British Studies* 1 (1961): 94–100; "Final Report of the Activities of the Medical Department," December 1, 1942, Box 258, Series I, PAWA, UM.

27. Letter from Harvey C. Orme, *New Horizons*, March 1944. On medical interventions and U.S. global power in the early twentieth century, see, for example, Warwick Anderson, *Colonial Pathologies: American Tropical Medicine, Race, and Hygiene in the Philippines* (Durham, NC: Duke University Press, 2006); Laura Briggs, *Reproducing Empire: Race, Sex, Science, and U.S. Imperialism in Puerto Rico* (Berkeley: University of California Press, 2002); Marcos Cueto, ed., *Missionaries of Science: The Rockefeller Foundation in Latin America* (Bloomington: Indiana University Press, 1994); and Julia F. Irwin, "Humanitarian Occupations: Foreign Relief and Assistance in the Formation of American International Identities, 1898–1928," Ph.D. dissertation, Yale University, 2009.

28. Gilmore, "African Report." On "contact zones," see Mary-Louise Pratt, *Imperial Eyes: Travel Writing and Transculturation* (New York: Routledge, 1992).

29. Duplex, "Americans Here Again," *African Morning Post*, November 8, 1941.

30. Rubbs, "Things as I See Them," *Spectator Daily*, July 17, 1942; "Americans Entertain Accra with Baseball," *Spectator Daily*, October 17, 1942.

31. Letter from Albert Ofori, "Africans Like Our Men: Natives Find Us Understanding and Ready to Help," *New York Times*, November 3, 1942.

32. British Secretariat to U.S. Consul Accra, August 25, 1942, Box 4, General Records, 1942–1949, Gold Coast Consulate, RG 84, NARA.

33. Harvey Neptune, *Caliban and the Yankees: Trinidad and the United States Occupation* (Chapel Hill: University of North Carolina Press, 2007), 52, 74.

34. "The African's Poor Lot," *Spectator Daily*, November 7, 1941; J. B. Wilson, "Between Employer and Employee," *Daily Echo*, November 13, 1941.

35. *Africa News Letter*, October 1, 1942, Box 258, Series I, PAWA, UM; "The African's Poor Lot," *Spectator Daily*, November 7, 1941; J. B. Wilson, "Between Employer and Employee"; Rubbs, "Things as I See Them: Letters of Note," *Spectator Daily*, November 7, 1941.

36. Whiteman to Van Dusen, November 14, 1941, Box 54, Series I, PAWA, UM. Such disputes over wage scales were not, of course, peculiar to Pan Am or

Africa but a common phenomenon with U.S. companies that employed local labor in foreign development projects. See, for example, Robert Vitalis, *America's Kingdom: Mythmaking on the Saudi Oil Frontier* (New York: Verso, 2009).

37. Alfred Lief, *The Firestone Story: A History of the Firestone Tire and Rubber Company* (New York: McGraw-Hill, 1951), 321–323; Greg Grandin, *Fordlandia: The Rise and Fall of Henry Ford's Forgotten Jungle City* (New York: Metropolitan Books, 2009), 141; Pan American Airways memorandum, August 28, 1941, Box 23, Series I, PAWA; Henry Stimson to Welles, October 1, 1941, CDF 882.7962/11, RG 59, NARA.

38. Roosevelt to Cordell Hull and Sumner Welles, June 11, 1941, Box 6, SF, PSF, PP, FDR, FDRPL; Lester A. Walton to Hull, September 6, 1941, CDF 882.7962/9, RG 59, NARA; memorandum of conversation, Harvey Firestone Jr. and Villard, November 3, 1941, CDF 882.7962/12, RG 59, NARA; Welles to Roosevelt, November 7, 1941, and Roosevelt to U.S. Legation Monrovia, November 6, 1941, OF 476, PP, FDR, FDRPL; Henry McBride to Welles, March 3, 1942, CDF 882.7961/51, RG 59, NARA; and Stimson to Hull, March 17, 1942, CDF 882.7962/70, RG 59, NARA.

39. "Address Delivered by Judge I. A. David, on board the Pan American Airways Plane on the Morning, 7th Oct. 1942, at Fisherman's Lake, Grand Cape Mount," *African Nationalist*, October 18, 1941; U.S. Embassy Monrovia to Secretary of State, July 24, 1942, CDF 882.796/10; November 4, 1942, CDF 992.796/12; and November 13, 1944, CDF 882.796/11–1344; all in RG 59, NARA; Earl P. Hanson to Samuel Pryor Jr., January 18, 1945, Box 25, Series II, PAWA, UM.

40. Letter from Mrs. J. Stuart Ramsey, *New Horizons*, November 1942.

41. Michael McCarthy, *Dark Continent: Africa as Seen by Americans* (Westport, CT: Greenwood Press, 1983), 125; "New Name," *New Horizons*, February 1942, 12.

42. Joseph Conrad, *Heart of Darkness and Selections from the Congo Diary* (New York: Modern Library, 1999), 17; archive.org/details/OrsonWelles-MercuryTheater-1938Recordings (accessed 3/12/13).

43. "Permanent Additions to Africa Left by PAA," Box 259, Series I, PAWA, UM.

44. Gilmore, "African Report"; "Permanent Additions to Africa Left by PAA"; Gilmore, undated memorandum, Box 23, Series I, PAWA, UM; "Description of PAA Camp in Africa," Box 259, Series I, PAWA, UM.

45. "America's New Lifeline to Africa," *Saturday Evening Post*, November 22, 1941; John O'Reilly, "U.S. Air-Route Bases in Africa Give American Tinge to Wilds," *New York Herald Tribune*, October 11, 1942; "Miracle in Africa," *Pittsburgh Sun-Telegraph*, June 26, 1942.

46. Letter from John Miller, *New Horizons*, May 1942.

47. Gilmore, "African Report"; letter from Hugh Ellis Jenkins, *New Horizons*, December 1942; Harold Whiteman to Van Dusen, January 2, 1942, Box 54, Series I, PAWA, UM.

48. Gilmore, "African Report"; Whiteman to Van Dusen, January 2, 1942; H. E. Baldwin, "Pan Africa and the Natives," *Africa News Letter*, August 1, 1942, Box 258, Series I, PAWA, UM. On paternalism, see J. Douglas Smith, *Managing White Supremacy: Race, Politics, and Citizenship in Jim Crow Virginia* (Chapel Hill:

University of North Carolina Press, 2002), 300; and Mary Renda, *Taking Haiti: Military Occupation and the Culture of U.S. Imperialism, 1915–1940* (Chapel Hill: University of North Carolina Press, 2001).

49. Baldwin, "Pan Africa and the Natives."

50. Pan American Airways advertisement illustrated by Stevan Dohanos, "America's New Lifeline to Africa," 1941, www.americanartarchives.com/dohanos .htm (accessed 8/21/12).

51. Geir Lundestad, "Empire by Invitation? The United States and Western Europe, 1945–1952," *Journal of Peace Research* 23 (1986): 263–277; Christina Klein, *Cold War Orientalism: Asia in the Middlebrow Imagination, 1945–1961* (Berkeley: University of California Press, 2003), 13.

52. Whiteman to Van Dusen, December 17, 1941, Box 54, Series I, PAWA, UM; E. Robin Little, "Transatlantic Circuit," *New Horizons*, December 1942; "Miracle in Africa," *Pittsburgh Sun-Telegraph*, June 26, 1942; Raymond Clapper, "Tells of U.S. Blazing Sky Trails across Africa," March 16, 1942, Folder 7, Box 259, Series I, PAWA, UM; E. R. Littler and W. A. Creamer, "American Airbases across Africa," *Engineering News Record*, February 11, 1943, 133–134. On the "swarm/horde" trope, see, for example, Nayan Shah, *Contagious Divides: Epidemics and Race in San Francisco's Chinatown* (Berkeley: University of California Press, 2001), 1–76.

53. See Benjamin L. Alpers, "This Is the Army: Imagining a Democratic Military in World War II," *Journal of American History* 85 (June 1988): 129–163.

54. "New Name," *New Horizons*, February 1941; letter from Charles H. Sharp, *New Horizons*, June 1942.

55. W. Banks Tobey, "Random Route Reflections," *New Horizons*, May 1941; letters from Tom McMillan, *New Horizons*, May and June 1944; letter from Everett Fischer, *New Horizons*, March 1943; *Africa News Letter*, July 15, 1942, and August 1, 1941, Box 258, Series I, PAWA, UM.

56. Letter from William Stempel, *New Horizons*, May 1942.

57. John O'Reilly, "African Chiefs See U.S. Airport, Chuckle over American Ways," *New York Herald Tribune*, October 18, 1942.

58. Allan Raymond, "U.S. War Materials and American Experts to Speed Their Delivery, Moving across Africa to British," *St. Louis Post-Dispatch*, February 1, 1942; "Voodoo Today, Vot Tomorrow," *Skyways*, August 1943, 8. On sound as an instrument of conquest and domination, see Sarah Keyes, " 'Like a Roaring Lion': The Overland Trail as a Sonic Conquest," *Journal of American History* 96 (2009): 19–43.

59. Raymond Clapper, "Pan American Airways Has Done a Fine Job in Establishing Air Route to Africa; Operation of Line Should Be Continued by Civilian Hands," May 9, 1942, Folder 9, Box 23, Series I, PAWA; Sonia Tomara, "U.S. Air Line Built over Africa despite 150° Heat and Malaria," *New York Herald Tribune*, September 5, 1942; "Miracle in Africa," *New Horizons*, March 1942.

60. William M. Leary, *Fueling the Fires of Resistance: Army Air Forces Special Operations in the Balkans during World War II* (Washington, DC: Air Force History and Museums Program, 1995), 30; Thomas W. Ennis, "George Kraigher,

Pilot in Two Wars," *New York Times*, September 25, 1984; Banning, *Airlines of Pan American*, 531–532; George Kraigher obituary file, Box 29, Series I, PAWA, UM.

61. Philip Magnus, *Kitchener: Portrait of an Imperialist* (New York: E. P. Dutton, 1968).

62. "Misspelled Farewell," *New Horizons*, March 1943; letter from T. H. McMillan, *New Horizons*, April 1944; Parker C. Wiseman, "Life in Africa," *New Horizons*, May 1941.

63. Banning, *Airlines of Pan American*, 523.

64. Brock, *Flying the Oceans*, 181–183; Ray, "Pan American Airways and the Trans-African Air Base Program," 186–210; Culbert and Dawson, *Pan Africa*, 139–144, 151–153; Gilmore, "Chronicle of PAA-Africa Militarization," Box 74, Series II, PAWA, UM; Pan Am timetable, November 1, 1947, www.timetableimages .com (accessed 8/21/12).

65. Raymond Clapper, "Tells of U.S. Blazing Sky Trails across Africa."

5. From Open Door to Open Sky

1. "Memorandum on International Civil Aviation, as revised and adopted August 26, 1943 by the Interdepartmental Committee on International Aviation," CDF 811.796 Committee/158-1/4, RG 59, NARA.

2. Jordan A. Schwarz, *Liberal: Adolf A. Berle and the Vision of an American Era* (New York: Free Press, 1987), viii–x, 240; Adolf A. Berle Jr. and Gardiner Means, *The Modern Corporation and Private Property* (New York: Commerce Clearing House, 1932); Adolf Berle to Cordell Hull, September 9, 1942, in Adolf A. Berle Jr., *Navigating the Rapids, 1918–1971: From the Papers of Adolf A. Berle* (New York: Harcourt, Brace, Jovanovich, 1973), 481.

3. Schwarz, *Liberal*, 240; Berle to L. Welch Pogue, July 30, 1942, Box 45, AAB, FDRPL.

4. Lloyd C. Gardner, *Economic Aspects of New Deal Diplomacy* (Madison: University of Wisconsin Press, 1964), 263; Henry R. Luce, "The American Century," *Life*, February 1941, reprinted in *Diplomatic History* 23 (1999): 165.

5. Gardner, *Economic Aspects*, 39, 93.

6. R. E. G. Davies, *A History of the World's Airlines* (London: Oxford University Press, 1964), 139.

7. Alan P. Dobson, "The Other Air Battle: The American Pursuit of Post-War Civil Aviation Rights," *Historical Journal* 28, no. 2 (1985): 429–439. See also Alan P. Dobson, *Peaceful Air Warfare: The United States, Britain, and the Politics of International Aviation* (Oxford: Clarendon Press, 1991) and Jeffrey Engel, *Cold War at 30,000 Feet: The Anglo-American Fight for Aviation Supremacy* (Cambridge, MA: Harvard University Press, 2007), 17–52.

8. Berle to Franklin D. Roosevelt, "Recommendation of Air Policy to the Congress," undated [1944], Box 55, AAB, FDRPL.

9. J. M. Spaight, *Aircraft in Peace and the Law* (London: Macmillan, 1919), 106–107; Stuart Banner, *Who Owns the Sky? The Struggle to Control Airspace from the Wright Brothers On* (Cambridge, MA: Harvard University Press, 2008).

10. Spaight, *Aircraft in Peace and the Law*, 5–8, 45–47; Arthur K. Kuhn, "International Aerial Navigation and the Peace Conference," *American Journal of International Law*, 1920, 370.

11. Quoted in Spaight, *Aircraft in Peace and the Law*, 7.

12. Banner, *Who Owns the Sky?*, 1–68; Dobson, *Peaceful Air Warfare*, 4–9, 27–30; Convention Relating to International Air Navigation, 1919, in Spaight, *Aircraft in Peace and the Law*, 137–189; Wesley Phillips Newton, *The Perilous Sky: U.S. Aviation Diplomacy and Latin America, 1919–1931* (Coral Gables, FL: University of Miami Press, 1978), 78–82.

13. Marilyn Bender and Selig Altschul, *The Chosen Instrument: Pan Am, Juan Trippe, the Rise and Fall of an American Entrepreneur* (New York: Simon and Schuster, 1982), 261–262, 300.

14. Memorandum, Berle to Hull, April 30, 1943, and Berle, draft of aviation policy recommendations, April 26, 1943, Box 54, AAB, FDRPL; Berle to Roosevelt, "Recommendation of Air Policy to the Congress."

15. "Foreign Flying," *Wall Street Journal*, April 9, 1945; "The Future of International Airways: Has the United States a Policy?" *Harper's*, January 1944, 97–106; "The Postwar Air-Pulling," *Skyways*, July 1943, 22; "Aviation: All Dressed Up . . . ," *Time*, August 26, 1946, 81; "World Airways: Control of Bases Nub of Rivalry Shaping up as Powers Scan Postwar Prospects," *Newsweek*, March 1, 1943, 26; Joseph Kastner, "The Postwar Air," *Life*, November 1, 1943, 100, 112.

16. Airlines of the United States advertisement, "The Limitless Right of Way," *Newsweek*, April 10, 1944, 64.

17. Berle to Sumner Welles, January 9, 1943, Box 73, AAB, FDRPL; Hull to Jesse Jones, February 2, 1943, CDF 811.796/Committee/67A, RG 59, NARA; Pogue to Berle, February 23, 1943, CDF 811.796 Committee/88, RG 59, NARA. Minutes and other records from ICIA meetings can be found in the State Department's Central Decimal File, 1940–44, under the heading 811.796/Committee/, RG59, NARA.

18. Robert A. Lovett, "Memorandum for the Committee on International Civil Aviation," March 11, 1943, and memorandum of conversation, April 11, 1944, "Post-War International Civil Aviation," CDF 800.796/785, RG 59, NARA.

19. Adolf A. Berle, *New Directions in the New World* (New York: Harper and Brothers, 1940), 1–2, 6–7.

20. Memorandum, Berle to Hull, April 30, 1943, Box 54, AAB, FDRPL.

21. "Memorandum on International Civil Aviation, as Revised and Adopted August 26, 1943 by the Interdepartmental Committee on International Aviation," CDF 811.796 Committee/158-1/4, RG 59, NARA; "Airports Constructed Abroad in Whole or Substantial Part with United States Funds, according to United Nations Groupings," attachment in Paul T. David to Robert G. Hooker, March 10, 1943, CDF 811.796/Committee/86-1/2, RG 59, NARA.

22. Memorandum, "Meeting on Aviation Policy," November 11, 1943, CDF 811.796 Committee/182A, RG 59, NARA; "Policy of the War Department in Regard to Post-War International Civil Aviation," April 20, 1944, Box 93, PSF, FDR, FDRPL.

23. Memorandum, "International Aviation Conference," September 16, 1944, Frames 748–751, Reel 5, Berle Papers (microfilm); memorandum, "Aviation Policy," March 24, 1943, Box 54, AAB, FDRPL; memorandum, Adolf Berle to Sumner Welles, March 2, 1943, CDF 800.796/258, RG 59, NARA; Henry A. Wallace, "Freedom of the Air—A Momentous Issue," *New York Times Sunday Magazine*, June 27, 1943.

24. J. C. Hunsaker, "Research and Post War Aviation," Folder 19195, NASA.

25. L. Welch Pogue, "Memorandum of a Meeting on International Aviation Policy," June 8, 1943, CDF 811.796/Committee 141-1/2, RG 59, NARA; memorandum, Robert G. Hooker to Berle, May 12, 1943, CDF 800.796/324, RG 59, NARA.

26. Dobson, *Peaceful Air Warfare*, 136.

27. Alan P. Dobson, *Anglo-American Relations in the Twentieth Century: Of Friendship, Conflict, and the Rise and Decline of Superpowers* (New York: Routledge, 1995), 5, 109; Roderick Floud and Paul A. Johnson, eds., *The Cambridge Economic History of Modern Britain* (Cambridge: Cambridge University Press, 2003), 16; Gardner, *Economic Aspects*, 277–278. See also Randall Bennett Woods, *Changing of the Guard: Anglo-American Relations, 1941–1946* (Chapel Hill: University of North Carolina Press, 1990).

28. "Britain in the Postwar Air," *Fortune*, March 1944, 161.

29. On geopolitics and geoeconomics, see Neil Smith, *American Empire: Roosevelt's Geographer and the Prelude to Globalization* (Berkeley: University of California Press, 2003), 319.

30. Memorandum, "Projected Time Schedule for Preliminary Aviation Conferences," February 17, 1944, CDF 800.796/595, RG 59, NARA.

31. Memorandum, Hull to Roosevelt, March 11, 1944, CDF 800.796/673A, RG 59, NARA.

32. Telegram, John G. Winant to Hull, April 7, 1944, CDF 800.796/695, RG 59, NARA; memorandum, Morgan to Berle, June 15, 1944, CDF 800.796/6-1544, RG 59, NARA; Berle, "Report on Air Conversations," April 19, 1944, Box 63, AAB, FDRPL; Berle diary, July 29–30, 1944, in *Navigating the Rapids*, 492–494; Schwarz, *Liberal*, 233–235.

33. Berle diary, August 25, 1944, in *Navigating the Rapids*, 496; Circular Telegram to Certain Diplomatic Officers, September 11, 1944, CDF 800.796/9-1144, RG 59, NARA; "Delegation of the United States to the International Civil Aviation Conference Convening at Chicago Illinois, November 1, 1944," Document 17, *International Civil Aviation Conference, 1944, Documents* (Washington, DC: U.S. Government Printing Office, 1944).

34. Frank Hughes, "Air Rivalry? Forsooth! Says La Guardia Here," *Chicago Daily Tribune*, November 10, 1944; "America Last," *Chicago Daily Tribune*, November 28, 1944; "Sellout of U.S. Air Rights at Parley Feared," *Chicago Daily Tribune*, November 4, 1944; Frank Hughes, "Charge Berle Is Giving up U.S. Air Rights," *Chicago Daily Tribune*, November 27, 1944.

35. "Stage Is Set for Global Air Parley in City," *Chicago Daily Tribune*, October 25, 1944; Frank Hughes, "50 Nations Open Civil Aviation Sessions Today," *Chi-*

cago Daily Tribune, November 1, 1944; Ralph W. Cessna, "Control of World Aviation at Stake in Chicago Conference," *Christian Science Monitor*, November 1, 1944; "Verbatim Minutes of the Opening Plenary Session," November 1, 1944, Box 60, AAB, FDRPL; "President Urges Chicago Parley Bar 'Closed Air': Message to 50 Nations," *New York Times*, November 2, 1944; "Roosevelt's Aviation Plea," Associated Press, November 2, 1944; "Outline of the American Position," November 18, 1944, Frame 957, Reel 5, Berle Papers (microfilm).

36. Andrei Gromyko to Hull, October 26, 1944, and Edward Stettinius to Gromyko, October 27, 1944, CDF 800.796/10-2644, RG 59, NARA; Gromyko to Stettinius, October 30, 1944, CDF 800.796/10-3044, RG 59, NARA; U.S. Embassy London to Hull, November 2, 1944, CDF 800.796/11-244, RG 59, NARA.

37. "Seeks Seat at International Aviation Meet," *Chicago Defender*, November 11, 1944; George Merrell to Hull, August 7, 1944, CDF 800.796/8-744, RG 59, NARA; Syngman Rhee to Hull, October 9, 1944, and Donald Eddy to Rhee, November 10, 1944, CDF 800.796/10-944, RG 59, NARA.

38. George Padmore, "U.S., Britain Woo Ethiopia," *Chicago Defender*, November 11, 1944; Frank Hughes, "19 Latin American Nations Resist New Deal Air Plan," *Chicago Daily Tribune*, November 6, 1944.

39. Russell Porter, "Air Parley Agrees on American Plan," *New York Times*, November 11, 1944; George Messersmith to Hull, November 6, 1944, CDF 800.796/11-644, RG 59, NARA; Frederick P. Latimer Jr. to Hull, November 10, 1944, CDF 800.796/11-1044, RG 59, NARA.

40. Russell Porter, "British Fear Air in Wars of Future," *New York Times*, November 8, 1944; "Civil Airplanes Seen Carrying Army of Future," *Los Angeles Times*, November 8, 1944; "U.S. 'Fifth Freedom' of the Air Is Business Invasion of Small Nations, British Aviation Minister Charges," *Wall Street Journal*, November 27, 1944; *Daily Express*, November 16, 1944, enclosure in Herbert Pell to Berle, November 16, 1944, Box 59, AAB, FDRPL; Berle to Stettinius, November 18, 1944, CDF 800.796/11-1844, RG 59, NARA; William A. M. Burden, "Opening the Sky: American Proposals at Chicago," *The Atlantic*, March 1945, 51; Frank Hughes, "Small Nations Urged: 'Bow to U.S. Air Policy,'" *Chicago Daily Tribune*, December 4, 1944.

41. Henry Ladd Smith, *Airways Abroad: The Story of American World Air Routes* (Madison: University of Wisconsin Press, 1950), 171.

42. Berle to Stettinius, November 26, 1944, Frame 985, Reel 5, Berle Papers (microfilm).

43. See *Foreign Relations of the United States (FRUS)* 1944, Vol. 2, 604; Burden, "Opening the Sky," 53.

44. Berle, *Navigating the Rapids*, 503–504.

45. Roosevelt to Churchill, November 24, 1944; Churchill to Roosevelt, November 28, 1944; Roosevelt to Churchill, November 30, 1944; and Churchill to Roosevelt, December 1, 1944; all in Warren F. Kimball, ed., *Churchill and Roosevelt: The Complete Correspondence*, Vol. 3 (Princeton, NJ: Princeton University Press, 1984), 407–408, 419–427.

46. Smith, *Airways Abroad*, 187.

47. "Minutes of Meeting of the Joint Subcommittee of Committees I, III, and IV," November 27, 1944, Document 418 in *International Civil Aviation Conference, Chicago 1944;* Berle to Hull, November 27, 1944, CDF 800.796/11-2744, RG 59, NARA; Russell Porter, "More Nations Back American Air Plan," *New York Times,* November 29, 1944.

48. Berle, "Summary: The Work of the International Conference on Civil Aviation," December 6, 1944, Frames 1021–1025, Reel 5, Berle Papers (microfilm).

49. "Flushing Guest Says Atomic Bomb Produces One World—The Next," *New York Times,* October 25, 1946. On American attitudes toward the atomic bomb during the immediate postwar years, see Paul Boyer, *By the Bomb's Early Light: American Thought and Culture at the Dawn of the Atomic Age* (New York: Pantheon, 1985), 1–106.

50. Bender and Altschul, *Chosen Instrument,* 356; Dobson, *Peaceful Air Warfare,* 176.

51. Dobson, *Peaceful Air Warfare,* 184.

52. Pogue to Stettinius, July 16, 1945, *FRUS* 1945, Vol. 2, 1461–1463; memorandum, "Summary of Agreement Drafted in Bermuda Concerning Civil Use of United States 99-Year Leased Bases," January 23, 1946, *FRUS* 1946, Vol. 1, 1457–1458; Harry S. Truman, "Statement by the President on the Agreement Reached at the Civil Aviation Conference in Bermuda," February 26, 1946, in *Public Papers of the Presidents, Harry S. Truman, 1945–1953,* www.trumanlibrary.org/publicpapers/index.php?pid=1486&st=&st1= (accessed 10/19/12); Anthony Sampson, *Empires of the Sky: The Politics, Contests, and Cartels of World Airlines* (New York: Random House, 1984), 72; Dobson, *Peaceful Air Warfare,* 173–210. Other documents from the Bermuda conference can be found in *FRUS* 1946, Vol. 1, 1459–1481.

53. State Department press release, "Denunciation of IATA Agreements," July 25, 1946, CDF 800.796/7-2546, RG 59, NARA; Dean Acheson, Circular Telegram to Certain American Diplomatic Officers, August 14, 1946, CDF 800.796/8-146, RG 59, NARA.

54. Schwarz, *Liberal,* 247; Owen Brewster to Berle, November 26, 1944, Box 59, Berle Papers, FDRPL.

55. "Americans Fly Everywhere Now," *Fortune,* April 1945, 139–141.

56. Victoria de Grazia, *Irresistible Empire: America's Advance through Twentieth-Century Europe* (Cambridge, MA: Harvard University Press, 2005), 3.

6. Mass Air Travel and the Routes of the Cold War

1. Gene Banning, *Airlines of Pan American since 1927* (McLean, VA: Paladwr Press, 2001), 542.

2. Christopher Endy, *Cold War Holidays: American Tourism in France* (Chapel Hill: University of North Carolina Press, 2004).

3. The literature on Cold War modernization and development is extensive. See, especially, Michael Latham, *The Right Kind of Revolution: Modernization, Development, and U.S. Foreign Policy from the Cold War to the Present* (Ithaca, NY: Cornell University Press, 2011) and *Modernization as Ideology: American Social Sci-*

ence and "Nation Building" in the Kennedy Era (Chapel Hill: University of North Carolina Press, 2000); David Ekbladh, *The Great American Mission: Modernization and the Construction of an American World Order* (Princeton, NJ: Princeton University Press, 2011); Nils Gilman, *Mandarins of the Future: Modernization Theory in Cold War America* (Baltimore: Johns Hopkins University Press, 2003); and David C. Engerman et al., eds. *Staging Growth: Modernization, Development, and the Global Cold War* (Amherst: University of Massachusetts Press, 2003).

4. Jeffrey A. Engel, *Cold War at 30,000 Feet: The Anglo-American Fight for Aviation Supremacy* (Cambridge, MA: Harvard University Press, 2007).

5. Airlines of the United States and Canada advertisement, "Right in Your Own Back Yard!," *New Horizons*, January 1941, inside front cover; Ralph Thomas Walker, "Tomorrow We Live," *Skyways*, September 1943, 26–27, 80; "Biggest Airport in the World," *Fortune*, April 1945, 142–144; "*Skyways* Presents the Skytel," *Skyways*, November 1943, 24–25.

6. R. E. G. Davies, *A History of the World's Airlines* (London: Oxford University Press), 244; F. Barrows Colton, "Our Air Age Speeds Ahead," *National Geographic*, February 1948, 264.

7. R. E. G. Davies, *Pan Am: An Airline and Its Aircraft* (New York: Orion Books, 1987), 56–57; Roger Bilstein, *Flight in America: From the Wrights to the Astronauts* (Baltimore: Johns Hopkins University Press, 1994), 172; Frederick Graham, "By Plane—To the Four Corners of the Earth," *New York Times*, June 9, 1946; Pan American Airways advertisements, "Here Are Wings for Your Dreams" and "Pick Your Continent!," 1948, Box PA7, DA, JWT, JWHC.

8. Davies, *A History of the World's Airlines*, 256–257; Graham, "By Plane—To the Four Corners of the Earth"; "Rise in Vacation Travel Abroad," *U.S. News and World Report*, March 12, 1948, 23.

9. Office of International Trade, Department of Commerce, "Contribution of Travel Development to Closing the Dollar Gap," undated [1949], and "New European Travel Market Data Available for First Time," May 19, 1949, Box 9, Entry 56, RG 469, NARA; Wayne Parrish, "Aviation Notebook: A Progress Report," October 14, 1949, Box 20, WP, HHPL.

10. Frank Gannett, *Winging 'Round the World*, Box 319, Series I, PAWA, UM; Frederick Lewis Allen, "After Hours," *Harper's*, October 1953, 90–91; Bernard DeVoto, "Transcontinental Flight," *Harper's*, July 1952, 47–48; Wolfgang Langewiesche, "Look Down, Look Down," *Harper's*, November 1948, 64–69; "Eastward Bound," *Harper's*, December 1948, 61–67; and "The Middle East over the Wingtip," *Harper's*, January 1949, 85–93.

11. Letters from Verna Seabolt and Kathleen Barnes, *Pan American World Airways Teacher*, October 1954, Folder 14, Box 86, Series I, PAWA, UM.

12. "What's Wrong with the Airlines," *Fortune*, August 1946; Bernard DeVoto, "Transcontinental Flight," *Harper's*, July 1952, 47–48; Karl Shapiro, "Air Liner," *New Yorker*, May 10, 1947, 36.

13. "What's Wrong with the Airlines," *Fortune*; Wolfgang Langewiesche, "Thirty-Seven Frontiers," *Harper's*, June 1949, 51.

14. "Safety in the Air," *New York Times*, January 13, 1947; "Something Wrong?," *Christian Science Monitor*, January 15, 1947; "Safety Factors in Air Transportation," *Los Angeles Times*, January 26, 1947; "Troubles over Air Safety," *U.S. News and World Report*, January 31, 1947, 19–20; "Flaming Death," *Washington Post*, June 3, 1947.

15. David Bernstein, "Our Airsick Airlines," *Harper's*, May 1949, 68.

16. Karen Miller, " 'Air Power Is Peace Power': The Aircraft Industry's Campaign for Public and Political Support, 1943–1949," *Business History Review* 70 (1996): 297–327; John B. Rae, *Climb to Greatness: The American Aircraft Industry, 1920–1960* (Boston: Massachusetts Institute of Technology Press, 1968), 174; J. Fred Henry, "An American Tragedy," *Skyways*, November 1947, 12.

17. John F. Victory, "Keeping America First in the Air," speech before the National Press Club, March 23, 1945, Folder 19195, NASA; Gill Robb Wilson, "Air Power Is Uncle Sam's Fist," *Skyways*, February 1946, 12; J. Fred Henry, "A Strong Air Force," *Skyways*, December 1947, 18; Ivin Wise, "Our Defenses Are Down," *Skyways*, March 1948, 18; Steven Charles Call, "A People's Air Force: Air Power and Popular Culture," Ph.D. dissertation, Ohio State University, 1997, 176–227.

18. President's Air Policy Commission, *Survival in the Air Age: A Report by the President's Air Policy Commission* (Washington, DC: U.S. Government Printing Office, 1948), 7, 19–20, 25–26, 31–33, 133.

19. President's Air Policy Commission, *Survival in the Air Age*, 8; Thomas K. Finletter, "Responsibility of Air Power," *Skyways*, February 1951, 10–11, 45, 48. On antistatism and the history of American conceptions of "totalitarianism," see Aaron L. Friedberg, *In the Shadow of the Garrison State: America's Anti-Statism and Its Cold War Grand Strategy* (Princeton, NJ: Princeton University Press, 2000); Abbott Gleason, *Totalitarianism: The Inner History of the Cold War* (New York: Oxford University Press, 1995); and Benjamin L. Alpers, *Dictators, Democracy, and American Public Culture: Envisioning the Totalitarian Enemy, 1920s–1950s* (Chapel Hill: University of North Carolina Press, 2003).

20. John Lewis Gaddis, *Strategies of Containment: A Critical Appraisal of Postwar American National Security Policy* (New York: Oxford, 1982), 359; David T. Courtwright, *Sky as Frontier: Aviation, Adventure, and Empire* (College Station: Texas A&M University Press, 2004), 124; Aircraft Industries Association, *Aviation Facts and Figures, 1955* (Washington, DC: Lincoln Press, 1955), 33.

21. President's Air Policy Commission, *Digest of Public Discussions, September–December 1947, Part 1, Section 1—National Defense* (Washington, DC: U.S. Government Printing Office, 1947); President's Air Policy Commission, *Survival in the Air Age*, 5, 99–123, 127–129, 144; statement of David L. Behncke, President of the Air Line Pilots Association, before the President's Air Policy Commission, December 2, 1947, Folder FO 130040-14, ATC, NASM Library; Courtwright, *Sky as Frontier*, 125.

22. Roger Lewis, "TAP, Guided Missiles, and IHC," March 1958, Box 254, Series I, PAWA, UM; *Guided Missile Range Division Clipper*, Box 91, Series I, PAWA, UM; Allan C. Fisher Jr., "Cape Canaveral's 6,000-Mile Shooting Gal-

lery," *National Geographic*, October 1959; Pan Am Annual Report, 1968, Box 23, Series I, PAWA, UM. On the "military-industrial complex" and the emergence of the national security state, see, especially, Michael J. Hogan, *A Cross of Iron: Harry S. Truman and the Origins of the National Security State, 1945–1954* (New York: Cambridge University Press, 1998) and Julian E. Zelizer, *Arsenal of Democracy: The Politics of the National Security State, from World War II to the War on Terrorism* (New York: Basic Books, 2007).

23. Pan Am press release, "Trippe's Statement, Inaugural Ceremonies for First Round-the-World Air Service, June 17, 1947," Box 319, Series I, PAWA, UM.

24. State Department press release, June 1947, Box 319, Series I, PAWA, UM; Clementine Paddleford, "Globe-Girdling Clipper Plans Special Menus," *New York Herald Tribune*, June 17, 1947; "'RTW Scarf' Meets with Wide Approval as Record Sales Soar," *Sales Clipper*, May 1950, Box 14, AF, JWT, JWHC.

25. "50,000 to Circle the World by Air in '62, Pan Am Says," *New York Times*, September 22, 1962; Henry H. Arnold, "Air Power for Peace," *National Geographic*, February 1946, 193.

26. Christopher Endy, "Travel and World Power: Americans in Europe, 1890–1917," *Diplomatic History* 22 (1998): 565, 571.

27. See Michael J. Hogan, *Marshall Plan: America, Britain, and the Reconstruction of Western Europe, 1947–1952* (New York: Cambridge University Press, 1987).

28. Juan T. Trippe, "World Prosperity through Foreign Travel," October 27, 1948, Box 2, JTT, NASM.

29. Department of Commerce, "Contribution of Travel Development to Closing the Dollar Gap"; Endy, *Cold War Holidays*, 45–50, 81–99.

30. Theodore J. Pozzy, "Stronger Travel Development Campaign," March 9, 1949, Box 2, Entry 968, RG 469, NARA.

31. Endy, *Cold War Holidays*, 33–54; Henry Luce, "The American Century," *Life*, February 17, 1941, reprinted in *Diplomatic History* 23 (1999): 165.

32. Department of Commerce, "Contribution of Travel Development to Closing the Dollar Gap"; Department of Commerce, "New European Travel Market Data Available for First Time"; ECA Travel Development Section, "Summary Report: Survey among American Tourists Leaving Europe," undated [1950], Box 9, Entry 968, RG 469, NARA; Juan Trippe, testimony before the Senate Committee on Interstate and Foreign Commerce, June 28, 1949, Box 2, JTT, NASM.

33. "Tourist Air Travel," *New York Times*, September 6, 1948; "Historical Summary of PAA's Services to Puerto Rico," Exhibit No. PA-100, CAB Docket No. 2123, 1948, and "Pan American's Operations to Puerto Rico," 1953, Box 64, Series I, PAWA, UM; Carl Solberg, *Conquest of the Skies: A History of Commercial Aviation in America* (Boston: Little, Brown, 1979), 346.

34. Juan Trippe, speech before the Honolulu Chamber of Commerce, February 23, 1950, Box 2, JTT, NASM; Department of Commerce, "Summary Statement: Calendar Year 1948, Transatlantic Airline Passenger Capacity," Box 9,

Entry 56, RG 469, NARA; telegram, U.S. Embassy Paris to Secretary of State, May 3, 1949, Box 1, Entry 970, RG 469, NARA.

35. Pozzy to Hoffman, August 24, 1949, and Department of Commerce, "Why Air Fares to Europe Should Be Reduced," April 14, 1949, Box 1, Entry 970, RG 469, NARA; "New Travel Era Bekons" [sic], *Fort Wayne News Sentinel,* December 27, 1948; "Henry Ford of the Air," *New York Enquirer,* July 21, 1947; "Clipper Skipper," *Time,* March 28, 1949, 84–92; Juan Trippe, "Now You Can Take That Trip Abroad," *Reader's Digest,* January 1949, 69–72.

36. U.S. Consulate Nice to Secretary of State, December 7, 1951, CDF 900.52/12-751, RG 59, NARA; Lowell Thomas, "When Are You Going Abroad?" *Los Angeles Times,* April 27, 1952; Frederick Graham, "2 Planes Here Take off to Start New Low-Fare Flights to Europe," *New York Times,* May 1, 1952.

37. Banning, *Airlines of Pan American,* 582; Frank Cipriani, "Two-Thirds of Atlantic Flights Now Tourist Class," *Chicago Tribune,* September 26, 1954; Pan American Airways, "Pan Am will fly you around the World for only $135 down," 1955, Box PA21, DA, JWT, JWHC; Pan American Airways, "NOW—*prepaid* clipper tickets for immigrants," 1949, Box PA10, DA, JWT, JWHC, Duke.

38. Sherman Adams to John Foster Dulles, July 26, 1954, CDF 911.52/7-2654, RG 59, NARA; press release, June 12 White House Conference on People-to-People Partnership, May 31, 1956, www.eisenhower.archives.gov/research/online_documents/people_to_people/BinderV.pdf (accessed 8/27/12); "Development of International Travel, Its Present Increasing Volume, and Future Prospects," *Department of State Bulletin,* March 21, 1955, 491–495.

39. Elaine Tyler May, *Homeward Bound: American Families in the Cold War Era* (New York: Basic Books, 1988); Wayne Parrish, "Aviation Today and Tomorrow" speech, May 11, 1948, Box 20, WP, HHPL.

40. Clinton Golden and Bert Jewell to Enos Curtin, February 14, 1950, and U.S. Consulate Algiers to Secretary of State, Box 9, Entry 56, RG 469, NARA.

41. "Pan American World Airways (NY) Credits Advertising for Increased Air Travel," *J.W.T. News,* July 31, 1950, Box MN9, JWT Newsletters, JWT, JWHC; memorandum, W. E. Wyted to John DeVries et al., June 4, 1957, Box 11, WT, JWT, JWHC.

42. Minutes for meeting on Pan American World Airways, December 1956, and "Pan American's 'White Pea in a Pod,'" December 1956, Box 23, RB, JWT, JWHC.

43. "Pan American Advertising Geared to Attracting New Customers, Fighting Competition," *J. Walter Thompson Company News,* December 16, 1957, Box MN11, JWT Newsletters, JWT, JWHC; Script for System Sales Meeting, December 6, 1955, Box 23, RB, JWT, JWHC.

44. Pan American World Airways advertisement, "Can you name the homelands of these people?" *Life,* April 2, 1956; Pan Am Review Board Meeting, January 20, 1956, Box 42, DS, JWHC, Duke.

45. Pan American World Airways advertisements, "Do you collect ancient ruins?," etc., 1957, Box PA26, DA, JWT, JWHC.

46. Script for System Sales Meeting, December 6, 1955, Box 23, RB, JWT, JWHC; "Pan American Advertising Geared to Attracting New Customers, Fighting Competition," *J. Walter Thompson Company News*, December 16, 1957, Box MN11, JWT Newsletters, JWT, JWHC.

47. Script for System Sales Meeting, December 6, 1955; Pan American World Airways advertisement, "My sketch book proves you can see more of EUROPE when you fly Pan American," 1956, Oversize Box 4, WE, JWT, JWHC; "Rockwell Illustrations Depict Round-the-World Travel," *J. Walter Thompson Company News*, March 6, 1956, Box MN10, JWT Newsletters, JWT, JWHC.

48. Script for System Sales Meeting, December 6, 1955.

49. *Ladies' Home Journal* Research Department, "Opinions from 178 Travel Agents about Travel Copy Especially Designed to Appeal to Women," May 1952, Box 31, AVF, JWT, JWHC; TWA advertisement, "How to See More of Your Husband," 1954, Ad*Access Online Project, ad T2101, JWHC, library.duke.edu/digitalcollections/adaccess.

50. Pan Am advertisement, "This family of five will save $1,000," *Time*, September 19, 1955, and *Saturday Evening Post*, September 22, 1955; Pan Am advertisement, "Be there together," *Life*, May 13, 1957; American Airlines advertisements, "I'm Going with My Husband for Half Fare!," 1948, Ad*Access Project ad T0299; "Columbus may have discovered America, but *mother* discovered American!," 1949, #T0361; and "There's nothing like it on earth for traveling with a baby!," 1949, #T0350, JWHC, Duke, library.duke.edu/digitalcollections/adaccess.

51. *Ladies' Home Journal* Research Department, "Opinions from 178 Travel Agents"; Pan Am advertisement, "How business women can see the world on a secretary's pay check," *Charm*, January 1957; "New Horizons, Unique Benefits Extended to Business Girls," *J. Walter Thompson Company News*, December 12, 1955, Box MN10, JWT Newsletters, JWT, JWHC; TWA advertisement, "Who Says It's a Man's World?," 1953, Ad*Access Online Project, #T2098, JWHC, library.duke.edu/digitalcollections/adaccess; Pan Am advertisement, "When you are alone in a foreign land," *Foreign Affairs*, January 15, 1955.

52. "Opening the Cockpit Doors," *Time*, May 3, 1963; Organization of Black Airline Pilots, www.obap.org/obap-network/about-obap\ (accessed 10/19/12); Women in Aviation Resource Center, www.women-in-aviation.com/cgi-bin/links/detail.cgi?ID=387 (accessed 10/19/12).

53. American Airlines advertisement, "He Commands Your Flagship and Your Confidence," 1949, Ad*Access Online Project #T1778, JWHC, library.duke.edu/digitalcollections/adaccess; Pan Am, "Eyes That See around the World," 1956, Oversize Box 4, WE Papers, JWT, JWHC; Pan Am advertisement, "It's Nice to Know Uncle Sam's Your Skipper When You Fly to Faraway Places," 1956, Box PA23, DA, JWT, JWHC.

54. John B. Lansing and Dwight M. Blood, *The Changing Travel Market* (Ann Arbor: Survey Research Center, Institute for Social Research, University of Michigan, 1964), 84, 89.

55. Leland Stowe, "The Knack of Intelligent Travel," *Reader's Digest*, September 1952, 103–106; George Kent, "How to Be an American Abroad," *Reader's Digest*, June 1949; William J. Lederer and Eugene Burdick, *The Ugly American* (New York: Norton, 1958).

56. Endy, *Cold War Holidays*, 5.

57. Marilyn Bender and Selig Altschul, *The Chosen Instrument: Pan Am, Juan Trippe, the Rise and Fall of an American Entrepreneur* (New York: Simon and Schuster, 1982), 433–445; telegram, U.S. Embassy Paris to Secretary of State, July 24, 1950, CDF 911.5240/7-2450, RG 59, NARA; U.S. Embassy Paris to Secretary of State, November 15, 1950, CDF 911.5251/11-1550, RG 59, NARA; "L'Aviation Francaise devra-t-elle payer une dette de M. Truman?" *Le Rassemblement*, November 4, 1950; "Notre aviation menacée par un Trust américain," *Carrefour*, October 31, 1950; "Les ambitions de la 'Pan American' menacent l'existence d'Air-France," *Libération*, November 30, 1950; memorandum for the President, "Civil Aviation Consultations with France," February 13, 1951, CDF 911.5251/12-1351, RG 59, NARA.

58. Hans Heymann Jr., *The Soviet Role in International Civil Aviation* (Washington, DC: RAND Corporation, November 1957); Secretary of State to U.S. Embassies Athens and Ankara, July 31, 1953, CDF 961.5200/7-3153, RG 59, NARA; U.S. Embassy Ankara to Secretary of State, August 11, 1953, CDF 961.5282/8-1153, RG 59, NARA; Air Transport Association, "Red Star into the West," 1959, CDF 911.7261/1-458, RG 59, NARA. On the meanings of aviation in Russian and Soviet culture, see Scott W. Palmer, *Dictatorship of the Air: Aviation Culture and the Fate of Modern Russia* (New York: Cambridge University Press, 2006).

59. Telegram, Secretary of State to U.S. Embassies in New Delhi and other Asian capitals, February 5, 1958, CDF 961.7291/1-2158, RG 59, NARA; U.S. Embassy Delhi to Secretary of State, January 2, 1958, CDF 961.7291/1-258, RG 59, NARA; Heymann, *The Soviet Role in International Civil Aviation*; "Russia's Jet Marks Bid to Build World Airline," *Business Week*, September 14, 1957, 95. On the Cold War in the Third World, see Odd Arne Westad, *The Global Cold War: Third World Interventions and the Makings of Our Times* (New York: Cambridge University Press, 2005).

60. State Department Circular Airgram, March 11, 1948, CDF 800.796A/3-1148, RG 59, NARA; Harry S Truman inaugural address, January 20, 1949, www.trumanlibrary.org/whistlestop/50yr_archive/inagural20jan1949.htm (accessed 10/19/12).

61. Hans Heymann Jr., *Civil Aviation and U.S. Foreign Aid: Purposes, Pitfalls, and Problems for U.S. Policy* (Washington, DC: RAND Corporation, January 1964), 5–11.

62. Telegram, U.S. Consulate Jidda to Secretary of State, December 22, 1945, CDF 890F.796/12-2245, RG 59, NARA; TWA-Saudi Arabia Agreement, enclosure in U.S. Embassy Cairo to Secretary of State, July 12, 1946, CDF 890F.796/7-1246, RG 59, NARA.

63. U.S. Consulate Jidda to Secretary of State, July 8, 1946, CDF 890F.796/7-846, RG 59, NARA; Dean Acheson to Robert P. Patterson, August 9, 1946, CDF 890F.796/8-246, RG 59, NARA.

64. Press release, "Private Enterprise 'Point Four' Aids Underdeveloped Countries," October 28, 1957; "TAP Spurs Airlines' Growth," *The Clipper*, May 1958; "The Technical Assistance Program in the Air," undated [1958], Box 254, Series I, PAWA, UM.

65. Roger Lewis, "TAP, Guided Missiles, and IHC"; "TAP Spurs Airlines' Growth," *The Clipper*, May 1958, Box 254, Series I, PAWA, UM; Lewis to Coleman A. Harwell, May 12, 1959, Box 528, Series II, PAWA, UM.

66. Wolfgang Langewiesche, "The Middle East over the Wingtip," *Harper's*, January 1949, 85–93.

67. For further analysis, see Jenifer Van Vleck, "An Airline at the Crossroads of the World: Ariana Afghan Airlines, Modernization, and the Global Cold War," *History and Technology* 25 (2009): 3–24.

68. U.S. Embassy Kabul to Secretary of State, February 17, 1951, CDF 989.52/2-1751, March 25, 1952, CDF 989.52/3-2552, and November 22, 1952, CDF 989.52/11-2252, RG 59, NARA; U.S. Embassy Cairo to Secretary of State, March 31, 1952, CDF 989.52/3-3152, RG 59, NARA; Louis Dupree, "The Emergence of Technocrats in Modern Afghanistan," American Universities Fieldstaff Report, South Asia Series, Vol. XVII, no. 5, 1974; Louis Dupree, *Afghanistan* (Princeton, NJ: Princeton University Press, 1978), 485–494.

69. U.S. Embassy Kabul to Secretary of State, May 20, 1953, CDF 989.52/5-2053, RG 59, NARA; telegram, Secretary of State to U.S. Embassies Kabul and Karachi, March 3, 1955, CDF 982.72/30355, RG 59, NARA.

70. Telegrams, U.S. Embassy Kabul to Secretary of State, April 29, 1956, CDF 989.72/4-2856, and April 30, 1956, CDF 989.72/4-3056, RG 59, NARA; joint ICA-State Department cablegram to U.S. Embassy Kabul, May 8, 1956, CDF 989.72/5-856, RG 59, NARA; Prochnow to James R. Durfee, July 2, 1956, CDF 911.7200-PAA/7-256, and August 7, 1956, CDF 911.7200-PAA-7/256, RG 59, NARA.

71. U.S. Embassy Kabul, "U.S. News Bulletin," April 18, 1957 and June 5, 1957, Box 205, Series I, PAWA, UM; telegrams, U.S. Embassy Kabul to Secretary of State, January 26, 1957, CDF 989.72/1-2557, and February 1, 1957, CDF 989.72/1-3157, RG 59, NARA; "Review of Pan Am Status—Ariana," November 27, 1979, Box 678, Series II, PAWA, UM; Air Transport Development Project PPA 06-37-036, May 1956, Box 728, Series II, PAWA, UM.

72. "Ariana Afghan Airlines: A Short History," 1974, Folder F1A-612000-01, ATC, NASM Library; "TAP in Action: Profile of a Partnership—Wings over Afghanistan," undated [1962], Box 205, Series I, PAWA, UM.

73. U.S. Embassy Kabul to Secretary of State, September 17, 1957, CDF 989.72/9-1757, and November 26, 1957, CDF 989.72/11-2657, RG 59, NARA; R. E. G. Davies, *Airlines of Asia since 1920* (McLean, VA: Paladwr Press, 1997), 86; "Ariana Afghan Airlines: A Short History"; "TAP in Action"; "Technical Cooperation in the Field of Air Transportation," April 1967, Box 216, Series I, PAWA, UM.

74. "Dade Countians Helping Afghans Build an Airline" and "Camels to Airliners," Box 380, Series I, PAWA, UM. Additional clippings: Folder 8, Box 205, Series I, PAWA, UM.

75. Jane Kilbourne to Mrs. R. W. Beecher, October 29, 1959, and Roger Lewis to Faiz Mohamed Ahmedzai, November 10, 1959, Box 528, Series II, PAWA, UM; "Afghan Women Now Sew Western Styles," March 9, 1961, Folder 17, Box 380, Series I, PAWA, UM.

76. Abdul Karim Hakimi, address over Kabul Radio, August 29, 1958, enclosure in U.S. Embassy Kabul to Secretary of State, October 1, 1958, CDF 989.72/10-158, RG 59, NARA; *Bakhtar*, June 29, 1959, quoted in U.S. Embassy Kabul to Secretary of State, August 11, 1959, CDF 989.724/8-1159, RG 59, NARA; Davies, *Airlines of Asia*, 86.

77. Lewis to Norman P. Seagrave, April 17, 1959, Box 528, Series II, PAWA, UM; Dupree, *Afghanistan*, 511, 637–641.

78. C. J. Davies to Mehrabuddin Paktiawal, January 10, 1984, Box 205, Series I, PAWA, UM; "Many Workers for Afghan Airline Defect to the West," *New York Times*, September 16, 1980; Davies, *Airlines of Asia*, 88.

79. Heymann, *Civil Aviation and U.S. Foreign Aid*, 20, 26, 33–34.

80. Rae, *Climb to Greatness*, 215; Davies, *A History of the World's Airlines*, 174.

7. The Jet Age and the Limits of American Power

1. Pan Am press release, "Jet Age Pioneers Were a Cross Section of the Flying Public," October 26, 1983, Folder F1P-167000-70, ATC, NASM Library; "Off to Paris in Jet Time," *Life*, November 10, 1958, 113–117; "123 on Board First Regular Jet to Paris," *Chicago Daily Tribune*, October 27, 1958.

2. Mary V. R. Thayer, "Mamie Christens Jet Clipper 'America,'" *Washington Post*, October 17, 1958; Juan T. Trippe, "Address Given at the Christening of the Pan American Jet Clipper America," October 16, 1958, Box 1, JTT, NASM; public relations memo, "The Clipper Ships," August 11, 1961, Box 6, Series I, PAWA, UM; "Queen of the Ocean Air," *New York Herald Tribune*, October 19, 1958.

3. John McDonald, "Jet Airliners: Year of Decision," *Fortune*, April 1953, 246; "Transatlantic Jet Flights Start Today," *Washington Post*, October 4, 1958; Richard Witkin, "Idlewild Opened to Jet Airliners; Strict Rules Set," *New York Times*, October 4, 1958; Robert Alden, "Six-Hour Record Is Set," *New York Times*, October 5, 1958; "Jet Transports," *Washington Post*, May 28, 1950.

4. Mitchell Gordon, "British Start World's First Scheduled Jet Airliner Flights Today," *Wall Street Journal*, May 2, 1952; "Jets for Pan American," *New York Times*, October 22, 1952; "Jets for Pan Am," *Washington Post*, October 23, 1952; "Jet Transports: Why U.S. Holds Back," *Business Week*, April 18, 1953, 77–82.

5. Steve Budiansky, *Air Power: The Men, Machines, and Ideas That Revolutionized War, from Kitty Hawk to Gulf War II* (New York: Viking, 2004), 355–358; Margaret Connor, *Hans von Ohain: Elegance in Flight* (Reston, VA: American Institute of Aeronautics and Astronautics, 2002), 29–95.

6. Budiansky, *Air Power*, 354; Roger E. Bilstein, *Flight in America: From the Wrights to the Astronauts* (Baltimore: Johns Hopkins University Press, 1994), 179, 181–182.

7. Budiansky, *Air Power*, 361; Bilstein, *Flight in America*, 181–182, 223–224, 227.

8. "Air Force Position on Jet Transports," *Aviation Week*, June 2, 1952, 16; John McDonald, "Jet Airliners: Year of Decision," *Fortune*, April 1953, 246; John Stuart, "1952 Now Looms as Critical Year for U.S. Air Transport Industry," *New York Times*, May 4, 1952; "No Comets for Us," *Washington Post*, May 6, 1952; "Will Britain Rule the Jet Transport Air Waves?," *Newsweek*, September 15, 1952, 87; "Behind in Jets," *Washington Post*, August 30, 1952; "Jet Transports," *Washington Post*, December 27, 1952; "Air Force Opposes U.S. Jet Airliners," *New York Times*, May 13, 1952.

9. Harrison E. Salisbury, "Soviet Displays Big 4-Jet Bomber," *New York Times*, May 2, 1954; Joseph and Stewart Alsop, "Matter of Fact . . . ," *Washington Post*, May 26, 1954; Robert Hotz, "Russian Jet Airpower Gains Fast on U.S.," *Aviation Week*, May 23, 1955, 12; Clifton Daniel, "Moscow Air Show Unveils 4-Jet Transport-Airliner," *New York Times*, July 4, 1955; Stanley Johnson, "Soviet Bids for Airline Supremacy," *Washington Post*, March 25, 1956; "New Russian Jet Airliner Visits London," *Aviation Week*, March 26, 1959, 29; "Queen of the Ocean Air," *New York Herald Tribune*, October 19, 1958; "Tu-104 Interior: Damask for Jet Age," *Aviation Week*, April 16, 1956, 126–127; "Red Jet Service to London," *Aviation Week*, May 7, 1956, 34; "Jet Age—Red Style," *Newsweek*, April 2, 1956, 74; "Russia's Jet Marks Bid to Build World Airline," *Business Week*, September 14, 1957, 94–95.

10. John B. Rae, *Climb to Greatness: The American Aircraft Industry, 1920–1960* (Cambridge, MA: The MIT Press, 1968), 206; "Gamble in the Sky," *Time*, July 19, 1954; "Why Boeing Has Undertaken 'Project X,'" *Boeing Magazine*, November 1952; Wellwood Beall, "The Story behind Project X," *Boeing Magazine*, December 1952; Daniel L. Rust, *Flying across America: The Airline Passenger Experience* (Norman: University of Oklahoma Press, 2009), 189–192.

11. "Debut of America's Jet Transport," *Boeing Magazine*, June 1954, 8–9, NASM Library; "Twelve-Hour World: First American Jet Airliner," *Newsweek*, March 8, 1954, 51; "707 Is Longer than Distance Flown by Wright Brothers," *The Clipper*, November–December 1958, NASM Library; "The Jet Age Arrives: Billion Dollar Battle Turns Drawing-Board Dreams into Reality," *Newsweek*, March 5, 1956, 72–75; Boeing press release, "Boeing 707 Jet Stratoliner," September 30, 1955, Box 109, Series I, PAWA, UM.

12. Carl Solberg, *Conquest of the Skies: A History of Commercial Aviation in America* (Boston: Little, Brown, 1979), 387–395; Boeing press release, "Pan American 707 Purchase Marks New Era, Says Allen," October 13, 1955; Douglas Aircraft press release, October 13, 1955, Box 109, Series I, PAWA, UM; Ansel E. Talbert, "Pan Am Ordering 45 Jet Air Liners, Cost 269 Million," *New York Herald Tribune*, October 14, 1955; "Airlines of U.S. Stock up for Jet Age," *Life*, October 24, 1955, 32–37.

13. "One Million Passengers Have Now Flown the Boeing 707 Jetliner!" *Boeing Magazine*, November 1959; Richard Sweeney, "Boeing 707 Is 'Honest' Airplane,"

Aviation Week, October 6, 1958, 70–79; "What the Experts Say," *Boeing Magazine*, October 1958, 14–15; "The New Jet-Winged World," *Newsweek*, March 25, 1957, 87–89; Boeing, undated [1960], Box 109, Series I, PAWA, UM.

14. Bilstein, *Flight in America*, 228; Jeffrey Engel, *Cold War at 30,000 Feet: The Anglo-American Fight for Aviation Supremacy* (Cambridge, MA: Harvard University Press, 2007), 173–175.

15. R. E. G. Davies, *A History of the World's Airlines* (London: Oxford University Press, 1964), 488–490; Bilstein, *Flight in America*, 231; Solberg, *Conquest of the Skies*, 403.

16. "International Airlines: The Great Jet Gamble," *Fortune*, June 1958; "Gamble in the Sky," *Time*, July 19, 1954; Cecil Brownlow, "Jet Age, Growth Strain Airports, Airways," *Aviation Week*, February 25, 1957, 149–150; Robert Hotz, "Aviation's Modernization Problem," *Aviation Week*, October 20, 1958.

17. John B. Lansing and Dwight M. Blood, *The Changing Travel Market* (Ann Arbor: Survey Research Center, Institute for Social Research, University of Michigan, 1964), 89, 107–108; "Jets across the U.S.," *Time*, November 17, 1958.

18. L. M. Hughes, "Airlines Double Ad Spending in Five Years," *Air Transport World*, June 1965.

19. Murray Barnes to sales managers, July 6, 1958; "Jet Promotional Schedule," undated [1958]; Donald J. Dougherty to traffic and sales managers, June 4, 18, and 27, 1958, Box 27, Series II, PAWA, UM; internal memorandum, July 7, 1958, Box 109, Series I, PAWA, UM; "Advertising Carries a Heavy Load as Jet Planes Roar into a New Travel Era," *Advertising Week*, November 14, 1958.

20. John W. Ogilvie to sales managers, July 22, 1958, Box 109, Series I, PAWA, UM; Dougherty to Thomas M. Dixon, June 30, 1958; and industrial relations manager, Atlantic Division, to Robert P. Sheils, April 25, 1958, Box 27, Series II, PAWA, UM; "Your Jet Clipper," 1958, Box 110, Series I, PAWA, UM.

21. Dougherty to sales managers, June 27, 1958; Barnes to sales managers, July 6, 1958, Box 27, Series II, PAWA, UM; "Pan American Atlantic Division Announces First Jet Service to the U.S. on Oct. 27th," *J. Walter Thompson Company News*, September 22, 1958; "A Pan Am 'First'—Jet Clipper Service to Europe This Fall," *J. Walter Thompson Company News*, July 14, 1958, Box MN11, JWT Newsletters, JWT, JWHC; H. B. Miller to local PR reps and sales managers, August 21, 1958, Box 110, Series I, PAWA, UM; "How They're Promoting Travel on 707s," *American Aviation*, November 3, 1958, 26.

22. "Copy for Use in Jet Announcement," undated [1958], Box 27, Series II, PAWA, UM; Pan Am advertisement, "Good Afternoon, Ladies and Gentlemen, This Is the Captain Speaking," April 24, 1958, Box 11, Winfield Taylor Papers, JWT, JWHC; John E. Henry to Beverly Zimmerman, October 31, 1958; Henry to J. Ely Van Hart Jr., October 31, 1958; Henry to Dorothy Krever, November 3, 1958; Maurice F. Hanson to Samuel Meek and Winfield Taylor, November 3, 1958; all in Box 11, WT, JWT, JWHC. On Cold War–era ideologies of gender and domesticity, see Elaine Tyler May, *Homeward Bound: American Families in the Cold War Era* (New York: Basic Books, 1988).

23. Pan Am press release, "Jet Passengers to Travel in Living-Room Comfort," July 1958; Pan Am publicity brochure, Box 109, Series I, PAWA, UM; American Airlines publicity brochure, "Welcome Aboard Your American Flagship," 1959, Folder F1A-500000-15, ATC, NASM Library.

24. Gordon Williams, "Penthouse in the Sky," *Boeing Magazine*, October 1958; "Once Aboard, You Won't Believe Your Eyes . . . or Ears," *Boeing Magazine*, July 1958; "Research in Comfort," *Boeing Magazine*, June 1955; advertisements, *Boeing Magazine*, October 1958, November 1958, and May 1959.

25. L. M. Hughes, "Airlines Double Ad Spending in Five Years," *Air Transport World*, June 1965; "Taking Off on a New Sales Pitch," *Business Week*, August 3, 1968, 44–45; Pan Am press release, "Cookout above the Clouds," July 1958; "PAA Prepares First Jet Travel Book for Les Gals," *Atlantic Clipper*, spring 1959, Box 109, Series I, PAWA, UM.

26. "First 6 Weeks: 12,168 Jet Pax," *Atlantic Clipper*, December 1958, Box 109, Series I, PAWA, UM; Gerald J. Barry, "Jet Age: So Far, So Fast," *Newsweek*, September 14, 1959, 100; Anthony Sampson, *Empires of the Sky: The Politics, Contests, and Cartels of World Airlines* (New York: Random House, 1984), 110.

27. "Jet-Age Growing Pains," *Newsweek*, May 9, 1960, 97–98; "Jet Debt," *Time*, November 21, 1960; "Losing Altitude," *Time*, July 21, 1961; "World Airlines Race to Fill Jet Seats," *Aviation Week & Space Technology*, March 7, 1960; *Pan American World Airways Annual Report*, 1962–1965, Box 23, Series I, PAWA, UM; Solberg, *Conquest of the Skies*, 406.

28. "Jet Clippers Are Here," Folder 4, Box 109, Series I, PAWA.

29. Harrison E. Salisbury, "Nixon and Khrushchev Argue in Public as U.S. Exhibition Opens," *New York Times*, July 25, 1959.

30. Bob Considine, "All Aboard for Moscow," Hearst newspapers, July 7, 1959; Pan Am press release, undated [July 1959], Box 179, Series I, PAWA, UM; Associated Press, "U.S. Jet Sets NY-Moscow Record," July 24, 1959.

31. Trippe, speech before New York Board of Trade, October 8, 1959, Box 10, JTT, NASM.

32. American Airlines, "Welcome Aboard Your American Flagship," 1959, Folder F1A-5000000-15, ATC, NASM Library; *Pan American Airways Annual Report*, 1965, Box 23, Series I, PAWA, UM; "Military Airlifts to Saigon Are Up," *Pan American Clipper*, October 15, 1965, NASM Library.

33. "Era of the Seven-League Sell," *Time*, December 18, 1964; Pan Am World Wide Marketing Service materials, Box 71, Series I, PAWA, UM; Victoria de Grazia, *Irresistible Empire: America's Advance through Twentieth-Century Europe* (Cambridge, MA: Harvard University Press, 2006).

34. Reports on ICH hotels, *Overseas Clipper*, 1963–1965, NASM Library; "Air-Conditioned Luxury Takes Mystery out of Mysterious East," *The Clipper*, July 15, 1965, NASM Library; TWA, "Coming Home from Overseas—to Get the U.S. Kind of Care," 1962, Folder FIT-662000-10, ATC, NASM Library.

35. Clay Blair Jr., "So You're Flying Jet to Europe," *Saturday Evening Post*, January 10, 1959.

36. Alastair Gordon, *Naked Airport: A Cultural History of the World's Most Revolutionary Structure* (New York: Metropolitan Books, 2004), 184–188; "The Jet-Age Idlewild," *Fortune*, June 1958, 125–128; "Gateways to the Jet Age," *Time*, August 15, 1960. Idlewild was renamed John F. Kennedy International Airport in December 1963.

37. Gordon, *Naked Airport*, 202; Vincent Scully, "Rethinking Saarinen," in Eeva-Liisa Pelkonen and Donald Albrecht, eds., *Eero Saarinen: Shaping the Future* (New Haven, CT: Yale University Press, 2006), 24. On abstract expressionism and the Cold War, see Serge Guilbaut, *How New York Stole the Idea of Modern Art* (Chicago: University of Chicago Press, 1985).

38. Pelkonnen and Albrecht, "Project Portfolio," in *Eero Saarinen*, 214; Susan Santala, "Airports: Building for the Jet Age," in *Eero Saarinen*, 303; "New Airport for Jets, Dulles International Airport," *Architectural Record*, March 1960, 175–182; "Fine New Front Door," *Life*, December 16, 1957, 103–104; Ada Louise Huxtable, "Jet Age Triumph," *New York Times*, April 8, 1962.

39. Robert C. Ruark, "This Is Your Life in the Jet Age," *New York World and Sun*, August 29, 1958; *Boeing Magazine*, August 1952 and January 1954; *Overseas Division Clipper*, March–April 1963 and October–November 1964, NASM Library.

40. Braniff International advertisement, "Air Strip," 1965, and National Airlines advertisement, "Fly My 747s to Miami," 1971, www.youtube.com (accessed 7/28/12); "Fly Me," *Time*, November 15, 1971; TWA, "The Thrilling Threesome with TWA," Folder FIT-662000-07, ATC, NASM Library.

41. "Glamor Girls of the Air," *Life*, August 25, 1958, 68–76; Bowen Northrup, "I'm Natasha, Fly Me," *Wall Street Journal*, September 1972. On the "stewardess mystique," see Kathleen Barry, *Femininity in Flight: A History of Flight Attendants* (Durham, NC: Duke University Press, 2007).

42. *Pan American World Airways Annual Report*, 1960 and 1965, Box 23, Series I, PAWA, UM; TWA route maps, Folder FIT-662000-10, ATC, NASM Library; Thomas L. Friedman, *The World Is Flat: A Brief History of the Twenty-First Century* (New York: Picador/Farrar, Straus and Giroux, 2007).

43. Pan Am press release, "You'll Arrive before You Leave on New Pan Am Jet Route," August 7, 1959, Box 109, Series I, PAWA, UM.

44. Arthur Herzog, "It's the Innest, It's the Jet Set," *New York Times Magazine*, October 28, 1962, 32, 110–112; Peter Maas, "Boswell of the Jet Set," *Saturday Evening Post*, January 1963.

45. Malvina Lindsay, "We Won't Lose out on Jet Age Whirl," *Washington Post*, December 31, 1955; Mary Roblee, "In the Jet Set: New York to Paris in a Boeing 707," *Vogue*, December 1958, 79; George Sokolsky, "These Days," *Washington Post*, November 30, 1955; Herzog, "It's the Innest, It's the Jet Set."

46. Maas, "Boswell of the Jet Set," 31–33; Tom Hoctor, "Beatlemania," *The Clipper*, March 1964, NASM Library; Arthur M. Schlesinger, *Robert Kennedy and His Times* (New York: Mariner Books, 2002), 387.

47. "Paris Fashions Bear Pan Am's Trademark," *The Clipper*, July–August 1964, NASM Library.

48. Solberg, *Conquest of the Skies*, 407; *Pan American World Airways Annual Report*, 1960, Box 23, Series I, PAWA, UM; "Teenagers Fly PAA to Forum," *Overseas Clipper*, February–March 1964, NASM Library; Robert Mansfield, "Teen-Age Ambassadors of the Jet Age," *Boeing Magazine*, July 1958; Robert Mansfield, "Europe in Ten Days—By Air," *Boeing Magazine*, September 1958.

49. Robert J. Serling, *Howard Hughes' Airline: An Informal History of TWA* (New York: St. Martin's, 1983), 306–308; "Immigrants Can Ride Now on Pan Am Pay Later Plan," *Overseas Division Clipper*, November 1963; "First to Fly," *Overseas Division Clipper*, February–March 1964, NASM Library.

50. Richard Witkin, "Russian Jetliner Inaugurates Direct Soviet-U.S. Passenger Flights," *New York Times*, July 16, 1968; "Aboard the First Flights," *Life*, July 26, 1968; Jackson Heights Young Republican Club to Pan Am, undated [1967]; Norman P. Blake to Harold Gray, April 27, 1967; and Karlis Minka to John Shannon, November 7, 1966; all in Box 427, Series II, PAWA, UM; "Fair Warning, Friends," *New York Daily News*, November 22, 1967; "Another Poor Bargain," *Aviation Week & Space Technology*, November 7, 1966.

51. S. W. Bryant, "What Jet Travel Does to Your Metabolic Clock," *Fortune*, November 1963, 160–163; "Why Jet Travel Makes You Tired," *Science Digest*, October 1964, 25; "Those Circadian Rhythms," *Time*, December 17, 1965, 66; "FAA Plans to Conduct Survey of Jet Fatigue Effects on Crew," *Aviation Week & Space Technology*, January 11, 1960, 54.

52. Chester L. Cooper, "Fly Now, Pay Later," *Foreign Policy*, Fall 1975, 165; Bilstein, *Flight in America*, 237.

53. "Text of President's Statement on the Balance of Payments Problem and How to Meet It," *New York Times*, January 2, 1968; "Air Ticket Tax of 5% Proposed on Foreign Trips," *Wall Street Journal*, February 1, 1968. For further analysis of the travel deficit, see Christopher Endy, *Cold War Holidays: American Tourism in France* (Chapel Hill: University of North Carolina Press, 2004), 182–202.

54. William W. Yates, "'Visit U.S.A. Year' Ushers in a New Travel Decade," *Chicago Daily Tribune*, February 7, 1960; "Americans on the Go—'60," *Newsweek*, June 13, 1960, 81–84

55. Jerry Hulse, "Travelines," *Los Angeles Times*, January 21, 1968; "An Inauspicious Beginning," *Washington Post*, January 2, 1968; Robert A. Wright, "Proposals Shock Tourist Industry," *New York Times*, February 6, 1968; Ernest Dunbar, "Memo to LBJ: See America First . . . If You Can," *Look*, April 16, 1968, 48; Lansing and Blood, *The Changing Travel Market*, 10

56. "The Travel Deficit," January 12, 1968, Box 6, JTT, NASM; *Pan American World Airways Annual Report*, 1961, 1965, 1968, Box 23, Series I, PAWA, UM; Endy, *Cold War Holidays*, 200.

57. Richard Johnson, "Travel Notes," *Esquire*, October 1968, 20–22.

58. Art Seidenbaum, "Join the Jet Set—and Keep Running Aground," *Los Angeles Times*, December 4, 1968; "Air Traffic Jam Greets 1st Russ Jet," *Chicago Tribune*, July 16, 1968; "Airports at Sea," *Time*, May 30, 1969.

59. Bilstein, *Flight in America*, 262; Marilyn Bender and Selig Altschul, *The Chosen Instrument: Pan Am, Juan Trippe, the Rise and Fall of an American Entrepreneur*

(New York: Simon and Schuster, 1982), 501–504; Gene Banning, *Airlines of Pan American since 1927* (McLean, VA: Paladwr Press, 2001), 613, 621.

60. Warren Pfaff, "A Flying Leap," Box DG17, JWT Publications; "Preliminary Exploration of Consumer Perceptions of the 747 Plane," Box 15, AF, JWT, JWHC; "Up in the Air with the Jumbo Jet," *Sales Management*, December 15, 1969; ads by Hank Syverson, 1969, Box PA89, DA, JWT, JWHC.

61. Banning, *Airlines of Pan American*, 620–621; Steve Davis, "Pan Am Waitlists 2,500 Eager to Go 'Anywhere' When 747 Jumbo Jet Takes off for Destination X," *Travel Weekly*, November 12, 1968; Bender and Altschul, *Chosen Instrument*, 507, 517–519; Robert Gandt, *Skygods: The Fall of Pan Am* (New York: Morrow, 1995), 135.

62. Gandt, *Skygods*, 166–167, 173.

63. Bilstein, *Flight in America*, 264–266; Juan T. Trippe, address upon receiving the 1966 Wright Brothers Memorial Trophy, December 15, 1966, Box 5, JTT, NASM; Lieutenant Colonial Donald I. Hackney, "The Supersonic Transport as an Instrument of National Power," Army War College thesis, 1966, 55, www.dtic.mil/cgi-bin/GetTRDoc?AD=ADA510143 (accessed 10/24/12).

64. "Flight of the Tu-144," *New York Times*, January 2, 1969; Richard D. Lyons, "The Russians Lead with the SST," *New York Times*, January 5, 1969; Wayne Parrish, "Lessons of History," speech to Aero Club of Washington, DC, May 28, 1974, Box 21, WP, HHPL.

65. Rust, *Flying across America*, 206–208; David Gero, *Flights of Terror: Aerial Hijack and Sabotage since 1930* (Sparkford, UK: Patrick Stevens, 1997); Tim Naftali, *Blind Spot: The Secret History of American Counterterrorism* (New York: Basic Books, 2005), 19–53; William Lee Brent, *Long Time Gone: A Black Panther's True-Life Story of His Hijacking and Twenty-Five Years in Cuba* (iUniverse, 2000).

66. Paul Hoffman, "Arab Guerrillas Kill 31 in Rome during Attack on U.S. Airliner, Take Hostages to Athens, Fly On," *New York Times*, December 18, 1973; "Kissinger Tells 'Moral Outrage' at Hijacking," *New York Times*, December 19, 1973.

67. Gordon, *Naked Airport*, 236–237.

68. Arthur Hailey, *Airport* (Garden City, NY: Doubleday, 1968), 60.

69. Ibid., 178.

70. Ibid., 60–61, 438.

71. "Americans Fly Everywhere Now," *Fortune*, April 1945, 139–141.

Conclusion

1. Valerie Lester, *Fasten Your Seat Belts! History and Heroism in the Pan Am Cabin* (McLean, VA: Paladwr Press, 1995), 263–265.

2. Agis Salpukas, "Pan Am Loses Aid from Delta," *New York Times*, December 4, 1991; Martha M. Hamilton, "Aviation Pioneer Pan Am Folds Its Wings at Age 64," *Washington Post*, December 5, 1991; Carlyle C. Douglas, "A Symbol of a Different Era Collapses," *New York Times*, December 8, 1991; Lester, *Fasten Your Seat Belts!*, 266–268.

3. T. A. Heppenheimer, *Turbulent Skies: The History of Commercial Aviation* (New York: Wiley, 1995), 314–316; John R. Meyer and Clinton V. Oster Jr., eds., *Airline Deregulation: The Early Experience* (Boston: Auburn House, 1981), 1–12; Brian F. Havel, *Beyond Open Skies: A New Regime for International Aviation* (New York: Wolters Kluwer Law & Business, 2009), 264–266.

4. Robert Gandt, *Skygods: The Fall of Pan Am* (New York: William Morrow, 1995), 125; Heppenheimer, *Turbulent Skies*, 338–339.

5. Marilyn Bender and Selig Altschul, *The Chosen Instrument: Pan Am, Juan Trippe, the Rise and Fall of an American Entrepreneur* (New York: Simon and Schuster, 1982), 518; Gandt, *Skygods*, 94, 166.

6. Gene Banning, *Airlines of Pan American since 1927* (McLean, VA: Paladwr Press, 2001), 604; Gandt, *Skygods*, 94–222; Lester, *Fasten Your Seat Belts!*, 263.

7. Banning, *Airlines of Pan American*, 646–648; Gandt, *Skygods*, 198–199.

8. Banning, *Airlines of Pan American*, 649–661; Gandt, *Skygods*, 208–211; Heppenheimer, *Turbulent Skies*, 340, 344.

9. United Kingdom Air Accidents Investigation Branch, *Report on the Accident to Boeing 747-121, N739PA at Lockerbie, Dumfresshire, Scotland on 21 December 1988*, Aircraft Accident Report No 2/90 (EW/C1094), July 1990, www.aaib.gov.uk/cms_resources.cfm?file=/dft_avsafety_pdf_503158.pdf; Banning, *Airlines of Pan American*, 659–661; Heppenheimer, *Turbulent Skies*, 339; Gandt, *Skygods*, 271; Robert D. McFadden, "Megrahi, Convicted in 1988 Lockerbie Bombing, Dies at 60," *New York Times*, May 20, 2012.

10. Heppenheimer, *Turbulent Skies*, 332–337.

11. Jad Mouawad and Michael J. de la Merced, "United and Continental Said to Agree to Merger," *New York Times*, May 2, 2010; Hugo Martin, "Merged United and Continental Operations off to a Bumpy Start," *Los Angeles Times*, March 5, 2012; David Fagin, "United We Fall: The Merger between Continental and United off to a Shaky Start," March 9, 2012, www.huffingtonpost.com/david-fagin/united-and-continental-airline-merger_b_1379046.html, (accessed 10/24/12); "United's Profit Falls on Continental Merger," *New York Times*, July 26, 2012; Google search, "United Continental merger problems" and "United Continental merger," October 24, 2012; Matthew Yglesias, "American Fail Lines," October 8, 2012, www.slate.com/articles/business/moneybox/2012/10/american_airlines_delays_the_airline_that_can_t_get_its_passengers_to_their_destinations_on_time_manage_its_workers_or_even_keep_its_seats_bolted_down_.html (accessed 10/24/12).

12. Wayne Curtis, "The Golden Age of Air Travel," *ForbesLife*, September 26, 2011; Federal Aviation Administration, "Passenger Boarding (Emplanement) and All-Cargo Data for U.S. Airports," calendar year 2011, www.faa.gov/airports/planning_capacity/passenger_allcargo_stats/passenger/ (accessed 10/24/12); FAA press release, "Airline Passenger Travel to Nearly Double in Two Decades," March 8, 2012, www.faa.gov/news/press_releases/news_story.cfm?newsId=13394 (accessed 10/24/12); Seth Stevenson, "The Southwest Secret," June 12, 2012, www.slate.com/articles/business/operations/2012/06/southwest_airlines_profitability_how_the_company_uses_operations_theory_to_fuel_its_success_.html (accessed 10/24/12); Doug Cameron, "Southwest Airlines' Net Profit Soars on Strong

Growth," *Wall Street Journal*, July 19, 2012; Southwest Airlines Fact Sheet, www.swamedia.com/channels/Corporate-Fact-Sheet/pages/corporate-fact-sheet (accessed 10/24/12).

13. Roger Bilstein, *Flight in America: From the Wrights to the Astronauts* (Baltimore: Johns Hopkins University Press, 1994), 286; Daniel L. Rust, *Flying across America: The Airline Passenger Experience* (Norman: University of Oklahoma Press, 2009), 215–224; John Baldwin Seales, "Airline Ads: The Need to Appeal to Today's Market," *The Travel Agent*, December 2, 1976.

14. 2012 Skytrax World Airline Awards, www.worldairlineawards.com (accessed 10/24/12). The 2012 top ten list, in order, is as follows: Qatar Airways, Asiana Airlines (South Korea), Singapore Airlines, Cathay Pacific (Hong Kong), All Nippon Airways (Japan), Etihad Airways (United Arab Emirates), Turkish Airlines, Emirates, Thai Airways International, and Malaysia Airlines. Diana Gialo, "Where Luxury and Flying Go Hand-in-Hand," www.nbcnews.com/id/39639344/ns/travel-business_travel/ (accessed 10/25/12); Singapore Airlines advertisement, "Across the World with the Singapore Girl," www.youtube.com/watch?v=fNEJrd6GkSY (accessed 10/25/12).

15. Gary Shteyngart, "A Trans-Atlantic Trip Turns Kafkaesque," *New York Times*, September 29, 2012.

16. *The 9/11 Commission Report, Including Executive Summary: Final Report of the National Commission on Terrorist Attacks upon the United States* (Washington, DC: U.S. Government Printing Office, 2004), 168.

17. See Jane Mayer, "The Predator War," *New Yorker*, October 26, 2009; Peter W. Singer, *Wired for War: The Robotics Revolution and Conflict in the 21st Century* (New York: Penguin, 2009); and Nick Turse and David Englehardt, *Terminator Planet: The First History of Drone Warfare, 2001–2050* (CreateSpace Independent Publishing Platform, 2012).

18. See www.airandspace.si.edu/americabyair (accessed 10/27/12).

19. www.airlinesoriginals.com (accessed 10/27/12); "Second Exclusive Pan Am/Marc Jacobs Bag Launches," Business Wire, June 11, 2007; www.flight001.com/aboutus/(accessed 10/26/12).

20. Erik Lacitis, "Museum of Flight Showing 75 Years of Flying in Style," *Seattle Times*, February 10, 2008.

21. "For Those Who Think Young," *Mad Men*, AMC, written by Matthew Weiner and directed by Tim Hunter, original air date July 27, 2008; "Flight 1," *Mad Men*, AMC, written by Lisa Albert and Matthew Weiner and directed by Andrew Bernstein, original air date August 3, 2008; "The Jet Set," *Mad Men*, AMC, written by Matthew Weiner and directed by Phil Abraham, original air date October 12, 2008; "Out of Town," *Mad Men*, AMC, written by Matthew Weiner and directed by Phil Abraham, original air date August 16, 2009; "Souvenir," *Mad Men*, AMC, written by Lisa Albert and Matthew Weiner and directed by Phil Abraham, original air date October 4, 2009; "Tea Leaves," *Mad Men*, AMC, written by Erin Levy and Matthew Weiner and directed by Jon Hamm, original air date April 1, 2012.

22. "Wee Small Hours," *Mad Men*, AMC, written by Dahvi Waller and Matthew Weiner and directed by Scott Hornbacher, original air date October 11, 2009.

23. *"Pan Am* Canceled: ABC Series Officially Grounded," June 20, 2012, www. huffingtonpost.com/2012/06/20/pan-am-canceled-abc-series-dead_n_1612908. html (accessed 10/30/12); Brian Gallagher, "Christina Ricci and Jack Orman Talk *Pan Am* Season 1," September 25, 2011, www.movieweb.com/news/christina-ricci -and-jack-orman-talk-pan-am-season-1 (accessed 10/26/12); Brooks Barnes, "Prime Time Ambitions," *New York Times*, August 28, 2011; Angela Watercutter, "TV Fact-Checker: A Former Stewardess on *Pan Am's* Friendlier Skies," September 25, 2011, www.wired.com/underwire/2011/09/tv-fact-checker-pan-am/ (accessed 10/26/12).

24. "Pilot," *Pan Am*, ABC, written by Jack Orman and directed by Thomas Schlamme, original air date September 25, 2011; "Eastern Exposure," *Pan Am*, ABC, written by Jack Orman and Moira Walley-Beckett and directed by Thomas Schlamme, original air date October 16, 2011.

25. Susan J. Douglas, "If Only Sexism Could Be Cancelled," *In These Times*, November 2011, 35; Tom Chiarella, "Stewardesses," *Esquire*, September 2011, 60. The term "stewardess mystique" is from Kathleen Barry, *Femininity in Flight: A History of Flight Attendants* (Durham, NC: Duke University Press, 2007).

26. Cullen Murphy, *Are We Rome? The Fall of an Empire and the Fate of America* (New York: Mariner Books, 2008); Troy Patterson, "Pan Am Reviewed: ABC's New Jet-Age Drama Pays Delightful Attention to Detail," September 23, 2011, www.slate.com/articles/arts/television/2011/09/pan_am.html (accessed 10/26/12).

27. "What's Wrong with the Airlines," *Fortune*, August 1946; Arthur Hailey, *Airport* (Garden City, NY: Doubleday, 1968); Virginia Postrel, "Up, Up and Away," *Atlantic*, January/February 2007.

28. Henry Luce, "The American Century," *Life*, February 7, 1941, 61–65.

Acknowledgments

I will never cease to marvel that I ended up writing about aviation. In some ways, though, it seems uncannily destined. Like the Wright brothers, I was born and raised in Dayton, Ohio, the "birthplace of aviation." Both of my grandfathers served as fliers during World War II. My mother's father was an Army pilot who, after retiring, built an airplane by hand in his basement; my father's father was a bombardier with the 14th Air Force, also known as the Flying Tigers. Still, I vividly recall my childhood impatience during annual school and family trips to the U.S. Air Force Museum at Wright-Patterson Air Force Base. I remember once watching my maternal grandfather study a small display of aircraft spark plugs for a good twenty minutes, thinking: *Why on earth do people care so much about these old airplanes?*

Even I, however, was not immune to aviation's allure. As a child and teenager, I welcomed any opportunity to go to the local airport. Dayton International Airport, with its gray, low-ceilinged corridors and bunker-like motel, is no Dulles or Charles de Gaulle. Yet to me, those bland gray corridors were gateways to the world. I watched the jets ascend and wondered where they would land. Paris? Nairobi? Tokyo? More likely they were headed to Pittsburgh, Nashville, or Toledo. Nonetheless, wherever the jets from Dayton were flying, I wanted to go. Like my protagonists, I dreamed of seeing the world, and aviation made it seem possible for those dreams to become real.

This book was initially inspired by Wendell Willkie's evocative paean to aviation's ability to create "one world": "A navigable ocean of air blankets the whole surface of the globe. There are no distant places any longer: the world is small and the world is one." And indeed, the process of working on the book has proven to me that no person is an island—not even a historian immersed in the solitary grind of archival research and writing. Quite literally, I could not have completed this book without the help of many individuals and institutions along the way.

First of all, I would like to thank my editor at Harvard University Press, Joyce Seltzer, who understood and believed in this book from our very first conversation in the lounge of the Washington, DC, Marriott Wardman Park Hotel in 2008. The manuscript benefited greatly from her meticulous reading and analysis. I also thank Brian Distelberg for his assistance with my many inquiries about publishing logistics, and I thank Julie Palmer-Hoffman and Edward Wade for their careful attention to copyediting and production.

At Yale, I have been privileged to be part of a vibrant and engaging intellectual community—initially as a graduate student and now as a professor—for the past ten years. Jean-Christophe Agnew supported and guided this book in every way from its tentative beginnings as a paper in his graduate seminar "The American Century." His lengthy, richly erudite comments on my chapter drafts always pushed my thinking in new directions. Matthew Frye Jacobson not only challenged me to embolden my arguments about cultural history but also consistently encouraged me to have confidence in my project. (I will never forget one conversation with Matt, when I was experiencing anxiety about my project, in which he offered five key words of wisdom: "Okay. Take a deep breath." I am still trying to follow this advice.) Seth Fein has been a mentor in all aspects of my work—teaching, research, and writing. Seth's seminar on transnational history, which I took during my first semester of graduate school, was a transformative experience that inspired me to venture across geographical and methodological borders. Paul Kennedy and Beverly Gage graciously took the time to read the entire manuscript and offer insightful suggestions for revision; Paul and Bev have also been wonderful mentors and sources of professional guidance. Among the many colleagues whose intellectual camaraderie has sustained my spirits and enhanced my work, I would particularly like to thank Patrick Cohrs, Laura Engelstein, Joanne Freeman, Glenda Gilmore, Gilbert Joseph, Bill Rankin, and Adam Tooze. Weekly conversations in the Yale International History Workshop and Colloquium in International Security Studies have also greatly enriched my thinking. Last but certainly not least, I thank my students, graduate and undergraduate, from whom I have learned so much.

I am grateful to have spent my graduate school years in the company of an exceptionally brilliant, dynamic, and fun cohort. I particularly thank the members of my writing group—Adam Arenson, Gretchen Heefner, Theresa Runstedtler, Helen Veit, and Erin Wood—for their productively challenging critiques of my work. Aaron O'Connell and Katherine Unterman also offered insightful feedback during formative stages of my research and writing. Mary Greenfield's detailed and perceptive comments, informed by her own terrific project on the history of steamships in the nineteenth century, proved enormously generative during my first round of revisions. Kathleen Belew, Gerry Cadava, Lisa Pinley Covert, Jay Driskell, Faith Hillis, Charles Keith, Jana Lipman, and Andrew Sackett have all provided inspiration, wisdom, commiseration, and laughter throughout the years.

For their generous financial support of my research and writing, I thank the AHA/NASA Fellowship in Aerospace History, the Daniel P. Guggenheim Fellowship at the Smithsonian Institution's National Air and Space Museum, the George C. Marshall Foundation, the Hartman Center for Advertising, Sales, and Marketing History at Duke University, the Pan Am Historical Foundation, the Franklin and Eleanor Roosevelt Institute, the Herbert Hoover Presidential Library, and, at Yale, International Security Studies, the John Morton Blum Fellowship in American History and Culture, and the Frederick W. Hilles Publication Fund. I also could not have completed this book without frequent help from many archivists and librarians. For their assistance with research queries and image permissions, I would especially like to thank David Langbart at NARA; Maria Estorino, Rochelle Pienn, and Steve Hersh at the University of Miami; Lynn Eaton at Duke University; Jane Odom at the NASA History Office; Dawne Dewey and Gino Pasi at Wright State University; Matt Schaefer at the Hoover Presidential Library; and Judith Schiff and Christine Weideman of Manuscripts and Archives at Yale's Sterling Memorial Library.

The National Air and Space Museum gave me a scholarly home as I conducted my research in Washington, DC, and my work greatly benefited from its curators' expertise. Dominick Pisano has been a wise, generous, and unfailingly encouraging mentor from the day that I first e-mailed him. Martin Collins's vast knowledge of both historiographical and theoretical literature helped me to clarify, and amplify, my broader arguments about the relationship among technology, culture, and power. For sharing their knowledge of aerospace history in many enjoyable conversations, I also thank NASM curators Tom Crouch, Ron Davies, Bob van der Linden, and Margaret Weitekamp, as well as my fellow NASM fellows Aaron Alcorn, Matt Hersch, Willie Hiatt, Phil Tiemeyer, and especially Christine Yano—whose company made a research trip to the Pan Am archives feel like a vacation and whose perspective as an anthropologist broadened my intellectual horizons.

I have presented portions of this book at numerous conferences and talks, including the annual meetings of the American Historical Association, the Organization of American Historians, the Society for Historians of American Foreign Relations, and the Society for the History of Technology. Although it would be impossible to thank everyone who has offered useful comments and suggestions, I would like to acknowledge the contributions of Jeremy Adelman, Brooke Blower, Elizabeth Borgwardt, Joseph Corn, David Courtwright, Jeffrey Engel, Bruce Mazlish, Emily Rosenberg, and Daniel Sargent. Portions of Chapter 3 were previously published in "The 'Logic of the Air': Aviation and the Global 'American Century,'" *New Global Studies* 1, no. 1 (2007): 1–33, and are reprinted with permission from De Gruyter. Brief portions of one section of Chapter 6 are reprinted from "An Airline at the Crossroads of the World: Ariana Afghan Airlines, Modernization, and the Global Cold War," *History and Technology* 25, no. 1 (2009): 3–24.

My anonymous reviewers for Harvard University Press read the manuscript with great care and thoroughness, and their incisive critiques proved immensely helpful as I began the process of revision. I also received invaluable feedback from Michael Adas, Frank Costigliola, and Vanessa Schwartz, each of whom generously read the entire manuscript and offered detailed suggestions for revision. It has been a true privilege to discuss my ideas and arguments with historians whose own scholarship I so greatly admire.

While acknowledging all of the friends who supported and encouraged me throughout the process of writing this book would require several more pages, I would particularly like to thank Linda Amerighi, Linda Hase, Gretchen Heefner, Ryan Irwin, Aaron O'Connell, Maria Satterwhite, Taylor Spence, Kate Unterman, and Helen Veit. During the intense final months of revision, Bernard Buba sustained my energy and morale with love, patience, and phenomenal culinary talents. It has also been a privilege to get to know Hayley Buba, whose passion for reading reminds me of why I became a historian. I owe special thanks to Sahr Conway-Lanz, who lived with this project for five years and whose influence on my life and work runs through every one of these pages.

My deepest debts of gratitude are to my family. *Empire of the Air* has been written in the memory of my grandfathers, William Van Vleck and Aubrey Cadle, both of whom were decorated aviators during World War II. Especially after researching and writing this book, I am not only proud of but frankly in awe of their service and accomplishments. My grandmother, Ruth Van Vleck, has always been a lively fount of encouragement and good humor; her fascinating stories about traveling the world with my grandfather after the war did much to pique my initial interest in the history of international aviation. I thank my cousin Robin Romblad, a fellow teacher, for her friendship throughout the years. As a proud "cat lady," I must acknowledge my zany calico girl, Charlie, my sole companion during many hours of writing and a constant source of joy and consolation. Finally: my parents, Richard and Caren Van Vleck, have been my indefatigable champions, supporting me in every possible way. On a daily basis, their steadfast belief in me—and the example of their own lives—gives me hope, courage, and inspiration. I lovingly dedicate my first book to them.

Index